GASIFICATION FOR PRACTICAL APPLICATIONS

Edited by **Yongseung Yun**

Gasification for Practical Applications

http://dx.doi.org/10.5772/3132
Edited by Yongseung Yun

Contributors

Yongseung Yun, Seung Jong Lee, Seok Woo Chung, G.G. Fouga, G. De Micco, H.E. Nassini, A.E. Bohé, Yoshiaki Yamazaki, Younes Chhiti, Sylvain Salvador, A. Sanjeevi Gandhi, T. Kannadasan, R. Suresh, Jonathan Kamler, J. Andres Soria, Youngchul Byun, Moohyun Cho, Soon-Mo Hwang, Jaewoo Chung, Sevgihan Yildiz Bircan, Kozo Matsumoto, Kuniyuki Kitagawa, Yuli Grigoryev, Marek Sciazko, Tomasz Chmielniak, Pravin Kannan, Ahmed Al Shoaibi, C. Srinivasakannan, Maurício Bezerra de Souza Jr., Leonardo Couceiro Nemer, Amaro Gomes Barreto Jr., Cristina Pontes B. Quitete, L. Sivakumar, X. Anithamary

Published by InTech

Janeza Trdine 9, 51000 Rijeka, Croatia

Publishing Process Manager Oliver Kurelic
Technical Editor InTech Prepress, Novi Sad
Cover InTech Design team

Additional hard copies can be obtained from orders@intechopen.com

Gasification for Practical Applications, Edited by Yongseung Yun
p. cm.
ISBN 978-953-51-0818-4

For EU product safety concerns:
IN TECH d.o.o., Prolaz Marije Krucifikse Kozulić 3, 51000 Rijeka, Croatia,
info@intechopen.com or visit our website at intechopen.com.

Meet the editor

Dr. Yongseung Yun majored in chemical engineering, received Ph.D. at the University of Utah in 1990, MS from KAIST, Korea in 1981, and BS at Yonsei University, Korea in 1979. At KIST in Korea from 1981-1984, he researched on the oil agglomeration and FBC. For his Ph.D., he worked on the low temperature air oxidation of coal. At Brown University in 1991-1992, he researched about the pretreatment methods for coal liquefaction. He has concentrated on coal gasification for IGCC from 1993 until now. From 2000, he expands gasification research to wastes, and from 2007 on CO2 capture technologies. He works as the vice president in KAWET, KSWM, and in Korea DME Association. Also serves as the Editor for Korean Industrial Chemistry News of KSIEC.

Contents

Preface

Gasification, which produces gas for fuel or chemicals, remains a basic technology for society. Since current society has evolved to use gas and liquid forms of energy, the gasification technology, which converts low-grade feedstock to useful forms of gas or liquid, continues to be an important option, unless some revolutionary technological breakthroughs in renewable energies are made to replace it.

The most common question regarding gasification is whether there is still a need for more development and demonstration for the technology that has already been improved and implemented in many industries. In fact, gasification technology has been in the market for more than a century. Old gasification technology employed the atmospheric and air-using system, mainly for coal, while the recent gasification system uses a more compact, oxygen-using, fast-reacting system, with more complicated monitoring and control schemes than earlier technologies.

The atmospheric air-using system has been improved to be suitable for biomass and wastes as a system that exhibits moderate efficiency and minimizes byproducts. There does, however, exist significant room for upgrading the biomass and wastes gasification - in scaling up the capacity, in compactness for treating the same amount of feedstock, and in optimizing the reactor for the specific characteristics of biomass and wastes.

It is very pessimistic, however, to consider how much the gasification technology has improved and advanced during the last 30 years or so. Whereas industrial application of gasification has dramatically increased, fundamental technology was only focused on the segmental upgrading, rather than an innovative technology jump.

There are relatively new areas for biomass and wastes utilization, where gasification technology can play a major role. Supercritical and plasma gasification technologies are particularly noteworthy, and are included in two chapters in the book. Even though large scale plants are not yet fully developed, it can have a major impact on the future gasification process, depending on the progress of detailed technologies.

The Market for the gasification plants had remained as one that required major investment and heralded many major projects worth several billion dollars, but the actual market was not as big as anticipated, except in China. Accordingly, the gasification was focused on the market where the current technology could be utilized without technical risks and any significant investment could be reserved for new technology in gasification.

To reach a higher efficiency in gasification, and to reach a wide-spread commercialization stage, most immediate tasks in gasification plants reside in reduction of constructions costs, and in upgrading the availability of plants with innovative ideas in design and operations. Increased number of plants will surely reduce the cost and enhance availability, and increased experience from more plants will in turn also induce a wider application. At the current most probable estimate, renewable energy, which should replace the fossil fuel based technologies, appears to be fully competitive without government incentives around the year 2030. At least until year 2030, gasification will be one of the most practical technologies that can bridge the transition period, by providing clean liquid fuels, gas, and chemicals from low grade feedstock in an environmentally benign way.

Many earlier books on gasification dealt in depth with the thermodynamics, reaction kinetics, feedstock characteristics, and current commercially available gasifiers, etc. Although this is essential information for readers who want to study fundamentals first, in many cases readers want to see the practical application data first, and can study the fundamentals later for their specific needs.

Because the content of this book is compiled of chapters from individual authors, it might not be a concise setup, but it will be quite useful as material to review the status and the future direction of gasification technologies. In this regard, the book provides the practical information for researchers and graduate students who want to review the current situation, and to upgrade and bring in a new idea to the conventional gasification technologies.

The book is comprised of chapters of different aspects of gasification. First, the gasification of various feedstocks of coal/coal-related materials, biomass, and wastes. Second, the cost & economics of gasification, and last, the modeling and simulation of the gasification process. For the biomass and wastes gasification that has emerged as an important new energy source, examples of application on the bio-oil, supercritical and plasma gasification are given.

Each chapter allocates considerations for stranded resource development and cost determination. In addition, practical application examples regarding the recent modeling and simulation are provided, which have gained more importance with the advances in computational hardware and software capabilities.

I would like to thank the authors for contributing their chapters, and revising where needed. Finally, I should not omit my sincere thanks to Mr. Oliver Kurelic, who proposed the book concept on gasification in December 2011, and for his managerial support during the whole process. The work together, with his dedication on this endeavor, was an impressive experience, during the whole process, and in the finalization of the book - "Gasification for Practical Applications."

Yongseung Yun
Institute for Advanced Engineering, Suwon,
Republic of Korea

Section 1

Coal and Coal Related Materials Gasification

Chapter 1

Considerations for the Design and Operation of Pilot-Scale Coal Gasifiers

Yongseung Yun, Seung Jong Lee and Seok Woo Chung

Additional information is available at the end of the chapter

http://dx.doi.org/10.5772/49951

1. Introduction

Although there are many successful commercial coal gasifiers, the basic form and concept have not been improved for the last 20 years or so. Details on the design and operation for the commercial coal gasifiers are closely guarded as proprietary information. Considering the recent technology jump in CFD and monitoring systems, at least some coal gasifiers should come out as a more revolutionary style. Especially it's important to test the novel gasifier types when the gasification has widened the application scope in environmental and biomass areas. Many research ideas should have a chance to design and test in the more realistic conditions of high pressure and high temperature with molten slags. This chapter wants to give an introduction and practical considerations to design and operate the bench scale to pilot scale gasifiers at the actual coal gasification conditions.

The chapter consists of following sections. Each part will give a practical view point to build and test the gasifier at the actual gasification conditions, which are toxic and explosion-prone when the syngas is not trapped inside the gasifier. The scope of the chapter will be focused on the pilot-scale size since the purpose is to focus on the wide distribution of information on the coal gasifiers as well as to stimulate the more active involvement of research groups on the future coal gasifier development.

Key items are, currently known types of coal gasifiers, selection guidelines of coal gasifiers, comparison of slurry type vs. dry type gasifiers, and the discussion regarding the operating pressures and manufacturing limits, etc. Another aspects are the difference in slagging gasifiers and partial/non-slagging gasifiers, coal selection guidelines for gasification, application of CFD for the gasifier design, coal feeding methods, and in-situ estimation of gasification status inside the gasifier.

Other points are the choice in gasifier wall (refractory, membrane wall), slagging/fouling related problems, and finally the future direction of coal gasifiers.

Institute for Advanced Engineering (IAE) has worked in the pilot scale coal gasifiers from 1994. Figures 1-3 demonstrate the coal gasifiers of 1-3 ton/day scale at the operation range of 3-28 bar, 1,300-1,600°C [1-3]. Figure 1 shows two slagging coal gasifiers of 3 ton/day capacity. Left side gasifier was built in 1994 and operated since at the maximum pressure of 28 bar and 1,400-1,550°C. Right-hand side gasifier was mainly applied to the waste oil gasification and used as a test bed for the top-feeding coal gasifier.

Figure 1. Pilot-scale coal gasifiers of slagging type (Left: side-feeding/max. 28 bar, Right: top-feeding/max. 5 bar)

Figure 2 shows the 2 ton/day pilot-scale coal gasifier which chose the top-feeding, partial/non-slagging entrained-bed type and normally operated at 20 bar, 1,300-1,450°C range. Another type of gasifier which chose the membrane wall, top-feeding, slagging type is shown in Figure 3. Idea of applying membrane wall with a layer of refractory was applied to make a gasifier as small as possible.

2. Selection guidelines of coal gasifiers

History of coal gasification starts from early 20th century, but the real commercial size of gasifiers can be supplied from limited vendors. Table 1 shows the commercially available coal gasifiers that can treat coal over 1,500 ton/day. To reach this size of gasifiers, 3-4 steps of development are necessary: bench scale, 10-30 ton/day, 200-500 ton/day, and finally the 1,500-3,000 ton/day commercial size. Pilot coal gasifiers typically include bench to 30 ton/day scale.

Figure 2. Pilot-scale coal gasifier of top-feeding, partial/non-slagging entrained-bed type (max. 21 bar)

Figure 3. Pilot-scale coal gasifier of membrane wall, top-feeding, slagging type (max. 21 bar)

Key factors in deciding the suitable gasifier type will be discussed in this section. As shown in Table 1, currently known coal gasifiers can be classified with choices on the reactor type which will decide the residence time in gasifier, coal feeding method and location, gasifier stages and number of burner nozzles to supply reactants, gasifier wall type in protecting the metal gasifier wall, whether coal ash will be converted to slag or just fly-ash, and the oxidant whether to use oxygen or air.

Item	Shell	Uhde	Conoco-Phillips	Siemens	GE Energy	MHI	OMB	Lurgi
Country	Netherlands	Germany	USA	Germany	USA	Japan	China	Germany
Reactor Type	Entrained	Entrained	Entrained	Entrained	Entrained	Entrained	Entrained	Fixed
Feeding	Dry/Side	Dry/Side	Slurry/ Side	Dry/Top	Slurry/ Top	Dry/Side	Dry/Side	Dry/ Top
Stages	1	1	2	1	1	2	1	1
Wall	Membrane	Membrane	Refractory	Membrane	Refractory	Membrane	Membrane	-
Slagging	Yes	Yes	Yes	Yes	Yes	Yes	Yes	No
Oxidant	O_2	O_2	O_2	O_2	O_2	Air	O_2	Air/O_2
Burners	4	4	2+1	1	1	4+4	4	-

Table 1. Currently available commercial coal gasifiers

First of all, most important remark will be that there is no universal coal gasifier to meet all the different technical requirements. Each gasifier has developed to meet the specific needs from the customers and should see where the preferred gasifier type has the most proven experience in the industry. One of the most frequently asked question is that a specific gasifier can be utilized interchangeably both for the power generation and for the chemical production. If the plant size is small, this option might be possible with limited option. But most commercial gasification plants usually cost 10-200 million US$. With this high capital cost, the gasifier which is the core part of the plant should be designed to maximize the wanted final product with highest efficiency, along with minimum maintenance and without any accident.

Item	Option
Reactor type	Entrained, Fluidized, Fixed(Moving-bed)
Coal feeding	Dry, Wet(Slurry)
Feeding location	Top, Side
Gasifier wall	Refractory, Membrane wall
Ash treatment	Slagging, Non-slagging
Gasifier pressure	High. Medium, Atmospheric
Oxidizing agent	Oxygen, Enriched oxygen, Air
Syngas cooling	Quench, Radiant/Convective cooling
Gasifier stages	One, Two
Burner number	One, Multi

Table 2. Selection Items and Option for Coal Gasifier

2.1. Entrained-bed vs. fluidized-bed vs. fixed-bed

Currently available gasifiers can be classified basically as three reactor types. The processes that require a high throughput capacity in a single reactor generally employ entrained-bed type, as in IGCC, since the reactor size can be minimized by fast residence time (typically less than 5 sec) in the gasifier as well as by high pressure. Although large scale operation by entrained-bed type has successfully demonstrated and employed commercially, the experience is not long enough as fixed or fluidized-bed gasifiers. Also most prominent disadvantage of entrained-bed gasifier is in its high capital cost involved due to condensed configuration of parts.

Fluidized-bed has been developed basically for the application to low-grade fuels or feedstock, like a low-grade coal and wastes that contain various materials. After two oil shocks in the 1970's, many companies were interested in using low grade fuels which were not an interested material, mainly it was coal. Operating principle of fluidized bed involves even distribution of oxidizing agent through the distribution plate in bubbling type, or through the reactor in circulating type. Gas bubbles tend to flow via the less congested area, in turn result in dead zone inside the reactor. This causes the difficulty in scale-up design and operation. Most prominent fluidized-bed examples are FBC boiler and waste pyrolysis plants.

Fixed-bed has a long history of industrial experience as a so-called Lurgi type, which is still used in a large number in China. Due to its long industrial experience, it's reliable. But it's not suitable for the single large scale gasifier. Lurgi recently has achieved to make a gasifier of 1,600 ton/day capacity.

Item	Entrained-bed	Fluidized-bed	Fixed-bed
Residence time in reactor	3-5 sec	minutes	>30 min
Single unit size	Medium-Very large	Medium	Medium
Pressurized reactor	Easy	Not-easy	Not-easy
Complexity	Complex	Complex	Simple
Coal particle size	< 100 microns	6-10 mm	6-50 mm
Coal range	All ranks	Limit in agglomerating coals	Limit in agglomerating coals
Oxygen consumption (O_2/coal ratio)	Large (0.9-1.0)	Medium	Low (0.7-0.8)
Tar formation	None or Very little	Small	Many
Industrial experience	From 1980's	From 1970's	From 1930's
Advantages	Large scale operation	Suitable for low grade fuels	Reliable
Disadvantages	Expensive	Difficult in scale-up, Not suitable for fines	Limit in size

Table 3. Comparison of typical three gasifier types

2.2. Dry feeding vs. slurry feeding

Dry feeding gasifiers were developed mainly in Europe, while the gasifiers that had been developed in United States were slurry-feeding type. Table 4 summarized the key differences of dry and slurry feeding systems.

Maximum carbon conversion in the single-pass gasification without char-recycling could be obtained from the high-reactivity coals. The actual gasifier operation yielded nearly 100% carbon conversion for the high-reactivity coals. In general, dry-feeding entrained-bed gasifier can treat all ranks of coal while the slurry-feeding entrained-bed gasifier is suitable for bituminous coals of higher rank. However, unless the gasifier is designed to cover all different reactivity of coal in the reaction, even for the dry-feeding gasifier, low carbon conversion would result if the gasifier volume were not sufficient to sustain enough residence time of coal powder. In this case, the char-recycling process is required.

Item	Dry-feeding	Slurry-feeding
Coal type	All ranks	Not suitable for high moisture-containing low-rank coals
Efficiency	high	moderate
Carbon conversion	>99%	>99%
Capital cost	high	Moderate
Typical gasifier wall type	Membrane wall	Refractory
Cold gas efficiency	High	Moderate
Typical max. gasifier pressure	45 bar	80 bar
Key application area	Electricity generation	Chemical production
Commercial gasifiers	Shell, Uhde, Siemens, MHI	GE energy, Conoco-Phillips

Table 4. Comparison of dry and wet (slurry) feeding type gasifiers

Maximum gasifier pressure is limited to about 45 bar in the dry-feeding gasifier and to about 80 bar for the slurry feeding system. The bottleneck of the maximum available gasifier pressure is in the coal powder feeding system for the dry feeding type and in the economically manufacturable pressure vessel of large size which is more than few meters diameter in commercial applications.

2.3. Gasifier stages

Most coal gasifiers employ a single stage which is simple in design and less expensive with respect to manufacturing pressure vessel. When the feed coal is relatively uniform in quality and in other properties, the residence time inside the gasifier will be constant in theory if the constant feeding is guaranteed. When the coal and oxygen feeding is uniform, all the times, the performance of the gasifier will be satisfactory, although there would be some mechanical or components related problems. This point will be crucial in designing and operating the pilot coal gasifier. The most important factor in operating coal gasifiers should be the constant feeding of coal powder. Feeding of oxygen and steam is relatively easy since there are in gas states.

Unfortunately, coal is becoming more and more heterogeneous and lower quality. In many plants, feed coals are mixed from widely different origins. In this case, particle residence time inside the gasifer might not sufficient to guarantee the full conversion of all the input coals. Low reactivity or larger size coal particles that are contained in the input feed coal would pass through the gasifier without fully reacting.

Two stage design is introduced to accommodate the heterogeneous coal particles in a single reactor. Feeding amount of coal and oxygen can be manipulated in two separate positions at the gasifier. By adjusting the feeding amounts, hot local temperature is possible in the gasifier that will gasify even the least reactive particles coming with the coal feed. If the slagging is required, the temperature zone that is enough to melt all the inorganics should exist inside the gasifier.

One thing should be noted here. If one single pass through the gasifier is not sufficient to convert all organic components to syngas, unreacted char can be collected and recycled to make a carbon conversion above 99%. But recycling usually incorporates expensive additional feeding systems. If possible, it is the best to make a gasifier to fulfill 100% carbon conversion in a single pass through the gasifier.

2.4. Top-feeding vs. side feeding

Gasification produces gas and solid products as syngas and slag/fly-ash. Gas naturally tends to move upward and solid moves downward by gravity. If the properties of gas and solid apply just as they are, side feeding would be most natural. But side feeding produces operational problems in the areas of slag tap as well as in the syngas outlet which is located at the top section of the gasifier. In addition, slag temperature should be monitored and maintained at high enough temperature to ensure the smooth flow of molten slag.

Top feeding is injecting coal and oxygen, steam from the top side of the gasifier at the velocity above 20 m/s. Typical commercial top feeding coal gasifiers have a L/D ratio of about 1.5, in that the gasification flame might reach the slag tap area and can maintain the smooth passage of molten slag or ash with the fast flowing hot syngas through the slag tap. If the L/D ratio is higher than 2, careful arrangement to maintain the slag tap temperature should be employed like a slag tap burner.

Item	Top-feeding	Side-feeding
Advantages	Simple design (usually one feed nozzle)	Separate gas and solid flow direction
Disadvantages	Entrainment of fines	Complex design (2-12 feed nozzles)

Item	Top-feeding	Side-feeding
Main problem area	Nozzle erosion (Short life span)	Slag-tap plugging, Syngas exit line plugging
Design aspect	Simple	Complex

Table 5. Comparison of top-feeding and side-feeding methods

2.5. Refractory vs. membrane wall

Entrained-bed gasifiers run at 1,300-1,600°C, which requires a certain way of protecting the metal wall in the gasifier vessel. There are two ways to protect the vessel metal wall: by refractory or by membrane wall. Sometimes water jacket is used, but still requires the refractory protection.

Simply put, refractory system is cheap but bulky and heavy while the membrane wall is expensive and requires a good manufacturing skill. For the small pilot coal gasifier, using refractory of high chromium content (20-60%) is the cheapest way. Large gasifiers are using the brick refractory, but the pilot scale gasifier employs the mixture of refractory powder and water to fill the mold of the gasifier.

Refractory system is heavy and requires a long time (more than one day) of pre-heating before the gasification run. Membrane wall system is like an engine that is quick to ignite and run.

2.6. Slagging vs. non-slagging

Inorganics in coal should be treated to become a harmless material. Slagging gasifier converts inorganic parts to slag that is made by treating ash at the temperature above the ash fusion temperature. Non-slagging gasifier transforms the inorganics to ash form that is sometimes causing heavy metal leaching problem.

Ash that is made in the typical coal combustors like in coal fired boilers might leach heavy metals when stored outside. But, the intertwined structure in slag that is made during the melting in the gasifier prevents the heavy metals to come out at the normal environmental conditions unless the slag is meted again at high temperature above the melting temperature. In theory, slag should be the target to obtain, rather than ash that might cause a secondary environmental problem by heavy metal leaching.

But the problem is that utilization of slag is quite limited in current market although it is environmentally more benign, while fly-ash has many customers who want to buy. Slag can be used as a construction material or supplement for construction bricks, but the utilization record is not so bright. Fly-ash from the combustion processes has a well proven record in use during the last 5-8 years as cement fillers. When the fly-ash contains less than 5% carbon (preferably less than 3%), the ash is widely used as a supplement of cement filler.

Conventional non-slagging gasifiers adopt fluidized-bed type of reactor. Recent reports indicate that entrained-bed type of non-slagging gasifier might provide the advantages of fast reaction and the utilization of inorganics as a fly-ash form, or use the collected fly-ash as a low-grade fuel.

Item	Slagging	Non-slagging
Gasifier temperature	1,400-1,600°C	Less than 1,450°C (entrained-bed) 850-950°C (fluidized-bed)
Final type of inorganics(ash)	Slag	Ash
Utilization of slag/ash	Still not well accepted in industry	Well proven as cement filler

Table 6. Comparison of slagging and non-slagging types

Figure 4. Slag(left) from slagging gasifier and fly-ash(right) from non-slagging gasifier

2.7. Gasifier pressure

In the case of IGCC, gasifier pressure is typically determined by the gas turbine compressing pressure requirement. Operating pressure of commercial coal gasifiers are in the range of 22-28 bar in the IGCC plant using 7FA gas turbine. The 1.5th generation IGCC where using 7FB gas turbine requires a gasifier pressure at 41 bar to fulfill the inlet gas pressure for the 7FB machine. Higher gasifier pressure can push the gas turbine blades more strongly and thus can produce more power.

When the final product is chemical intermediates that should be used in the ensuing high pressure conversion process, high pressure operation is all the times more economical than the atmospheric or low pressure operation and the following syngas compression. Gas compression is one of the expensive processes and requires a heavy maintenance.

If the pressure of the chemical conversion process that is using the syngas from the coal gasifier requires higher than 50 bar, practically slurry feeding system is preferred over the dry-feeding. Dry feeding of coal powder above 50 bar is not practical by the currently available technologies till now.

Some people argue that the gasification pressure gives a profound variation in syngas composition. Gasification reaction itself would be dependent upon the pressure by thermodynamic principles. But in reality commercial gasifiers convert all carbon and hydrogen in coal to CO and H_2 at the optimal operating condition, and more H_2 is produced when steam is more added or slurry feeding is employed. If one pass of coal through the gasifier cannot reach >99% carbon conversion, the char or fines will be recycled to achieve the necessary conversion. Therefore when the gasifier is operating at the optimal condition which means that proper amount of oxygen and steam are supplied for more than 99% carbon conversion at all times, the gasifier pressure would not significantly influence the final syngas composition that will be used as a raw gas for power generation or manufacturing chemicals.

2.8. Oxidizing agent

In gasification, using oxygen is like driving a luxurious sports car whereas using air is like driving a small compact car. Pure oxygen pushes the gasification reaction with real fast response, while using air for the gasification responses rather slowly. Applying oxygen requires a heavy initial investment (notably ASU(air separation unit)) to gain fast response in controlling the gasifier temperature and not to worry about retaining high temperature to melt the ash components in coal. Using air will significantly simplify the gasification system and reduce the capital cost, but keeping the gasifier temperature above the ash fusion temperature is challenging. Especially small scale gasifiers could not maintain the gasifier temperature due to its inherent higher heat loss through the gasifier wall compared to large scale gasifiers.

If we consider the future gasifier plant that is to connect to CO_2 capture equipment, oxygen is the general trend. When air is used as an oxidizing agent, nitrogen is diluting the flue gas stream and will cost more in the downstream of CO_2 capture and separation.

Oxidizing agent	Oxygen	Air
Capital cost	High (ASU: about 15% of IGCC plant cost)	Moderate
Typical O_2%	95	21-24
CO_2 capture aspect	Competitive	Unfavorable
Heating Value of syngas	-	1/3 of O_2 case
Commercial gasifiers	All other coal gasifiers	Mitsubishi Heavy Industries, Japan

Table 7. Comparison of using oxygen and air for coal gasification

2.9. Power generation vs. chemical feedstock generation

The choice of coal gasifier could be different whether the final product is for electricity generation or for chemical product. Chemical product inherently requires more hydrogen in the molecular structure to be a higher value fuel like CH_4. Stable chemicals need to stabilize the structure as the $-CH_2-$ form which requires also more hydrogen.

Purpose	Power generation	Chemical feedstock
Target	Maximize total CO/H_2 amount Minimize heat loss Maximize efficiency	Maximize total H_2/CO ratio (Maximize H_2 content) Allow some heat loss Maximize high profit end-product
Gasifier material	High grade (expensive)	Not necessarily high grade
Gasifier size	Big (2,000-3,000 ton/day)	Moderate-Big (few hundreds - 3,000 ton/day)
Spare gasifier	Generally not in use	Usually use
Syngas cooling	Radiant syngas cooler	Quick quenching - moderate heat recovery
Typical gasifier type	Entrained-bed	Entrained, Fluidized, Fixed
Pressure range	22-28 bar (1st generation IGCC) 42 bar (1.5th generation IGCC)	Depend on the syngas conversion process pressure

Table 8. Choice of gasifier by the final product

Key question is whether one single gasifier can be utilized both as a power generating and also as a chemical feedstock producing gasifier. The answer is simply NO. Because plants that employ coal gasifier need 30-100 million US$ for the construction in general, the gasification plant should be designed and operated to optimize for the specific products unless the plant is designed as such from the very beginning.

2.10. Manufacturing limits

Manufacturing limit in the coal gasifier should be evaluated in terms of pressure, gasifier diameter, and manufacturing equipments. Coal gasifier is basically a pressure vessel which has a practical manufacturing limit simply by available steel rolling machine and by economics of manufacturing cost. Manufacturing a pressure vessel above 100 bar would not be practical purely due to the manufacturing ability of 3,000 ton/day scale gasifier as a single vessel, and it is never be economical since the wall thickness of large coal gasifier might be too large.

Pilot scale coal gasifiers are treating the coal in 1-30 ton/day range, in that no practical problem exists in manufacturing unless the size is too compact so that space for nozzles and cooling pipes is simply not available.

3. Coal selection guidelines for gasification [4]

The main content of this section had been published in the earlier paper in 2007[4]. Key parts are illustrated here. Table 9 illustrates what would be the most suitable coal for pilot-scale and commercial gasifiers. Pilot gasifier has a much smaller diameter in slag tap and gasifier exit line than the commercial size gasifier. If the ash content in feed coal exceeds 10%, simply small slag tap cannot pass through the molten slag even the slag viscosity is as low as liquid. Because slag flow viscosity in many cases stays at the few hundreds of centipoise range even above 1,400°C, smooth discharge of slag cannot happen, which results in plugging the slag discharge port.

Item	Pilot-scale gasifier	Commercial size gasifier
Coal rank	subbituminous	subbituminous, bituminous
Ash content	less than 5%, max. ~10%	8-12%, max. 25%
Volatile content	>30% (preferable)	No limit
Coal reactivity	high (preferable)	moderate-high
Ash viscosity	less than 250 poise at operating temperature	less than 250 poise at operating temperature

Table 9. Suitable coal for pilot and commercial scale gasifiers

The important indices for selecting the coal are ash melting temperature, slag viscosity, ash content, and the fuel ratio (or gasification reactivity). The suitable coal should contain the following properties. First, the approximate criteria for the ash melting temperature would be at the range of 1300-1400°C. If the ash melting temperature is below 1,260°C in particular, more precaution should be exercised to prevent the increased possibility of plugging by fly-slag. When the ash melting temperature is above 1,500°C, adding the fluxing agent would be required, or the gasifier temperature should be increased with the anticipated problems in the refractory life. Second, low-enough slag viscosity at the gasifier operating temperature must be guaranteed where slag would flow freely along the gasifier inner wall. Third, ash and sulfur contents should be at the lowest level if possible, and a certain amount of ash needs to be present in coal to protect the gasifier wall by thin-layer coating.

Coal reactivity is definitely an important parameter in coal selection for the gasification, probably next to the proper ash melting behavior. For the fixed gasifier volume, more reactive coal would complete the reaction within the available residence time. Before performing the actual gasification tests, coal reactivity should be studied by several ways. The most simple and intuitive way is to compare the fuel ratio of the proximate analysis data. Fuel ratio is defined as the weight ratio of fixed carbon to volatile matter contents in coal. A lower fuel ratio means more reactivity in general, such that lower rank coals are more reactive. The most simple and intuitive selection guideline that has been reported seems to be the plot between the fuel ratio that represents the coal reactivity versus the ash fusion temperature representing the slag viscosity. It can give the idea regarding the possibility in gasifier plugging [12,13].

Coals with the low fuel ratio would be a better choice if the gasifier would run without the char-recycling process. That means higher volatile content coals that normally exhibit a higher reactivity. To verify the suitable coal reactivity, TGA analysis under the inert gas environment would be sufficient to differentiate the relative reactivity of candidate coals in selecting the suitable coal. Figures 5-6 illustrate examples of applying TGA data to estimate the indirect reactivity by comparing with some reference coal that showed a good performance in gasification.

It has been reported that coal reactivity measured by TGA under an inert gas correlates with the inverse of the fuel ratio [7]. Although most accurate analysis data would be obtained under the identical gasification conditions, reactivity itself could be obtained from an analysis under inert environment. Here, reactivity was simply defined as the ratio of weight change over the specified reaction time.

In the dry-feeding gasifier, the surface moisture content of dried coal is more important than the total moisture data because of the pneumatic feeding requirement of the coal powder into the gasifier. Since the moisture content does not present any technical problems after coal is dried to less than 3 wt%, moisture content would not be a discerning factor in feeding ability. But the drying cost could reach too high to impact the total plant operating cost.

Slags obtained from the gasification at slagging temperature conditions leach heavy metal compounds far less than the environmental regulations, with no noticeable differences among the slag samples from different coal samples, and thus leaching test for slag would not be a precise criterion in determining the coal suitability for gasification.

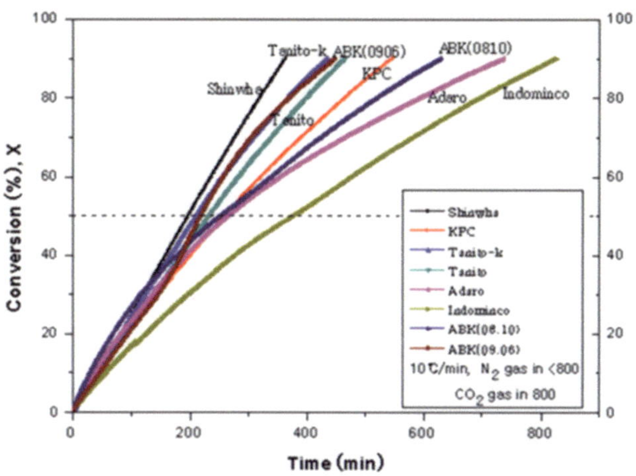

Figure 5. Rough comparison of reactivity for tested coals (TGA at Heating rate 10K/min till 800°C, 800°C isothermal, N_2 gas flow)

From the reactivity (indirect) point of view in Figure 6, Curragh and Denisovsky coals need a different gasifier design to account for longer reaction time.

Moisture content affects the operability of dry-feeding gasification system as well as the gasification efficiencies. Although moisture content of less than 2 wt% was used as a guideline in a dry-feeding commercial-scale coal gasifier [6], the moisture content of below 3 wt% demonstrated acceptable pneumatically conveying characteristics. In selecting the suitable coal for dry-feeding type gasifier, moisture content does not present any technical problems. It should rather be decided by economic consideration for drying and coal price.

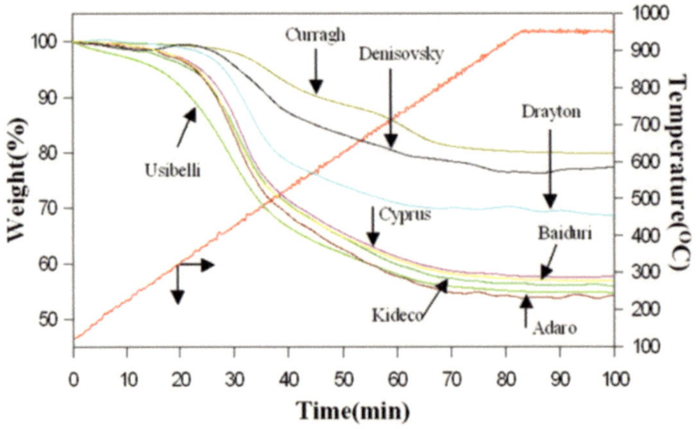

Figure 6. Indirect estimation of coal reactivity by TGA at 25 psig [4]

In gasifiers that require long-term continuous operation, low ash containing coals might be a better candidate since they produce a minimal fly-slag and bottom-slag that can act as a possible plugging material in exit-gas pipes or in the slag-tap. Judging from the operation results, the low ash containing coals showed significantly lower plugging problems by fly-slag in heat exchanging equipment like gas cooler after the gasifier.

On the other hand, because a certain level of ash in coal demonstrates a protecting function of the refractory as well as a function of heat loss minimization by coating the inner gasifier wall [8,9], an optimal ash content of the candidate coal should be judged on the basis of several interrelated parameters of coal price and ash-melting temperature. Since one of the many reasons for shutdowns in the demonstration IGCC plants of U.S.A., Europe, and Japan was slag and ash accumulation that can eventually develop to plugging and accompanying erosion, minimizing the fly-slag amount transported to the gasifier outlet is an area that should be scrutinized from the viewpoint of selecting the suitable coal. Coals of high ash content would definitely enhance the possibility of slag and ash accumulation.

Thereby, a preferable IGCC coal would possess only a reasonable amount of ash enough to coat the gasifier inner wall. The suitable ash content appears to be 1-6 wt% when there is a choice to select coal for the gasification system. For reference, a similar type of large-scale dry-feeding gasification indicated that coals containing less than 8 wt% ash content were recommended to recycle fly ash to coat the gasifier inner wall for insulating purpose, and the operating costs would increase from some 15% ash in coal[9]. Another reference reported that at least 0.5% ash is required to protect the gasifier inner wall when the wall is made of cooling tubes [10]. In addition, if coal is being imported or moved a long distance from the mine, higher ash content would only increase the cost for transportation and enhance the possibility of operational problems in gasifiers.

When the candidate coal meets the condition of ash melting temperature, another condition such as slag viscosity has to be considered. Suggested minimum gasifier operating temperature applicable in the dry-feeding gasifier was reported to be 50°C above the crystalline temperature of molten slag or 50°C above the temperature that corresponds to the 1,000 poise of slag viscosity for glassy slags [11]. Crystalline temperature is defined as the point where slag viscosity commences to increase sharply with decreasing temperature. Typically for the best performance, the gasifier is operated while maintaining the slag viscosity at the below 250 poise level. However, for practical applications, it would be better to maintain the gasifier temperature at about 100°C above the measured ash fluid temperature. All in all, slag viscosities of coals showing the glassy slag behavior were higher than those of molten slags above the crystalline temperature, signifying that more operational plugging problems by slag might occur for the coals of glassy slag.

Gasification temperature has a range for the proper conversion efficiencies. Typically, it is between 1,300-1,600°C. Oil gasification temperature is in the range of 1,300°C while the solid gasification operates at the higher temperature range. If the operating temperature is too low, carbon conversion gets lower mainly by insufficient reaction.

Coal selection can be summarized as follows. Coal properties of ash melting temperature, slag viscosity, ash content, and fuel ratio can be used as guides for estimating the plugging probability and gasification reactivity. First of all, the ash melting temperature and corresponding slag viscosity were used as a guide data for suitable coals. Next, low-rank coals of high reactivity were selected as the best candidate coals for dry-feeding entrained-bed coal gasification operation. Then, low ash coal would be chosen for the possibility of reduced operational problems related to slag and ash. Although the drying process would increase the cost for the subbituminous coals, more reactive coals with appropriate ash melting temperature should be the choice for dry-feeding entrained-bed gasification.

4. Application of CFD for gasifier design [5]

Although there have been several successful coal gasifiers that were commercially proven, many different design configurations are still possible for simple and reliable gasifier operation. As can be expected, tests of coal gasifiers at the actual high pressure and temperature conditions cost a lot of time and fund. Powerful simulation tools have made a major progress in computer simulation for the detailed analysis in reactors. It became a normal procedure to check the details in reactor design by CFD (Computational Fluid Dynamics). There are many limitations in applying CFD method in gasifier design, particularly in estimating slag behavior and slag-tap design. However, the CFD analysis proved to be useful in comparing the widely different design concepts as a pre-selection tool.

First, cold-flow simulation is applied to pre-select the configuration concepts, and the hot-flow simulation including chemical reactions follows to compare the concepts at more similar actual gasifier operation situation.

In designing a gasifier, many design parameters should be compared to obtain the optimal performance. Among design parameters for the entrained-bed gasifier, syngas flow direction, expected temperatures exiting the gasifier, size of any dead volume, L/D ratio, residence time inside the gasifier, and number and location of burner nozzles are most important.

From the relative evaluation of this preliminary analysis, most promising type and shape of the gasifier can be selected, after which more detailed CFD analysis including chemical reactions follows in order to obtain profiles of temperature, gas compositions, and particle flow path, etc.

As an example of CFD illustration, four cases of gasifier configuration of dry-feeding were first selected with two up-flow designs and two down-flow designs, as illustrated in Figure 7. In all cases, the feeding nozzles were positions to form a cyclonic swirl inside the gasifier with the purpose of increasing residence time. Case 1 is a reference design that is similar to the 3 ton/day coal gasification pilot plant at IAE in Korea. Thus, actual coal gasification database with more than ten different coals is available to verify the results in Case 1.

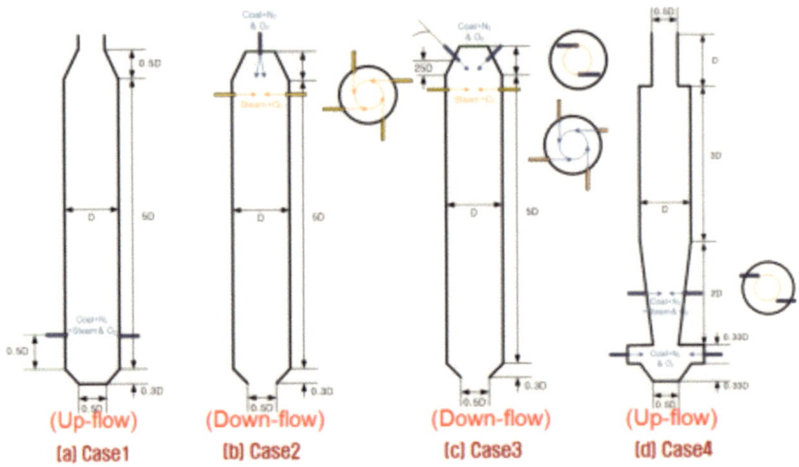

Figure 7. Four coal gasifier configurations compared in the CFD analysis [5]

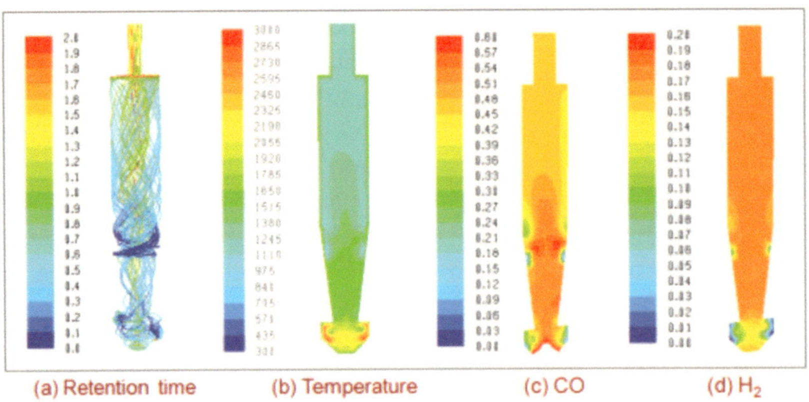

Figure 8. Hot-flow simulation result for up-flow Case 4 [5]

Table 10 summarized the hot-flow analysis results. Gas-phase residence time in Case 4 shows the highest value as 1.43 sec, while the down-flow Case 2 exhibited lowest as 1.03 sec. Residence time in reference Case 1 was 1.17 sec.

The pilot-plant gasification data in Case 1 configuration showed above 98% carbon conversion for the highly reactive Indonesian subbituminous coals [3]. For some un-reactive bituminous coals at the pilot gasifier of Case 1 configuration, residence time was not sufficient to guarantee the full carbon conversion in one pass through the gasifier. Recycling of un-reacted char particles to the gasifier, which means several passes through the gasifier, is one option to cope with this kind of low conversion efficiency in one pass, although more capital investment is required for additional equipments. In short, CFD analysis will be supplemented with actual pilot test results for the final design of the coal gasifier.

Case		1	2	3	4
Gas residence time (sec)		1.17	1.03	1.26	1.43
Gasifier exit gas temperature (°C)		1,202	1,081	1,065	1,021
Gasifier exit gas Comp. (vol %)	CO	54.13	52.81	52.70	51.46
Gasifier exit gas Comp. (vol %)	H_2	16.37	17.09	17.25	18.12

Table 10. Hot-flow gasifier CFD simulation result [5]

5. In-situ estimation of gasification status inside gasifier

Operating pilot coal gasifier produces profiles as in Figure 9. Gasifier temperature, pressure, and syngas composition are most basic data that are measured. In the pilot gasifier, inside temperature is measured directly by thermocouples in order to know the actual gasification condition. Syngas composition is readily measured by on-line GC or dedicated on-line gas analyzers.

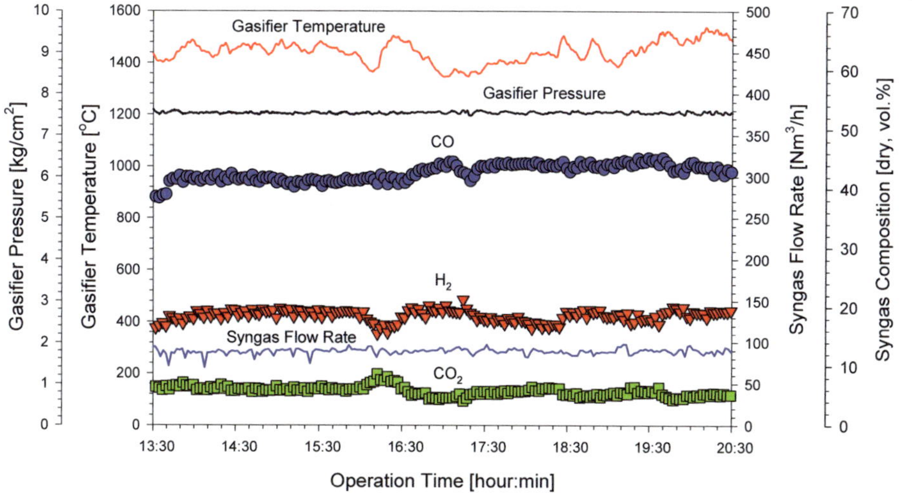

Figure 9. Typical gasification profiles at pilot scale dry-feeding coal gasifier (8 bar, Indonesian KPC coal)

If the gasification temperature is higher than 1,400°C where the chemical reaction is so fast that mass transfer limitation prevails, syngas composition can be reliably approximated by the thermodynamic equilibrium calculation which is readily available in most commercial process simulation softwares like ASPEN.

Examples of estimating the syngas composition by thermodynamic equilibrium calculation are shown in Figures 10-11. Both figures illustrate estimated syngas composition is satisfactory in engineering sense. In pilot plant, a notebook computer is used to calculate the expected syngas composition at the certain carbon conversion and reaction temperature while the gasifier is operated. In opposite way, from the known information on syngas composition, temperature, and coal property during the gasifier test, carbon conversion at that time can be calculated to verify how the gasifier is being operated.

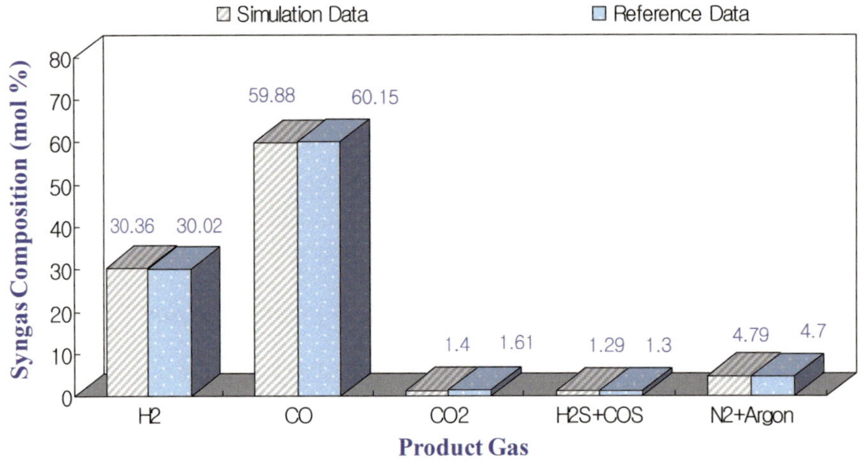

Figure 10. Comparison of syngas composition between simulated and actual commercial-scale plant data for Illinois No. 6 coal

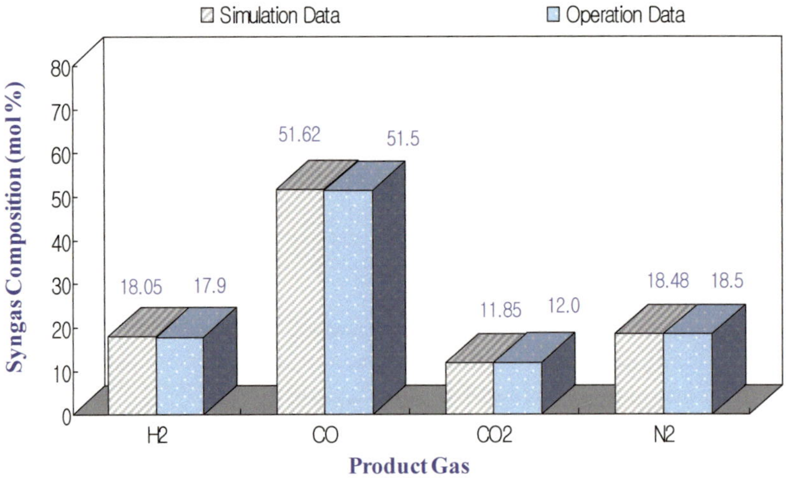

Figure 11. Comparison of syngas composition between simulated and actual pilot plant data for Indonesian subbituminous coal

Because the coal gasifier is normally under the pressure, direct looking into the gasifier is impossible. While we operate the gasifier, there are important variables to know in-situ, if possible, such as reaction temperature (typically 1,400-1,600°C), pressure, gas composition, and slag flow.

Gasifier temperature measurement by R-type thermocouple is a normal method in pilot plants, but in commercial gasifiers where at least several months of continuous operation is required thermocouple proved to be unreliable due to frequent wire disconnection under hot corrosive environment. Most commercial plants acquire temperature information indirectly by measuring such as steam production amount from the gasifier wall or methane content. Methane content in syngas has exhibited a reliable indirect information on temperature high or low limit, which is a very important data to prevent significant gasifier damage. If the gasifier temperature is too high, gasifier wall might be damaged, and if the temperature is too low, then the slag tap would face a plugging by re-solidified slags.

Figure 12 show the increase of CH_4 % from about 0 to 6,000 ppm by the drop of 100°C in gasifier temperature from 1,450°C to 1,350°C. Typical slagging coal gasifiers operate at temperatures where CH_4 content is maintained below the certain guideline value.

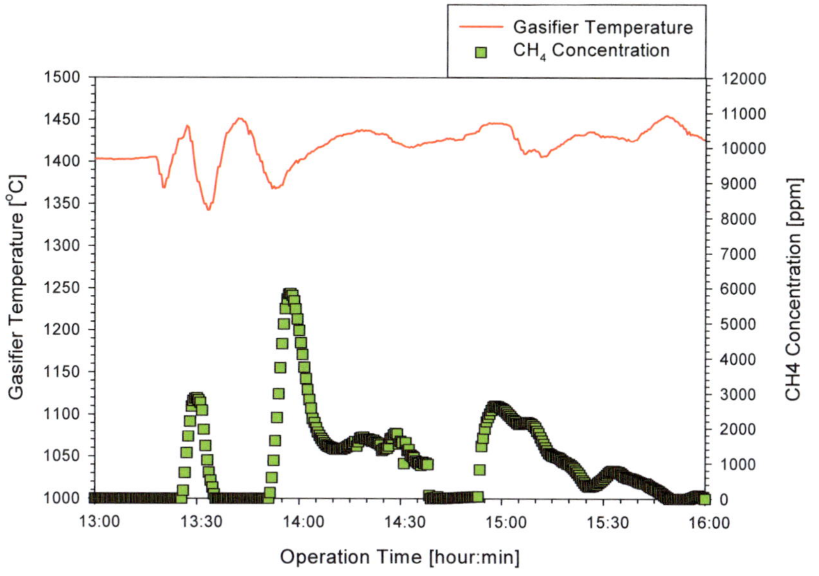

Figure 12. Relationship between gasifier temperature and CH_4 content (10 bar, Indonesian KPC coal)

6. Key areas of operation problems

There are key problematic areas that should pay attention in design and during operation. Main gasifier body would not explode unless a really bad manufacturer was chosen. There are weak points in gasifiers, which are slag tap, syngas exit line, and feed nozzles. Pilot plant requires frequent disassembling and reassembling to see the inside part and take samples for analysis after the test, which would increase the risk by many joint areas.

Gasifier problems basically reside in uncontrolled fluctuation of coal/oxygen, slag behavior, syngas leakage, and nozzle area. Smooth feeding is an essential part in all chemical reactions. In coal gasification, it is more important. A small sudden increase of oxygen while the coal feed is same can increase the gasifier temperature above 1,600°C in 10-30 seconds. Slag and molten fly-slag plug the slag tap and exit pipes or syngas cooling zone, if not properly monitored and operated. Many joint areas that are frequently reassembled inherently possess the possibility of loosening and eventually leakage with time. In the pressurized coal gasifier containing hot syngas whose components CO and H_2 are all easy to ignite with atmospheric oxygen, loosening joints definitely lead to syngas leakage, and surely a noisy explosion of that area.

6.1. Slag tap

The biggest operational problem identified during the pilot-scale gasification tests were the plugging in the slag discharge port by the bottom slag and the plugging in the syngas outlet area of the gasifier by the fly-slag, with the possibility of backfire explosion in the area of feed-lance nozzles. From the aspect of plugging by slag, slag viscosity with the gasifier temperature is an important index as described in the previous section for selecting the suitable coal. From the viscosity point of view, all subbituminous and most bituminous coals have shown the low enough slag viscosity among the tested coals, and thus it seems that they would not cause any operational problems by slag flow at the proper operation temperature, whereas a Russian coal yielded the highest slag viscosity that had caused an operational problem in slag discharge even under the gasifier temperature above 1,500°C. Higher ash content in coal increased the possibility of slag-related operational problems.

6.2. Syngas exit line

The most troublesome coal with plugging by fly-slag at the syngas outlet was Alaskan Usibelli coal from USA that showed an ash fluid temperature of 1,257°C. Figure 13 shows Alsakan Usibelli coal case of exit line plugging by fly-slag. Contrary to the case of Russian coal where slag viscosity values were more representing the actual behavior of slag in the gasifier, Usibelli coal demonstrated that ash fluid temperature for the raw coal was more representing the actual behavior of slag viscosity in the gasifier than the viscosity measurement for the gasified slag. Viscosity in the fly-slag of Usibelli coal exhibited at least

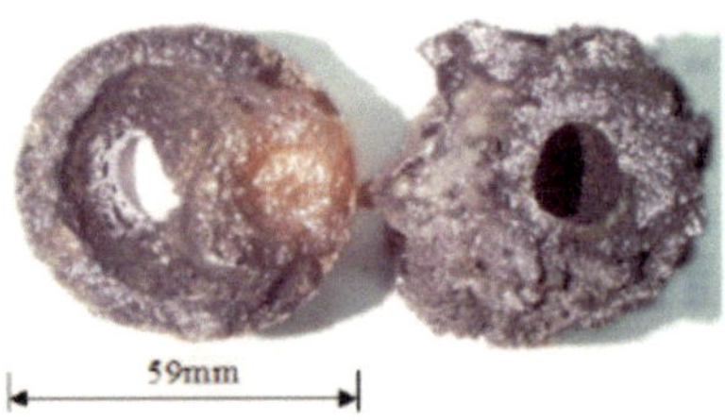

Figure 13. Deposited ash/slag at the exit port of pilot-scale coal gasifier (Alaskan Usibelli coal, 8 bar, 1,450°C)

a similar melting behavior that could be represented by the ash fluid temperature. The result till now signifies the importance of actual testing under the gasification conditions to confirm the gasification characteristics including the slag behavior.

Caution should be exercised when the candidate coal shows very low ash fusion temperature below 1,260°C with high ash content because the heat recovery system attached to the gasifier might show a higher plugging tendency.

6.3. Feed nozzle area

In the feed nozzle area, coal powder or coal slurry, oxygen, steam, hot syngas all meet at the small space. Moreover many joints exist, and mechanically nozzle itself contains many layers of metal tubes that expose to hot corrosive syngas. Welding points must meet the stringent specification to guarantee the long operation, and thus most gasifier vendors still supply the feed nozzles under their quality control.

If the welding joint in the feed nozzle break, syngas can pass though the hole and make the metal weak to break in sequence, which eventually ends up in explosion of feed nozzle area. More detailed discussion follows in the next section.

7. Safety consideration in coal gasification pilot plants

Institute for Advance Engineering in Korea has operated the pilot coal gasifiers from 1994, and has experienced several safety issues. During the design of the coal gasifier and the preparation of the constructed gasifier operation, items that need most careful concentration are,

- Maintain the enough higher pressure difference all the time at the coal feeding equipment over the gasifier
- Make sure that connected lines would not leak
- Welded area that would be exposed to hot syngas should be minimized
- Weakest and most dangerous area is the coal/oxygen feeding nozzle lines
- Toxicity of CO
- Any slightest possibility of contacting CO and Ni-based catalysts to produce nickel tetracarbonyl ($Ni(CO)_4$) which is one of the most fatal compound, more hazardous than CO

Coal gasifier deals with the syngas that consists of mainly CO and hydrogen at the high pressure and high temperature. Gasification also involves the pure oxygen with the coal powder or coal slurry. Under the normal operating situation in that reactive coal and oxygen are moving to the lower pressure region, coal and oxygen are reacting on the way through the gasifier and syngas are formed. Pressure at the coal feeding vessel remains at the higher pressure than the gasifier, so that hot syngas is not damaging the feeding lines. At any time, this pressure difference must be guaranteed, otherwise hot (1,300-1,600°C) syngas will flow backward through the coal powder and oxygen lines that will surely make an explosion.

Figure 14 shows the syngas flame along with the ignited coal particles that are flying around the flame at the leaked feed nozzle area. The accident occurred by the loosened ferrule at the coal feeding nozzle of the dry-feeding pilot coal gasifier that operated at 8 bar and around 1,450°C conditions. This flame looks similar to the flame of welding torch.

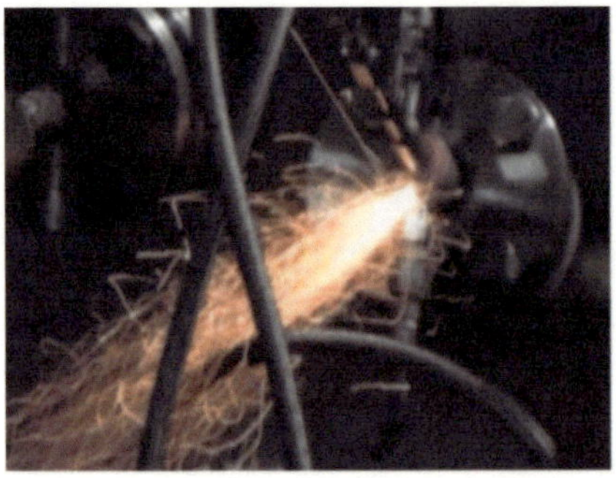

Figure 14. Picture showing the syngas flame caused by syngas leakage at the feed nozzle area

Figure 15. Damaged valve main body by the syngas explosion occurred during the 10 bar and around 1,500°C gasification pilot plant test

The force by the syngas explosion that occurs typically by the backward pressure to the feeding line amounts to tear out instantaneously the SUS metal of the value that should withstand 1,500 psi. Figure 15 demonstrates the damage to the valve main body by the syngas explosion occurred at the 10 bar and around 1,500°C conditions. The explosion should be avoided, but if it happens the damage area should be minimized. Best routine is to prevent any personnel who goes near the nozzle area during the hot gasification test. The explosion happens with a very short loud blast and will hiss out the syngas until the majority of syngas is vented out. Normal emergency routine involves the pushing the syngas out of the gasifier with nitrogen which is all the time maintained at the higher pressure than the gasifier and the oxygen line.

Figure 16 also exhibits the force of the syngas explosion. In the Figure, right-hand side is the gasifier (not shown in the figure) and the coal feeding vessel (not shown in the figure) is located at the left side of the Figure. There was a leak in the connecting tubes on the left side of the Figure. Then pressure of the feeding line suddenly drops to atmoshperic pressure and the hot syngas gushed to the feeding lines. Hot syngas reacts with coal powder and pure oxygen existing in the feeding line, resulting in the very explosive gas and push directly from the gasifier through the feeding line. Damaged shape in the Figure clearly illustrates the direction of the syngas explosion which is not following the curved SUS pipe, rather moves in direct line and tear the pipe in that direction.

Figure 16. Damaged SUS coal powder feeding pipe occurred during the 8 bar and around 1,500°C gasification pilot plant test

Figure 17 shows the importance of the welding quality in the feeding nozzle area. The accident occurred during the pilot coal gasifier operation with a subbituminous coal at 20 bar, 1,400°C. After the accident the nozzle parts were scrutinized and revealed that the vertical welding on the water cooling zone was an initial starting point and the hot syngas moved through the cooling water zone, after which the nozzle itself was damaged and finally the syngas with pure oxygen resulted in explosion. In the commercial system, water cooling system is operated with higher pressure than the gasifier pressure, but in the pilot system that might not use the high pressure water facility, the nozzle area should be monitored carefully and should make a way to prevent the possibility of syngas leakage through the cooling zone.

Carbon monoxide in syngas is typically 20-60% in the pilot coal gasifiers. Considering the allowable limit of CO concentration is 50 ppm and exposure to 0.1% CO can lead to fatality, the concentration of 20-60% which amounts to 20,000-60,000 ppm can lead to extreme safety hazards. Just one inhaling of syngas is enough to make a person to serious dizziness and vomiting.

Figure 17. Explosion accident at the coal feeding nozzle during the pilot gasifier operation at 20 bar, 1,400°C (Left: picture at normal operation, Right: picture at explosion time)

Syngas is widely in demand for manufacturing chemicals or synthetic fuels, which normally involves catalytic reactions. Extreme caution should be exercised when any nickel containing catalysts are employed with syngas. Although the chance is slim and little amount is used just as a test, any possibility inducing the formation of Nickel tetracarbonyl ($Ni(CO)_4$) should be checked and even the slightest inhaling by personnel should be avoided. Nickel tetracarbonyl is one of the most fatal compound, more hazardous than CO.

8. Future direction of coal gasifiers

If the commercially available coal gasifiers have reached already the best efficiency and satisfied all the industrial requirements, there would be no need to design and construct the pilot-scale gasifiers. Current coal gasifiers are still too expensive and too small in terms of coal-fired power plant. Coal price generally linkages with the oil price. Since the high oil price prompts to use more coal and pushes the coal price accordingly, low grade coal would be utilized more widely in the near future. Also there is a CO_2 issue that will impact the gasifier technology more suited in the CO_2 capture.

The future direction of R&D for coal gasifiers can be summarized as follows:

- Bigger capacity in a single gasifier
- Simplification of gasifier design
- Compactness
- Use of cheap low-grade coal
- Reduction of construction cost
- Increase in plant availability
- Response to CO_2 issue

9. Conclusions

Purpose of testing with the pilot-scale coal gasifier is to confirm the design concept before going to the commercial scale. In a sense, pilot gasifier is more dangerous than the big scale gasifier because the pilot gasifier requires frequent disassembling and contains more joint parts with smaller slag passage hole, which will increase the possibility in syngas backflow with eventual explosion. With knowing what is going on in the gasifier with the specific choice of design options, the best selection and design for the gasifier would possible.

Even with the long history of developing and commercial use of coal gasifiers, there is still a room in upgrading to a more efficient and cheaper version of coal gasifier and the pilot scale gasifier should follow to confirm the design logic and practical applicability. On the way to make a next generation coal gasifier, fundamental issues and experience from the past should be used as a cornerstone. Although it is not a vast experience compared to the almost century-old gasification system as in the fixed-bed type, the pilot-scale experience at IAE for the entrained-bed type gasifiers during the last 18 years or so might be useful for providing as guidelines which can act at least as a blocking block in preventing the worst case and act as a new starting point.

Author details

Yongseung Yun*, Seung Jong Lee and Seok Woo Chung
Institute for Advanced Engineering, Suwon, Republic of Korea

Acknowledgement

This work was supported by the Development of 300 MW class Korean IGCC demonstration plant technology of the Korea Institute of Energy Technology Evaluation and Planning(KETEP) grant funded by the Korea government Ministry of Knowledge Economy (*No. 2011951010001B*).

10. References

[1] Yun Y, Yoo Y.D. (2005) Comaprison of Syngas and Slag from Three Different Scale Gasifiers using Australian Drayton Coal. J. Ind. Eng. Chem. 11: 228-234.
[2] Yun Y, Lee S.J, Hong J. (2011) Operation Characteristics of 1 Ton/day-Scale Coal Gasifier with Additional Stage. Korean J. Chem. Eng. 28: 1188-1195.
[3] Yun Y, Chung S.W. (2007) Gasification of an Indonesian Subbituminous Coal in a Pilot-Scale Coal Gasification System. Korean J. Chem. Eng. 24: 628-632.
[4] Yun Y, Yoo Y.D, Chung S.W (2007) Selection of IGCC Candidate Coals by Pilot-scale Gasifier Operation. Fuel Processing Tech. 88:107-116.
[5] Yun Y, Ju J, Lee S.J (2011) Comparison of Design Concepts for Four Different Entrained-Bed Coal Gasifier Types with CFD Analysis. Appl. Chem. Eng. 22: 566-574 (in Korean).
[6] Yun Y, Lee G.B, Chung S.W (2003) Charactistics of Trace Gas and Solid Fines Produced from the Dry-Feeding Coal Gasifier. J. Korean Ind. Eng. Chem. 14: 511-518 (in Korean).
[7] Zaidi S (1995) Coal Reactivity: Correlation with Fuel Ratio and NMR Data. Fuel Processing Tech. 41: 253-259.
[8] Collot A. (2003) Matching Gasifiers to Coals. Proceedings 2003 Pittsburgh Coal Conference.
[9] Ploeg J. (2000) Gasification Performance of the Demkolec IGCC. Proc. of Gasification for the Future, 4th European Gasification Conference.
[10] Mehlhose F, Schingnitz M. (2003) Experience of Lignite and Hard Coal Gasification Gained in the Freiberg Test Facilities of Future Energy Gmbh. Proceedngs 2003 Pittsburgh Coal Conference.
[11] Moon, I, Cho C, Oh M. (2002) Viscosity of Coal Slags under Gasification Conditions. Energy Eng. J. 11: 149-159 (in Korean).

* Corresponding Author

[12] Moritsuka H et al. (1991) Design and Evaluation Study on a Demonstration Plant of an IGCC Power generation System. CRIEPI Report W90051.
[13] Moritsuka H. (1992) A Study of Thermal Efficiency and Load Control of an Air Blow IGCC Power Generation System. CRIEPI Report.

Chapter 2

Gasification Studies on Argentine Solid Fuels

G.G. Fouga, G. De Micco, H.E. Nassini and A.E. Bohé

Additional information is available at the end of the chapter

http://dx.doi.org/10.5772/48500

1. Introduction

As the global population growth and energy demand are steadily raising and the industry is forced to reduce the greenhouse gas emissions due to the global warming, there is an increasing pressure to improve the overall efficiency of the energy production systems. In this challenging framework, a renewed interest on coal gasification technologies has recently emerged worldwide, since they offer the potential of clean and efficient energy. One attractive characteristic of coal gasification technology is the possibility of co-production of electricity, hydrogen, liquid fuels and high-value chemicals that contributes to the improvement of power generation efficiency compared with conventional pulverised coal fired plants as well as the reduction of emissions of greenhouse gases and particulates to the atmosphere (Minchener, 2005). Gasification has also the additional advantage of accommodating a wide range of feed stocks, including low-cost fuels like petroleum coke, biomass, and municipal wastes (Higman & Van der Burgt, 2003).

As it will be explained in Section 2, Argentina is presently investigating the application of the concept of co-production for the integral exploitation of its coal reserves. Co-production of power, fuels and chemicals offers an innovative, economically advantageous mean of achieving the long-term energy goals of our country since it involves the integration in a single energy complex of three major building blocks: (1) gasification of coal to produce synthesis gas; (2) conversion of a portion of the synthesis gas to high-value products, such as high-purity hydrogen and liquid fuels; and (3) combustion of the remaining synthesis gas and unreacted gas from the conversion processes to produce electric power in a combined-cycle system. In the co-production concept, an energy complex produces not only power, but also fuels and/or chemicals. This concept greatly increases the flexibility of the complex and offers economic advantages compared with separate plants, one producing only power and the other only fuels or chemicals.

Following this objective, an extensive research and development program is being implemented in our country on solid fuel gasification technologies, beginning with both theoretical and experimental studies for understanding the mechanisms of the gasification reactions, in order to determine the optimum parameter conditions for the synthesis gas production and the further cleanup steps for the harmful contaminants removal. For providing indirect heating to the gasification reactors, replacing the partial combustion of the feed material that is needed to drive the endothermic gasification reactions, the alternative of using a nuclear high temperature gas reactor is being also evaluated (Nassini et al., 2011).

It is well-known that the chemical composition, the heating value and, then, the future use of the synthesis gas produced by solid fuel gasification is variable with the gasification technology employed, depending on a lot of factors such as solid fuel composition and rank; pre-processing and feeding procedures; gasification agents; operational conditions in the gasification reactor, i.e. temperature, pressure, heating rate, and residence time; and plant configuration characteristics like the flow geometry, ash removal method and gas cleaning system. There is a large number of gasification processes implemented at commercial level and the choice of a given gasification technology is difficult because it depends on diverse factors such as solid fuel availability, type and cost; size constraints; and production rate of energy. Even, in principle, all types of solid fuels can be gasified, the properties of the material to be processed are the least flexible factor to be considered in the analysis and, then, the gasification technology should be primarily matched to the properties of the solid fuels available for gasification (Collot, 2006).

According to that, a theoretical and experimental study is being now performed at laboratory scale, addressed to characterize the behaviour of Argentine solid fuels under typical gasification conditions and to identify the most suitable gasification process for the production of hydrogen and liquid fuels, respectively. The research program that is described below was designed to simulate in laboratory, as close as possible, the operational conditions of large-scale gasification plants and to provide the necessary information about fundamental mechanisms and kinetics of the gasification reactions for a further scaling up of experimental facilities.

2. Argentine energy situation and scientific background

The current energy matrix of Argentina is largely based on fossil fuels, i.e. petroleum oil and natural gas, but the preservation of non-renewable resources and the minimization of pollution are goals which today determine decisively further development of fossil fuel-fired power stations. In this sense, a so-called Hydrogen Law was dictated by the Argentine Congress in 2006 declaring of national interest the development of technologies needed for the progressive introduction of hydrogen as a clean energy carrier that can be used to meet the increasing residential, transportation and industrial demands. According to that, the national government is promoting all scientific activities related with the production, purification, safe storage and applications of hydrogen, as well as the development of more efficient energy production systems (Bohe & Nassini, 2011).

In order to bring together the requirements of a sustainable economic growth with the environmental protection, our country is then encouraging strategies for the rational and integral utilization of domestic coal reserves and this tendency is expected to increase with time, as natural gas and petroleum resources are becoming exhausted. The main domestic coal reserve accounting more than 7% of conventional energy resources of Argentina is a high-volatile sub bituminous coal that is extracted from the Río Turbio minefield, located in Santa Cruz province, in the south of the country (Carrizo, 2002). Another materials containing carbon and amenable to be gasified are asphaltites arising from minefields located in Mendoza (Beloff, 1972) and Neuquen (Savelev et al, 2008). Asphaltites would be an excellent raw material for the production of synthesis gas through solid fuel gasification due to their low content of ashes and high percentage of elemental carbon (Fouga et al., 2011).

It is well-understood that solid fuel gasification is a two-step process. In the first step, pyrolysis, volatile components of feed material are rapidly released at temperatures between 300 and 500 ºC, leaving residual char and mineral matter as by-products. The second step, char conversion, involves the gasification of residual char and it is much slower than devolatilization step, becoming then the rate-limiting step of the overall process. Even gasification reactions have been extensively studied during years worldwide, a better understanding of the fundamental reaction mechanisms and kinetics is still required for optimizing the design and operation of large-scale gasifiers in order to maximize the efficiency and economics of the overall gasification process.

Earlier studies demonstrated that the reactivity of chars to gasifying agents is very dependent on their formation conditions, particularly temperature, pressure, heating rate, time at peak temperature, and the gaseous environment. When volatile matter is generated, the physical structure of char changes significantly and swelling of fuel particles may occur. The complexity of char structure lies in the facts that the structure of a char itself is highly heterogeneous inside an individual particle and between different particles and the chemistry of a char is strongly dependent on the raw material properties. Then, a good understanding of the swelling of particles and the formation of the char pore structure during the devolatilization step, as well as the further evolution of the released volatile matter is essential to the development of advanced gasification technologies (Yu et al., 2007).

On the other hand, even coal is generally classified by its rank with fixed carbon content and calorific value as the major indicators, coal rank related parameters do not always provide adequate predictors for gasification reactivity since coals of similar rank may undergo quite different extents of reaction when they are gasified at a particular condition. Additionally to coal rank, reaction conditions and sample preparation procedures, several other factors are thought to influence the coal gasification reactivity such as the mineral matter content of coals which is known to influence the gasification reactivity because of the presence of reportedly catalytically active components (Domazetis et al., 2005).

3. Experimental approach

When introduced into a high-temperature atmosphere in a gasification reactor, solid fuel particles are heated at high heating rates (above 10^3 ºC/sec) and they undergo devolatilization and gasification simultaneously under more or less the same condition. In spite of this evidence, most of the char reactivity data reported in literature was obtained under gasification conditions that were different from the devolatilization conditions under which the chars were prepared. According to that, a high spread in char reactivity measurements is found, even for chars prepared from the same parent coal but under different pyrolysis conditions (Peng et al., 1995).

As earlier studies have demonstrated that the reactivity of chars to gasifying agents is very dependent on their formation conditions, to get meaningful data about kinetics of gasification reactions it is essential, at least, to produce chars in laboratory that replicate, as close as possible, the real conditions of char formation in large-scale gasifiers, i.e. high heating rates and intense gas convection around individual char particles.

The experimental approach followed to achieve both objectives in the char preparation is the so-called "two-stage" experiments in which the gasification reactivities are determined on char samples prepared in a previous pyrolysis step where parent coal particles are heated in an inert atmosphere at high heating rates and short residence times at high temperatures (Megaritis et al., 1998). A drop tube furnace was designed and built up for producing chars in laboratory at temperatures up to 1100 ºC and heating rates in the order of 10^3 ºC/sec, while the CO_2 and steam gasification reactivities of these ex-situ chars were measured in a thermo-gravimetric system adapted to work with corrosive gases and in tubular reactors coupled with gas chromatography. The experimental setups used for pyrolysis and gasification experiments are described in more detail in the following section.

4. Experimental procedures and methods

4.1. Characterization of solid fuels for gasification experiments

The first step of the experimental program consisted of a detailed physical and chemical characterization of the Río Turbio coal and several asphaltites called Emanuel, Susanita, Fortuna 4 and Toribia, and the main results are summarized in Table 1. It can be seen that Toribia and Fortuna 4 asphaltites have the highest volatile content (above 50 wt%) while Emanuel asphaltite has the highest fixed carbon content. Furthermore, the Río Turbio coal has the greatest ash content and porosity. BET areas were measured by N_2 adsorption/desorption according to Barrett–Joyner–Halenda (BJH) method (Barrett, 1951), and using Digisorb 2600 equipment (Micrometrics Ins. Corporation). The analysis of elemental composition indicates the presence of nickel and vanadium in three of the asphaltites, and the recovery of theses valuable metals could be of economical interest. Calcium and sodium, silicon and iron are present in most of the samples. The XRD measurements indicate that those elements are forming the following majority phases: quartz, calcium sulfate, hematite, and aluminum silicates.

Determination	Solid Fuels				
	Coal	Asphaltites			
	Río Turbio	Emanuel	Susanita	Fortuna 4	Toribia
Moisture	3.5 (wt %)	11.47 (wt %)	10.93 (wt %)	0.26 (wt %)	0.58 (wt %)
Volatile matter[a]	36.4 (wt %)	26.18 (wt %)	33.18 (wt %)	58.97 (wt %)	56.06 (wt %)
Fixed carbon[a]	51.2 (wt %)	68.67 (wt %)	55.50 (wt %)	40.57 (wt %)	43.25 (wt %)
Ash[a]	12.3 (wt %)	5.13 (wt %)	11.32 (wt %)	0.46 (wt %)	0.69 (wt %)
Density	1.107($g \cdot cm^3$)	0.679 ($g \cdot cm^3$)	0.642 ($g \cdot cm^3$)	0.412 ($g \cdot cm^3$)	0.427 ($g \cdot cm^3$)
C_T	59.8	64.3	63.6	78.0	75.3
N_T	2.78	3.27	3.24	2.92	3.14
S_T	0.86	2.36	0.7	4.5	4.40
Determination	**Char**				
BET area	96 ($m^2 \cdot g^{-1}$)	3.5 ($m^2 \cdot g^{-1}$)	3.17 (m^2g^{-1})	0.44 ($m^2 \cdot g^{-1}$)	na
Pore volume	0.064319	0.01495 (cm^3g^{-1})	na	na	na
Total porosity, ε_0	12 %	1.6 %	na	na	na
Ash content	20 %	6.95 %	16.94 %	1.12 %	1.57 %
Determination	**Ash**				
Elements present in Ash[b]	Na, Mg, Al, Si, K, S, Ca, Ti, Fe.	S, Ca, V, Fe, Si, Al, Ba, Ni, K, Sr, Mo, P, Cu.	Mg, Al, Si, S, Ca, V, Fe, Ni, Zn.	Na, Mg, Al, Si, K, Ca, V, Fe, Ni, Cu.	Na, Al, Si, K, Ca, V, Fe, Ni.
Main phases in Ash[c]	Fe_2O_3, SiO_2	SiO_2; Fe_2O_3; $CaSO_4$; $Ca_3V_2O_8$; $CaSiO_3$; $(Na,Ca)Al(Si,Al)_3O_8$	$CaSO_4$; SiO_2; $Ca_2Al_2SiO_7$.	SiO_2; NaV_6O_{15}	SiO_2; CaV_2O_6; $Al_6Si_2O_{13}$

[a] Moisture free
[b] Energy dispersive spectroscopy (EDS) and Energy dispersive X-Ray fluorescence spectroscopy (ED-XRF).
[c] X-Ray diffraction (XRD).
[na] not available

Table 1. Physical and chemical characterization of the Río Turbio coal and asphaltites.

4.2. Drop tube furnace for pyrolysis experiments

The drop tube furnace (DTF) that is shown in Figure 1 was used for preparing chars at high heating rates and short residence times at high temperatures from the Río Turbio coal and asphaltites. The reactor has a three-zone electric furnace able to operate up to 1100 ºC, which surrounds two concentric quartz tubes of 41 and 26 mm inner diameter, 1.30 and 1.20 m long, respectively. Primary nitrogen gas is injected at the bottom of the outer tube and is preheated while flowing upwards. When at the top of the outer tube, the gas is forced onto the inner tube through a flow rectifier and the gas flows downwards and leaves the reactor through a water-cooled collection probe. The solid fuel particles are entrained by a non-preheated secondary nitrogen gas jet to a water-cooled injection probe placed on top of the inner tube. The heating rate is estimated to be higher than 10^3 ºC/sec and the residence time of particles in the reactor less than 0.3 sec. The chars leave the reactor through the collection probe, and an extra nitrogen flow is added to the exhausted gases in order to quench the reaction and improve the collection efficiency in the cyclone.

The major operating parameters in the reactor were: (1) temperature of pyrolysis, ranging between 700 to 1100 ºC; (2) mass flow of solid fuel particles, through variations in the secondary nitrogen gas flow; and (3) particle residence time at high temperature, derived from the heated tube length which can be varied since the three axial zones have independent electric power supply.

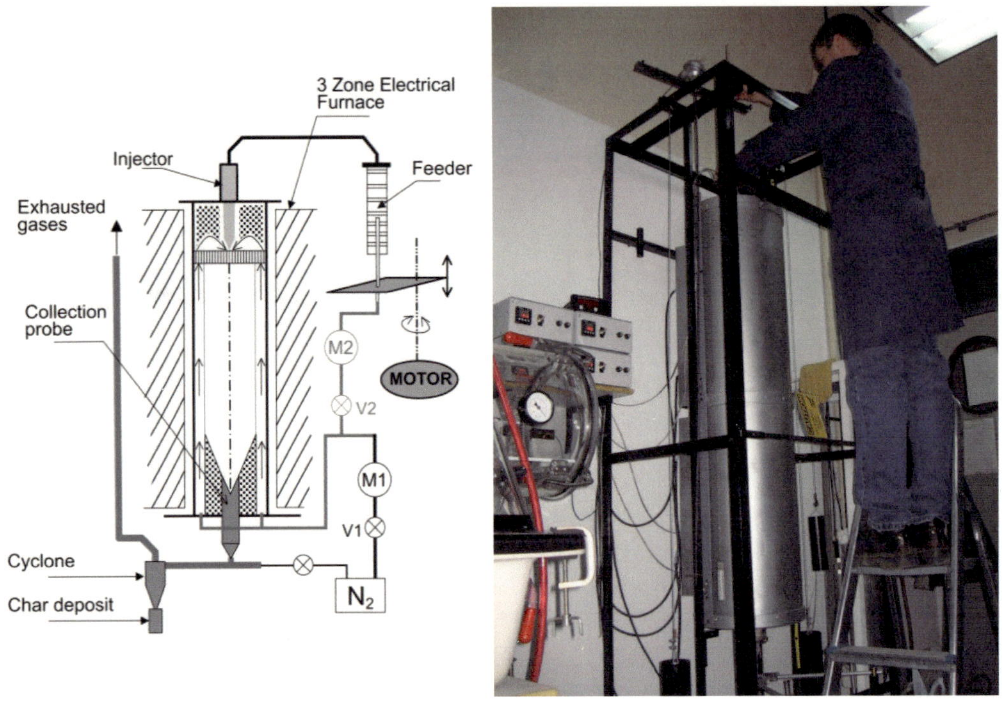

Figure 1. Drop tube furnace for pyrolysis experiments: (a) schematic view; (b) photograph taken during assembly.

4.3. Thermo-gravimetric system for gasification experiments

Gasification experiments using carbon dioxide and steam as gasifying agents were carried out in a thermo-gravimetric analyzer (TGA) that is schematically shown in Figure 2. This experimental setup consists of an electro-balance (Model 2000, Cahn Instruments, Inc.), a gas line, and a data acquisition system, having a sensitivity of ± 5 μg while operating at 950 ºC under a flow of 8 L/h. In a typical TGA run, the weight of the char sample is measured as a function of time and temperature as it is subjected to a controlled temperature program. TGA tests are usually carried out in two ways: (i) isothermal, where the sample is heated at a constant temperature, and (ii) non-isothermal with linear heating, where the sample is heated at a constant temperature rate.

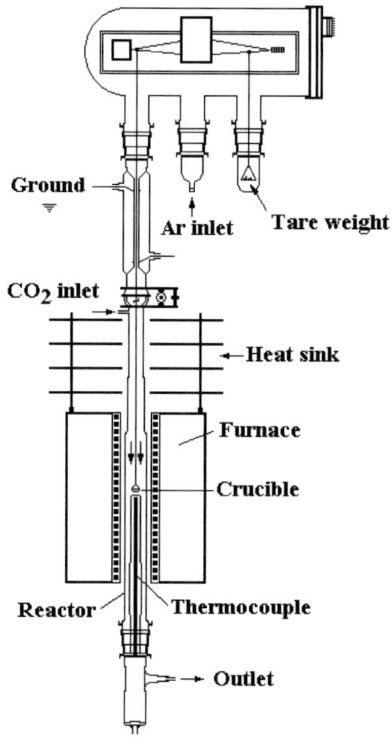

Figure 2. Thermo-gravimetric system for gasification experiments with carbon dioxide and steam.

The gasification rate under several experimental conditions of temperature, partial pressure of gasifying agent and sample mass, can be obtained from the temporal evolution of relative mass loss of char, as follows:

$$\alpha = \frac{m_0 - m}{m_0 - m_{ash}} \tag{1}$$

where m_0 is the initial mass of char, m is the mass of char at time t, and m_{ash} is the mass at the end of the gasification reaction when there is no more fixed carbon and corresponds to the ash content. According to equation (1) α takes values in the range between 0 and 1 and, hence, the gasification rate, R, can be expressed as:

$$R = \frac{d\alpha}{dt} = -\frac{1}{\left(m_0 - m_{ash}\right)} \frac{dm}{dt} \tag{2}$$

In mathematical form, R is expressed as a function of temperature (T), partial pressure of gas (p_{gas}) and reaction degree (α), as follows:

$$R = \frac{d\alpha}{dt} = K(T)F(p_{gas})G(\alpha) \tag{3}$$

where K(T) refers to an Arrhenius type equation, F(p_{gas}) expresses the dependence of R with the partial pressure of gasifying agent, and G(α) is a function that describes the geometric evolution of the reacting solid. This procedure allows to exclude mass sample effects and represents an appropriate approach for the analysis of gas-solid heterogeneous reactions (De Micco et al., 2010).

4.4. Tubular reactors coupled with gas chromatography

Gasification experiments using carbon dioxide and steam as gasifying agents were also carried out in tubular reactors coupled with gas chromatography. The experimental setup for gasification experiments with carbon dioxide is shown in Figure 3 and consists of a horizontal quartz tube surrounded by an electrical furnace, a gas control panel, and a gas chromatograph (SRI 8610 C) with a packed column Alltech CTR I and helium as carrier gas. Solid char samples of 10 mg were placed on a flat quartz crucible forming a loose packed bed and inside the tubular reactor where an argon flow of 3.5 L/h was maintained. For the isothermal experiments, char samples were heated at the working temperature for about 1 hour after which carbon dioxide was introduced into the reactor. At the same time, the exhausted gases were injected in a gas chromatograph every 5 minutes. To inject the gases into the chromatograph, the exhausted gas stream was connected to a 1 ml loop and, according to the gaseous flow used, the time required to fill the loop was 0.86 seconds. The Reynolds number corresponding to the experimental conditions indicates that the gaseous flow inside the reactor is laminar (De Micco et al., 2012).

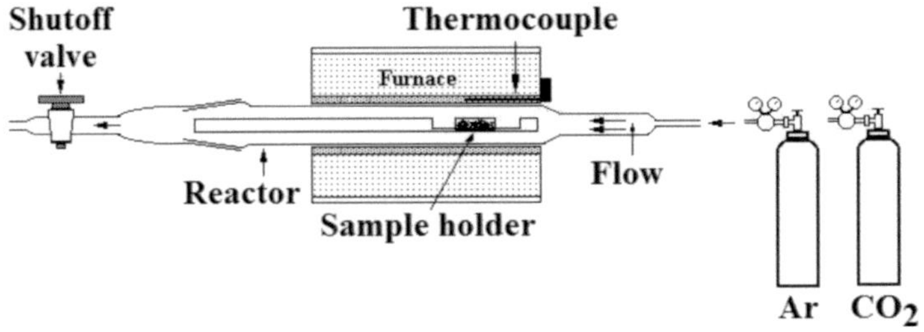

Figure 3. Tubular reactor coupled with gas chromatography for CO_2 gasification experiments.

The gasification rate is determined by monitoring the evolution of the concentration of reaction product, i.e. carbon monoxide (CO(g)), as a function of time. To follow the gasification kinetics, the peak areas corresponding to CO(g) concentration from the chromatograms registered every 5 minutes during the reactions are used. These areas are proportional to the amount of CO(g) moles formed during the time interval required to fill the loop. Since this time interval of 0.86 s is very small compared to the total time needed to achieve the complete reaction (more than 3000 sec), and assuming that no significant axial mixing occurs under laminar flow conditions, it can be considered that the peak areas are proportional to the instantaneous gasification rate. Plots of *CO-Area vs. time* were constructed for each gasification reaction and these experimental data were fitted with appropriated curves.

The number of moles formed at the time t can be calculated by integrating the curves from $t = 0$ to t, and the degree of reaction at time t can be obtained from the ratio of the previous result and the value of integrating the whole *CO-Area* vs. *time* for the complete gasification reaction, according to the following equation:

$$X(t) = \frac{n_{CO}(t)}{n_{CO}(t_f)} \tag{4}$$

where $X(t)$ is the degree of reaction at time t, $n_{CO}(t)$ is the number of moles of CO(g) formed from the beginning of reaction until time t, and $n_{CO}(t_f)$ is the total number of moles of CO(g) formed during the whole reaction.

The experimental setup for gasification experiments with steam is shown in Figure 4 and consists of a horizontal quartz tube surrounded by an electrical furnace, a gas control panel, a steam generator, a set of thermal and chemical traps for retaining the water molecules from the gaseous stream, and a gas chromatograph for analysing the gasification products. The kinetics of gasification reactions can be followed either by measuring the concentration of reaction products with the chromatograph or by gravimetric measurements in which case the gasification reaction is stopped at different reaction degrees.

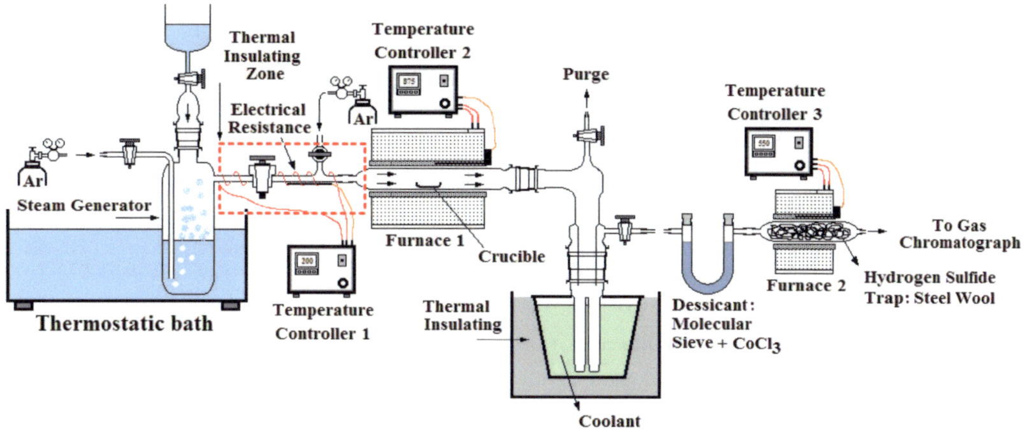

Figure 4. Experimental setup for gasification experiments with steam.

5. Modelling of gasification reactions

A good understanding of solid fuel reactivity and reaction kinetics with carbon dioxide and steam is required for careful optimization of gasification processes. For this reason, numerous studies are being performed worldwide in order to determine the kinetic parameters and reaction mechanisms of the gasification reactions, for each type of parent coal and char. In general, the gasification reaction is a heterogeneous gas-solid reaction where a porous solid is consumed leading to the formation of gaseous products such as carbon dioxide, carbon monoxide and hydrogen, and ash as a solid residue.

In order to obtain the kinetic parameters of gasification reactions, it is useful to measure the reaction rate under chemical control regime. To do that, it is necessary to find the experimental conditions under which mass-transfer resistance is absent. This is accomplished by changing the experimental parameters that influence the rate of the mass transfer processes occurring during the reaction.

Three main kinds of mass transport processes involving different physical phenomenon can be distinguished in this type of reaction: (a) transport of the gaseous reactant by bulk motion (mass convection); (b) transport of the gaseous reactant and products through the gaseous boundary layer (mass gaseous diffusion); and (c) transport of gaseous reactant and products within the solid pores (ordinary or Knudsen gaseous diffusion). The experimental parameters that can be systematically modified in order to make the mass transfer rate faster than the chemical reaction rate are the gaseous flow rate and the initial amount of solid reactant. Once the conditions are achieved to measure the reaction rate under chemical control in the selected range of temperatures, it is possible to apply different reaction models for describing the solid evolution during the reaction, and for obtaining the mathematical expression for the reaction rate.

In general, porosity, surface area and particle size of the solid fuel may vary during the reaction. There are many models that consider the effect of these changes to a different extent and, depending on the hypotheses the models can face various degrees of complexity. Furthermore, due to the porous nature of coal and char, it is not always possible to achieve complete chemical control of the reaction because the diffusion within the pores limits the overall rate of reaction. When this happens, both processes, chemical reaction and pore diffusion, exert an influence on the progress of reaction. Consequently, it is necessary to take into account mass transfer effects in the reaction rate expression.

Many models were developed and published for modeling the coal gasification reactions since the 1950s up to now. One of first approaches was done by Pettersen (Pettersen, 1957) who presented a method for a linear kinetic expression in the concentration and where appreciable concentration gradients were established in the pore system. He assumed uniform cylindrical pores with random intersections. The grain model was further developed by Szekely et al. (Szekely et al, 1976), representing the diffuse reaction zone of reacting porous solids and considering a solid made up of individual grains of equal size which could be spheres, long cylinders or flat plates. In this model, the solid surface area decreases nonlinearly with increasing the reaction degree. On the other hand, Bhatia & Perlmutter (Bhatia & Perlmutter, 1980) presented the random pore model which allows for arbitrary pore size distributions. In this model, the reaction surface changes due to two competing processes: (1) the effect of pore growth during gasification; and (2) the destruction of pores due to coalescence of neighboring pores. The model subsumes several earlier treatments as special cases. Other published models are the random capillary model (Gavalas, 1980), the discrete random pore model (Bhatia & Vartak, 1996), and the modified discrete random pore model (Srinivasalu et al., 2000).

In the analysis of gasification experiments with Argentine solid fuels, two different models were applied: (1) the grain model; and (2) the random pore model, and the mathematical formalisms are described briefly below.

Assuming separation of variables, the kinetic expression for the reaction rate is given by:

$$\frac{dX}{dt} = k(T)G(C_g)f(X) \tag{5}$$

being $X(t)$ the degree of reaction; $k(T)$ and $G(C_g)$ include the effects of temperature and gaseous reactant concentration in the reaction rate, respectively; and $f(X)$ accounts for the changes in physical or chemical properties of reacting solid with reaction degree.

The temperature dependence, i.e. the apparent reaction constant $k(T)$, is given by an Arrhenius equation with k_0 being the pre-exponentional factor and E_a the activation energy. The gas concentration dependence, $G(C_g)$, is given by a power law expression being n the reaction order with respect to gaseous reactant concentration, resulting :

$$k(T)G(C_g) = k_0 e^{-E_a/RT} C_g^n \tag{6}$$

Replacing $f(X)$ by the grain model for spherical grains (GM) and the random pore model (RPM) the expressions for the reaction rate and reaction degree vs. time are the following:

- Grain model:

$$\frac{dX}{dt} = k_0 e^{-\frac{E_a}{RT}} (1-X)^{\frac{2}{3}} \tag{7}$$

$$3\left[1-(1-X)^{1/3}\right] = k_{GM} t \tag{8}$$

- Random pore model:

$$\frac{dX}{dt} = k_0 e^{-\frac{E_a}{RT}} (1-X)\sqrt{\left[1-\psi \ln(1-X)\right]} \tag{9}$$

$$(2/\psi)\left[\sqrt{(1-\psi \ln(1-X))}-1\right] = k_{RPM} t \tag{10}$$

In the random pore model, in addition to the apparent reaction constant k_{RPM} there is another parameter, Ψ, which is related with the pore structure of the initial sample, and can be calculated from the experimental results with the following equation:

$$\psi = \frac{2}{2\ln(1-X_{max})+1} \tag{11}$$

where X_{max} is the value of reaction degree where the reaction rate is maximum.

6. Main results and discussion

6.1. Experimental design

Laboratory research activities on gasification reactions of Argentine solid fuels in presence of carbon dioxide and steam comprised a comprehensive theoretical and experimental study on the following two chemical reactions:

$$C + CO_2 \leftrightarrow 2CO \quad \Delta H = 159.7 \text{ kJ/mol} \tag{12}$$

$$C + H_2O \leftrightarrow CO + H_2 \quad \Delta H = 118.9 \text{ kJ/mol} \tag{13}$$

After determining the experimental conditions to get the chemical control regime of gasification reactions in the different experimental setups through the variation of the gaseous flow, sample mass and char particle size, which are detailed in Table 2, the effects of the following parameters were investigated:

1. Composition and rank of feed material, i.e. comparative behaviour of subbituminous coal and asphaltites;
2. Reaction temperature, in the range between 800 and 950 ºC;
3. Partial pressure of gasifying agent, between 30 and 80 %v/v;
4. Conditions of char formation, using chars prepared in the DTF at 850 and 950 ºC, respectively, and chars prepared in a fixed bed reactor at 950 ºC.

Gasifying agent	Río Turbio Coal		Emanuel Asphaltite	
	TG system	GC system	TG system	GC system
CO_2	Gaseous flow above 7.3 L/h. Sample mass below 10 mg.	Gaseous flow above 4.2 L/h. Sample mass below 16 mg.	Gaseous flow above 5 L/h. Sample mass below 2.5 mg.	na
H_2O	na	na	na	Gaseous flow above 2 L/h. Sample mass below 25 mg.

Table 2. Experimental conditions to get chemical regime in gasification reactions.

The main results of the theoretical and experimental research program are given in the following sections.

6.2. Effect of solid fuel composition and rank

Argentine solid fuels were pyrolysed in inert atmosphere (argon) using non-isothermal TGA runs and the resulting TGA curves are presented in Figure 5. The mass losses observed in all cases are due to a mixture of vapors and gases which are released during heating, including CO_2, CO, hydrocarbon species, tars, and so on, and they are in agreement with the corresponding values of volatile matter content given in Table 1.

Figure 5 shows that Fortuna 4 and Toribia asphaltites behaved similarly during pyrolysis, exhibiting a fast volatilization rate between 300 and 500 ºC where about 70 wt% of the total volatile matter was released, replicating the results of the proximate analyses where both materials showed very similar values of fixed carbon, ash and volatile matter. The same agreement between proximate analyses and non-isothermal TGA curves was detected for Susanita asphaltite and Río Turbio coal; in this case, two different portions can be distinguished in TGA curves: (1) a fast volatilization rate between 300 and 500 ºC where about 60 wt% of the volatile matter was released, and (2) a slow volatilization rate above 500 ºC where about 25 wt% of the volatile matter was further released. Finally, Emanuel asphaltite showed a singular behavior with a nearly constant volatilization rate between 300 and 900 ºC.

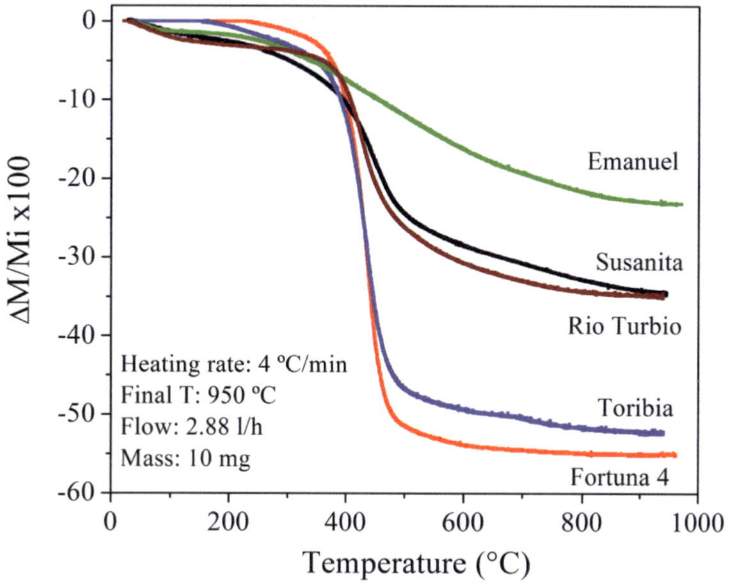

Figure 5. Thermo-gravimetric curves of non-isothermal pyrolysis tests with argon.

The comparative behaviour of Río Turbio coal and asphaltites under CO_2 gasification conditions was studied by performing non-isothermal and isothermal TGA measurements. The non-isothermal TGA curves are presented in Figure 6 and the experimental conditions were: temperature range between room temperature and 950 °C; heating rate of 4 °C/minute; partial pressure of CO_2: 80kPa; and sample mass: 10 mg.

It can be observed that the mass losses measured are due to the release of adsorbed water at low temperature (about 100 ºC) while, at higher temperatures (above 600 ºC), the mass losses corresponded to the gasification reaction of chars with CO_2, producing mainly CO(g). The initial reaction temperatures were: 630, 650, 680, 700, and 730 °C for Susanita, Emanuel, Río Turbio, Toribia, and Fortuna 4, respectively. These temperatures are indicative of the reactivity of chars, meaning that Susanita asphaltite has the highest reactivity and Fortuna 4 asphaltite has the lowest reactivity. Moreover, Toribia and Fortuna 4 asphaltites did not achieve the complete gasification when the temperature reached 950 ºC, showing a reduced reaction rate for the two samples. The other three chars presented a similar mass loss.

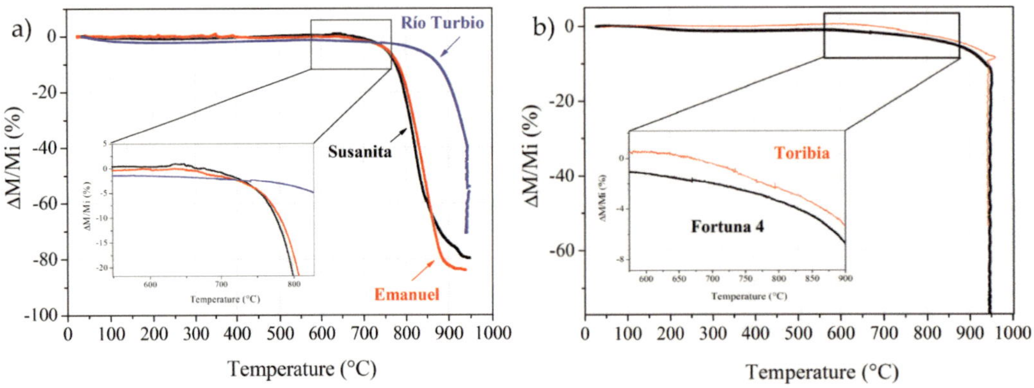

Figure 6. Thermo-gravimetric curves of non-isothermal gasification experiments with CO_2.

In order to study the kinetics of the gasification process, isothermal TGA curves must be obtained. Figure 7 shows the isothermal TGA curves corresponding to the CO_2 gasification of chars obtained in the non-isothermal pyrolysis tests. The experimental conditions were: temperature: 875 ºC; partial pressure of CO_2: 80 kPa; total gaseous flow rate [Ar - CO_2] :10 l/min, and initial sample mass: 25 mg.

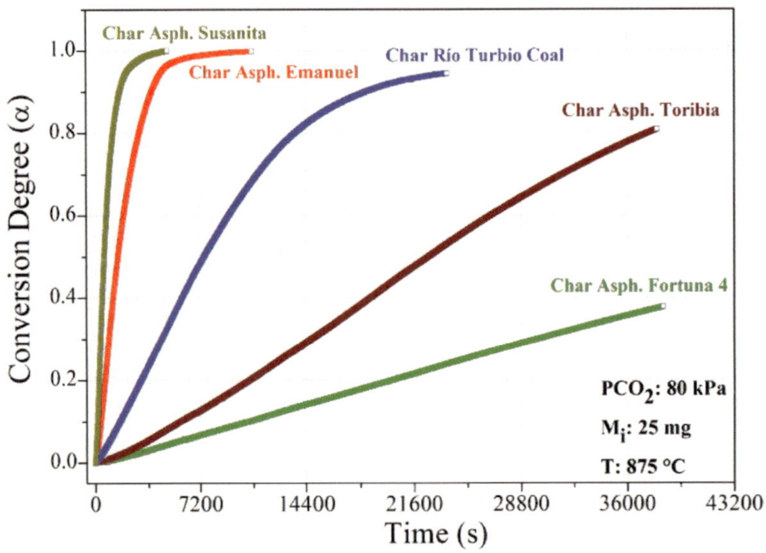

Figure 7. Thermo-gravimetric curves of isothermal gasification experiments with CO_2.

The wide range of reaction rates observed for these solid fuels (almost two orders of magnitude between Fortuna 4 and Susanita asphaltites) is indicating that they have different reactivities in presence of CO_2, and the difference may be attributed to the content of fixed carbon and mineral matter. As can be seen in Figure 7, Susanita and Emanuel asphaltites have a similar fixed carbon content, the same happens with Toribia and Fortuna with a lower fixed carbon content, while Río Turbio coal has an intermediate value. Related with the fixed carbon content, the most important effect that can be taken into account is the presence of more amounts of C-C bonds that may signify a greater net energy bond due to the absence or low content of impurities that can produce defects in the solid fuel matrix, increasing the reactivity with the CO_2.

In order to compare the relative influence between the fixed carbon content and the surface area of chars on the gasification rate, the BET areas were determined on chars obtained from Emanuel asphaltite and Río Turbio coal. Even the exposed surface area of Río Turbio char is almost 30 times higher than the Emanuel asphaltite one, the complete gasification reaction occurred after about 6 hours and 1 hour, respectively. This result indicates that the surface area has less influence on the reaction rate compared to the fixed carbon content of the feed material.

Other important characteristics which have a remarkable effect on the gasification rate are the elemental mineral matter and the mineral phases contained in the chars, since it is well-known that mineral matter naturally present in the carbonaceous matrix may act as a catalyst for the gasification reactions. As shown in Table 1, Fortuna 4 and Toribia asphaltites have very small amounts of mineral matter (ash content below 2 wt%), and they presented the lowest reaction rates. Emanuel asphaltite, even having a relatively low mineral matter content and a low BET area (7 wt% of ash and 3.5 m^2/g), has the highest diversity of metals such as V, Mo, Sr, Ni, and Cu among others, and then the catalytic effect of these metals could lead to the high reactivity observed in experiments. The same explanation can be applied to Susanita asphaltite which has a high ash content and also a fairly diversity of metals, showing the highest reactivity in presence of of CO_2. Finally, the reactivity of Río Turbio coal is intermediate between asphaltites of high and low mineral contents. A further catalytic effect that could be observed from the XRD measurements is that those chars containing calcium sulfate show higher reaction rates than those which have calcium forming other compounds.

6.3. Effect of gasification temperature and partial pressure of gasifying agent

From TGA and GC isothermal measurements obtained at several temperatures, partial pressures of reactants, and using CO_2 and steam as gasifying agents, the kinetic parameters of the rate equations of the gasification reactions could be determined.

In the method proposed by Flynn (Basan S. 1986), the activation energy can be determined from equation (5), although $G(C_g)$ and $f(X)$ are unknown functions. Replacing $K(T)$ by an Arrhenius equation and rearranging equation (5):

$$\int_0^{\infty} \frac{dX}{f(X)} = \int_0^{t} G(C_g) A \exp(-E/RT)\, dt \tag{14}$$

after taking the integral, we have:

$$F(X) = G(C_g) A \exp(-E/RT)\, t \tag{15}$$

and by taking the logarithm of both sides of equation (15):

$$\ln t = \ln\left[\frac{F(X)}{G(C_g)A}\right] + \frac{E}{RT} \tag{16}$$

The first term in the right hand side of equation (16) is a function of degree of reaction and partial pressure of carbon dioxide. Therefore, if partial pressure of carbon dioxide is keep constant, and the time to attain a certain reaction degree is determined as a function of temperature, equation (16) allows to obtain the activation energy from the slope of the plot ln t vs. T^{-1}. Analogous procedure can be applied to obtain the reaction order with respect to gaseous reactant replacing $G(C_g)$ by a power law expression. The reaction order can be obtained from the slope of the plot of ln t vs. C_g

By this method (also known as model-free method or iso-convertional method), the activation energies of gasification reactions were obtained and results are shown in Table 3, while the ln t vs. T^{-1} plots are shown in Figure 8.

The activation energies calculated are consistent with the fact that the Río Turbio coal showed a lower reaction rate compared to Emanuel asphaltite. Furthermore, the similar values of the activation energies for gasification reactions with CO_2 and steam show that the determining step in the mechanism of these reactions is independent of the gasifying agent used, and it can be associated more with the restructuring of carbon surface than with the gasifying agent accommodation.

Solid fuel	Activation energy	
	Gasification with CO_2(g)	Gasification with steam
Río Turbio coal	190 ± 10 kJ/mol	na
Emanuel asphaltite	185 ± 10 kJ /mol	186 ± 10 kJ/mol

na: not available

Table 3. Activation energies of gasification reactions with CO_2 and steam.

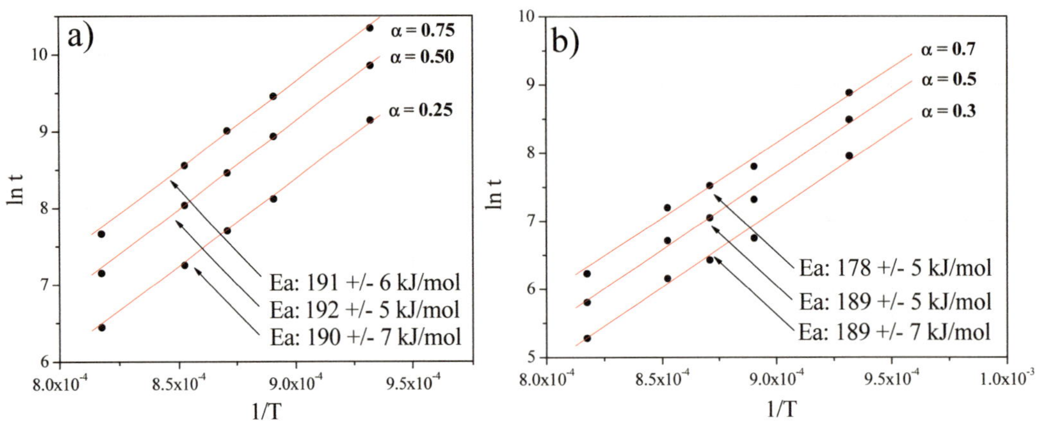

Figure 8. ln t vs. 1/T plot for the calculation of Ea with the Flynn method. (a) Río Turbio coal; (b) Emanuel asphaltite.

The reaction order (n) with respect to the gasifying agent was only determined for CO_2 gasification reactions and results are presented in Table 4 and Figure 9. The reaction order is 1 for the Río Turbio coal and 0.5 for the Emanuel Asphaltite. The first value may indicate that the reaction with CO_2 is produced by occupying only one active site on the surface while, on the opposite, the second value is indicating that a dissociate step on the surface of the particle may be occurring, with the CO_2 molecule being adsorbed and occupying two active sites. The last mechanism requires less activation energy for the breakdown of the C-O chemical bond than the first one, so the gasification reaction is expected to be faster, as it was already shown in the isothermal TGA curves

Solid fuel	Reaction order with respect to reactant	
	Gasification with CO_2(g)	Gasification with steam
Río Turbio coal	1	na
Emanuel asphaltite	0.5	1

na: not available

Table 4. Reaction order of gasification reactions with CO_2 and steam.

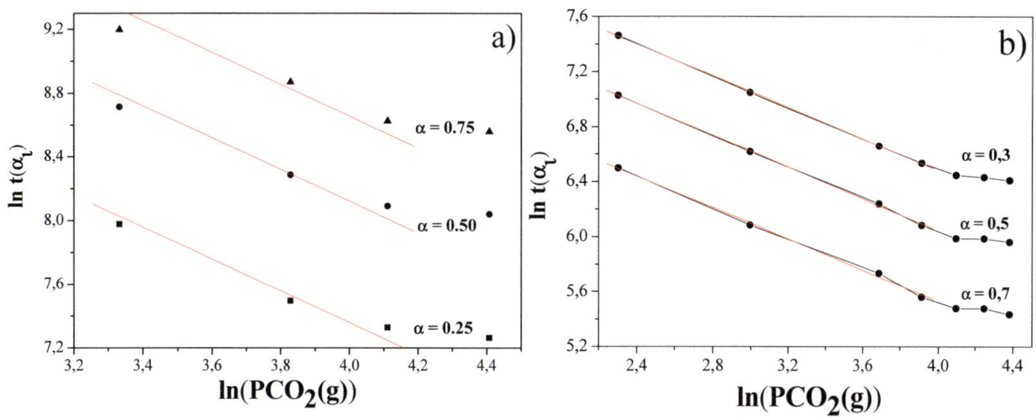

Figure 9. ln t(αi) vs. ln(PCO_2) plot for the calculation of the reaction order with respect to reactant. (a) Río Turbio coal; (b) Emanuel asphaltite.

6.4. Effect of gasifying agent

A comparative analysis of reactivity of Argentine solid fuels in presence of CO_2 and steam was also performed and main results are summarized in Figure 10. Figure 10(a) shows the conversion degree vs. time in TGA curves obtained from Emanuel asphaltite chars gasified at 875 ºC with CO_2 (at a partial pressure of 80 kPa) and steam (at a partial pressure of 20 kPa), respectively. It can be appreciated that the steam reactivity of Emanuel asphaltite char is a little bit higher than CO_2 reactivity even the steam partial pressure used in experiments was lower than the CO_2 partial pressure.

On the other hand, Figure 10(b) shows the conversion degree vs. time in curves obtained from Río Turbio coal chars gasified at 875 ºC in the tubular reactors shown in Figures 3 and

4. In this case, the difference in gasification reactivity between steam and CO_2 is more significant and matches better with the results found in literature (Roberts, D.G. 2000 and Messenbock, R.C. 1999).

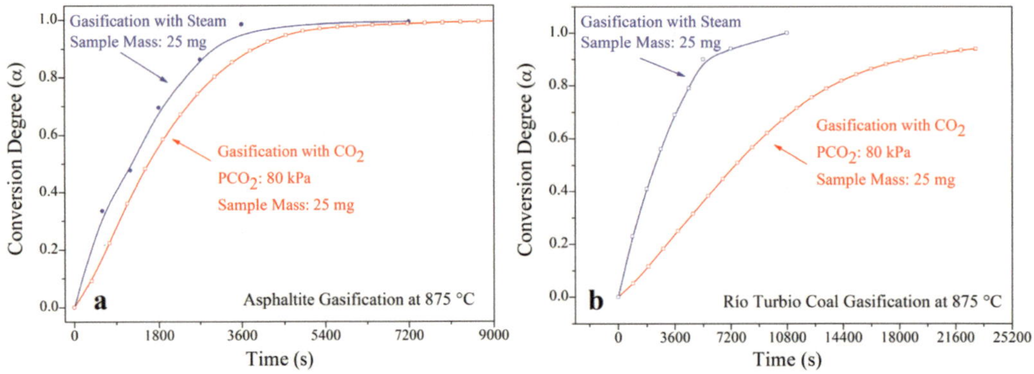

Figure 10. (a) TGA curves of Emanuel asphaltite char gasified with steam and CO_2; (b) gasification curves of Río Turbio coal char gasified with steam and CO_2 in tubular reactors.

6.5. Effect of char formation conditions

Chars from Río Turbio coal were prepared in three different conditions: (1) in the DTF at 850 ºC; (2) in the DTF at 950 ºC; and (3) in a Fixed Bed Reactor (FBR) at 950 ºC. Following, 10 mg of each char sample was gasified at 900 ºC in the TGA system with 30% CO_2 partial pressure at the same flow rate, in order to compare their gasification reactivities.

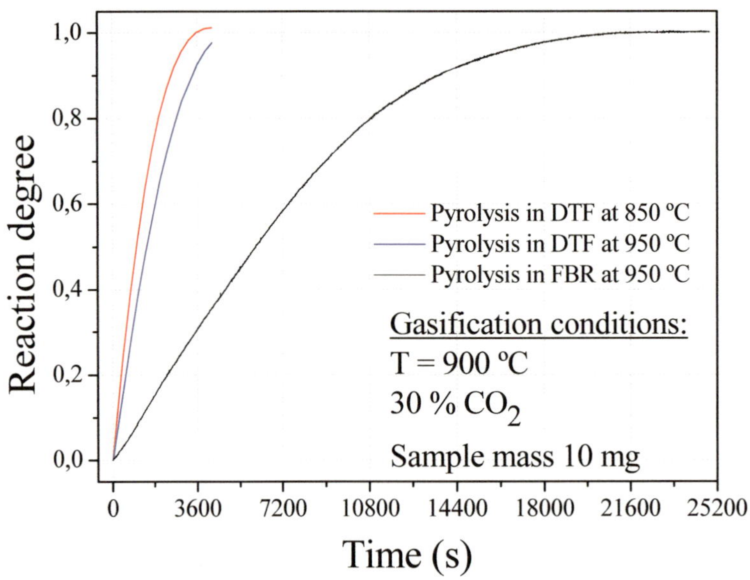

Figure 11. TGA isothermal CO_2 gasification curves of Río Turbio chars prepared in different pyrolysis conditions.

Figure 11 shows the TGA isothermal curves corresponding to the three different chars. It can be observed that both chars pyrolysed at high heating rates in the DTF followed a similar behavior and the gasification reactivity is comparable to reactivities of low-rank coals used in large-scale gasifiers. On the opposite, the TGA curve corresponding to the char prepared at low heating rate in the FBR shows that the gasification reaction progressed much slowly and it was completed after a long time, indicating a very low reactivity in presence of the gasifying agent.

These experimental results are demonstrating that the reactivity of chars to gasifying agents is very dependent on their formation conditions and, then, to get meaningful data about kinetics of gasification reactions, it is very important to produce chars in laboratory at high heating rates and intense gas convection around individual char particles, replicating the real operating conditions of commercial gasification reactors.

Finally, the reaction rate as a function of the conversion degree for those gasification experiments is presented in Figure 12, which also shows the predicted values by the grain model superimposed to the experimental measurements. It can be appreciated that the grain model is expected to well-simulate the gasification behavior of the Río Turbio coal in the temperature range used in experiments. From these fittings, the kinetic parameters of the theoretical models were calculated and they are given in Table 5.

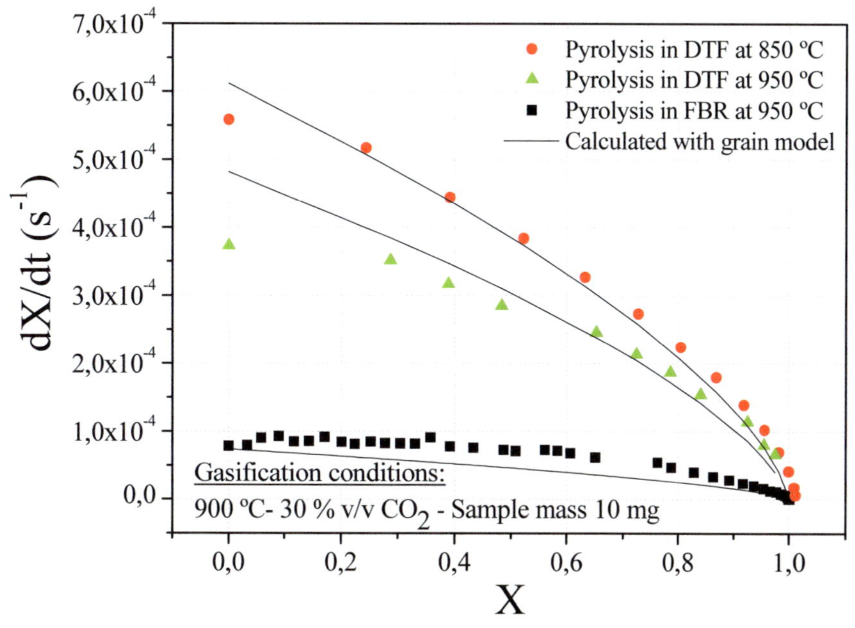

Figure 12. Theoretical and experimental CO_2 gasification rates of Río Turbio chars prepared in different conditions.

Char preparation	Model free method	RPM $\frac{dX}{dt}=k_0e^{-\frac{E_a}{RT}}(1-X)\sqrt{[1-\psi\ln(1-X)]}$			GM $\frac{dX}{dt}=k_0e^{-\frac{E_a}{RT}}(1-X)^{\frac{2}{3}}$	
	Ea kJ/mol	Ea kJ/mol	ko s^{-1}	Ψ	Ea kJ/mol	ko s^{-1}
Pyrolysis in DTF at 850 ºC	171 ± 10	165 ± 11	1.15 10^4	2	166 ± 11	1.51 10^4
Pyrolysis in DTF at 950 ºC	159 ± 22	158 ± 2	4.07 10^3	2	158 ± 2	5.24 10^3
Pyrolysis in FBR at 950 ºC	190 ± 10	na	na	na	195 ± 12	3.6 10^4

Table 5. Kinetic parameters of theoretical models used for simulating the gasification reactions.

It can be observed that the activation energy values of the gasification of chars prepared in the DTF are lower than the value corresponding to the char prepared in the FBR. Meanwhile, the values of the reaction rate constant are similar in all cases, independently of the char preparation method. These values are in agreement with the fact that the gasification reactions of chars prepared in DTFs are faster than the gasification reactions of chars which are pyrolysed in FBRs. Another important aspect is that the activation energy has the most significant effect on the char gasification rate in the present experimental conditions.

7. Conclusions and future works

A comprehensive theoretical and experimental research program is being implemented in Argentina at laboratory scale in the framework of a national strategy for the integral utilization of its domestic coal reserves, addressed to bring together the requirements of a sustainable economic growth with the environmental protection. The research program was designed to simulate in laboratory, as close as possible, the operational conditions of large-scale gasification plants and, then, to provide the necessary information about fundamental mechanisms and kinetics of the gasification reactions for a further scaling up of experimental facilities. For this purpose, specially-designed experimental equipment and test procedures were implemented for gasification experiments using carbon dioxide and steam as gasifying agents.

Experimental program on gasification with carbon dioxide is almost finished and experimental results show that all the Argentine solid fuels studied are amenable to be gasified since their gasification reactivities at high heating rates are comparable with those of low-rank coals used in large-scale gasifiers. Experimental program on steam gasification is just beginning but preliminary experimental results show that the reaction rate is higher than the reaction rate corresponding to the gasification with carbon dioxide.

As it was detected that some mineral phases present in the ashes may have a catalytic effect in gasification reactions, further studies to elucidate this influence are planned for the future, along with the construction of experimental setups for carbon dioxide and steam gasification experiments at higher pressures.

Author details

G. G. Fouga[1,2], G. De Micco[1,2], H. E. Nassini[1] and A. E. Bohé[1,2]
[1]*Comisión Nacional de Energía Atómica (CNEA), Republic of Argentina,*
[2]*Consejo Nacional de Investigaciones Científicas y Técnicas (CONICET), Republic of Argentina*

Acknowledgement

The authors would like to thank to the Agencia Nacional de Promoción Científica y Tecnológica (ANPCyT), Consejo Nacional de Investigaciones Científicas y Técnicas (CONICET), and Universidad Nacional del Comahue (UNComa) for the financial support of this work.

8. References

Barrett EP, Soguer LG, Halenda PP. (1951). The determination of pore volume and area distributions in porous substances. I. Computations from nitrogen isotherms. J. Am. Soc.;73: 373–80.

Basan S., Güven O., (1986). A comparison of various isothermal thermogravimetric methods applied to the degradation of PVC, Thermochimica Acta.106 169-178

Beloff, E.J. (1972). Minacar asphaltite: an asphaltite from Argentina, Fuel, Volume 51, Nº 2, pp. 156-159.

Bhatia S.K. & Perlmutter D.D. (1980). A random pore model for fluid-solid reactions: I. Isothermal, kinetic control, AIChE Journal, Volume 26, Nº 3, pp.379-385.

Bhatia S.K.& Vartak B.J. (1996). Reaction of microporous solids: the discrete random pore model, Carbon, Volume 34, Nº 11, pp.1383-1391.

Bohe, A.E. & Nassini, H.E. (2011). Hydrogen production and applications program in Argentina, *Nuclear Hydrogen Prodution Handbook,* edited by X. Yan & R. Hino, CRC Press, Taylor & Francis Group, pp. 695-723.

Carrizo, R.N. (2002). Mineral solid fuels. Geology and Natural Resources of Santa Cruz, Proceedings of XV Argentine Geological Congress, Santa Cruz, June 2002, pp. 759-771 (*in spanish*).

Collot, A. (2006). Matching gasification technologies to coal properties, International Journal of Coal Geology, Volume 65, pp. 191-212.

De Micco, G., Fouga, G. & Bohe, A. (2010). Coal gasification studies applied to hydrogen production, International Journal of Hydrogen Energy, Volume 35, pp. 6012-6018.

De Micco, G., Nasjleti, A. & Bohe, A. (2012). Kinetics of the gasification of Río Turbio coal under different pyrolysis temperatures, Fuel, Volume 95, pp. 537-543.

Domazetis, G., Liesegang, J. & James, B.D. (2005). Studies of inorganics added to low-rank coals for catalytic gasification, Fuel Processing Technology, Volume 86, pp. 463-486.

Fouga G. G., De Micco G. & Bohé A. E. (2011). Kinetic Study of Argentinean Asphaltite Gasification Using Carbon Dioxide as Gasifying Agent, Fuel, Volume 90, pp. 674-680.

Gavalas, G.R.(1980). A random capillary model with application to char gasification at chemically controlled rates, AIChE Journal, Volume 26, Nº 4, pp 577-585.

Higman, C. and Van der Burgt, M. (2003). Gasification, Gulf Professional Publishing, Elsevier, New York.

Megaritis, A., Messenbock, R., Collot, A., Zhuo, Y., Dugwell, D. & Kandiyoti, R. (1998). Internal consistency of coal gasification reactivities determined in bench-scale reactors: Effect of pyrolysis conditions on char reactivity under high-pressure CO_2, Fuel, Vol. 77, Nº 13, pp. 1411-1420.

Messenbock, R.C., Dugwell, D.R. & Kandiyoti, R. (1999). CO_2 and steam-gasification in a high-pressure wire-mesh reactor: reactivity of Daw Mill coal and combustion reactivity of its char, Fuel, Volume 78, pp. 781-793.

Minchener, A.J. (2005). Coal gasification for advanced power generation, Fuel, Volume 84, pp. 2222-2235.

Nassini, H., De Micco, G., Fouga, G. & Bohe, A. (2011). Kinetics studies on nuclear-assisted gasification of argentine coals, Proceedings of IAEA´s Technical Meeting/Workshop on Non-Electric Applications of Nuclear Energy, Prague, Czech Republic, 3-6 Ocober, 2011.

Peng, F.F., Lee, I.C. & Yang, R.Y. (1995). Reactivities of in-situ and ex-situ coal chars during gasification in steam at 1000-1400ºC, Fuel Processing Technology, Volume 41, pp. 233-251.

Pettersen E.E. (1957). Reaction of porous solids, AIChE Journal, Volume 3, Nº 4, pp 443-448.

Roberts, D.G. & Harris D.J. (2000). Char Gasification with O_2, CO_2, and H_2O: Effects of Pressure on Intrinsic Reaction Kinetics, Energy & Fuels, Volume 14, pp. 483-489.

Savelev, V., Golovko, A., Gorbunoa, L., Kamyanov, V. & Galvalizi, C. (2008). High-sulphurous Argentinean asphaltites and their thermal liquefaction products, Oil Gas Science Technology, Volume 63, Nº 1, pp. 57-67.

Srinivasalu, G.J. & Bhatia S.K. (2000). A modified discrete random pore model allowing for different initial surface reactivity, Carbon, Volume 38, pp. 47-58.

Szekely J., Evans, J.W.& Sohn H.Y. (1976). *Gas-solid reactions*, Academic Press, New York.

Yu, J., Lucas, J. & Wall, T. (2007). Formation of the structure of chars during devolatilization of pulverized coal and its thermo-properties: A review, Progress in Energy and Combustion Science, Vol. 33, pp. 135-170.

Chapter 3

Gasification Reactions of Metallurgical Coke and Its Application – Improvement of Carbon Use Efficiency in Blast Furnace

Yoshiaki Yamazaki

Additional information is available at the end of the chapter

http://dx.doi.org/10.5772/45732

1. Introduction

1.1. Feature of metallurgical coke and the role in blast furnace

Metallurgical coke is made from coal that is an organic compound, but is inorganic material composed of graphite. Metallurgical coke is porous media that contains pore of 50% in porosity. The size of metallurgical coke lump is from 25 mm to 50 mm (Fig. 1). In modern iron making process, coke has very important roles in iron making process because coke is, at the same time, used as reducing agent of ore, heat source of blast furnace, carburizing source of pig iron and spacer of gas and liquid transport through blast furnace. Metallurgical coke is charged from the top of blast furnace at first and moves to the bottom part. Reducing agents derived from coke are generated by following two reactions: (i) coke reacts with oxygen at the bottom part of blast furnace, and one carbon monoxide molecule is generated, (ii) coke reacts with carbon dioxide at middle part of blast furnace, and two carbon monoxide molecules is generated. Former reaction is *combustion* and latter reaction is named *carbon* (or *coke*) *solution-loss reaction*. Firstly, carbon monoxide generated from combustion reaction reduces ore (FeO_x) and becomes carbon dioxide. Then, carbon dioxide reacts with coke and two carbon monoxide molecules is generated.

1.2. Social background

Blast furnace operation consumes huge amount of carbon that finally becomes carbon dioxide. In recent years, worldwide, iron making materials (i.e. coal) are draining and soaring. So, improvement of carbon use efficiency to curtail carbon consumption is increasingly important issue from the viewpoint of material, energy resource and cost.

Combustion of coke takes the role of primary carbon monoxide generation. So, this gasification reaction is carbon consumption reaction. Carbon solution-loss reaction, also, consumes carbon, but is, on the other hand, gasified carbon recycle reaction (from carbon dioxide to carbon monoxide). Reactivity of these gasification reactions directly affects carbon use efficiency of iron making process. Gasified carbon produced by combustion reaction is finally emitted as carbon dioxide due to oxidization reaction of ore. Thus, control of coke solution-loss reactivity is important in order to improve of the carbon use efficiency. Both practical approaches and fundamental investigation are desired to this.

(a) A photograph of coke lump

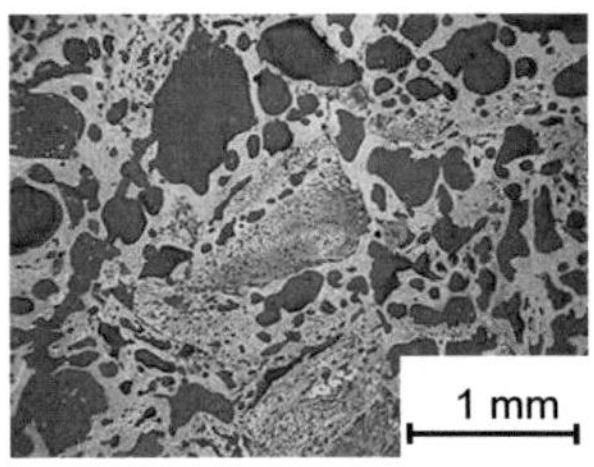

(b)cross-sectional image
of coke microstructure

Figure 1. Photographs of (a) coke lump and (b) cross-sectional image of coke microstructure

1.3. Purpose and outline of this chapter

As above, promoting and controlling solution-reactivity of metallurgical coke is very important in order to improve the carbon use efficiency. To realize the blast furnace operation in high carbon use efficiency, making of the coke which is satisfying four roles described in *1.1* and is solution-loss reactivity-promoted is required.

The practical purpose of the chapter is to propose the design guide of solution-loss reactivity-promoted (so-called *"highly reactive"*) coke from the viewpoint of use in blast furnace. There are many conditions (e.g. thermal, chemical or mechanical condition) which affect descending and reacting metallurgical coke in blast furnace. For this, the *proper* metallurgical coke should be made with considering the effects of these conditions. The chapter, hence, focuses on the fundamental knowledge and research about metallurgical coke gasification in the effect of thermal, chemical and mechanical condition. At first, the situation in blast furnace and the role of coke gasification reactions in blast furnace are introduced *in section 2*. The effect of catalyst, as useful way to promote the solution-loss reactivity, on solution-loss reactivity is discussed *in section 3*. Then, *in section 4*, the problem from the viewpoint of the strength caused by promotion of solution-loss reactivity is noted, and the phenomena of highly reactive and normal coke are investigated in order to provide the solution of the issue. To discuss about both of the reactivity and the strength of highly reactive coke, the reaction mechanism and phenomena of highly reactive coke before and after the gasification reaction is investigated in this section. Furthermore, we propose a proper method to make highly reactive coke catalyzed by metals.

2. Situation in blast furnace and role of metallurgical coke gasification

Situation in blast furnace and role of coke gasification can be discussed from two viewpoints. One is *chemical reaction* and *thermodynamic equilibrium state*. Metallurgical coke is gasified with oxidation product such as carbon dioxide, and reductive gas such as carbon monoxide is generated. In blast furnace, any fraction of the components is governed by thermodynamic equilibrium state in C-O-Fe system and reaction kinetics caused by the difference between actual and equilibrium state. The other is *the effect of coke solution-loss reactivity on carbon use efficiency* in blast furnace. Using coke with high solution-loss reactivity, equilibrium state changes because solution-loss reaction is endothermic and the temperature at TRZ decreases. As a result, necessary quantity of carbon (coke) decreases.

2.1. Chemical reaction and the thermodynamic equilibrium state in C-O-Fe system (Bannya, 2000)

2.1.1. Combustion (reducing gas generation)

In blast furnace, carbon atom of coke reacts with oxygen molecule from tuyere

$$\begin{aligned} &C(s) + O_2(g) = CO_2(g) \\ &\Delta H^\circ_{298} = -393.5 \text{ kJ/mol} \\ &\Delta G^\circ_{298} = -393500 - 2.99T \text{ J/mol}, \end{aligned} \tag{1}$$

where ΔH°_{298} is standard enthalpy change of formation, ΔG°_{298} is standard free energy and T is absolute temperature. Generated carbon dioxide by reaction of Eq. (1) reacts because there is much solid carbon as coke,

$$\begin{aligned} &C(s) + CO_2(g) = 2CO(g) \\ &\Delta H^\circ_{298} = 172.4 \text{ kJ/mol} \\ &\Delta G^\circ_{298} = -171660 - 175.02T \text{ J/mol}. \end{aligned} \tag{2}$$

As a result, following reaction occurs near the bottom part of blast furnace

$$\begin{aligned} &2C(s) + O_2(g) = 2CO(g) \\ &\Delta H^\circ_{298} = -221.1 \text{ kJ/mol} \\ &\Delta G^\circ_{298} = -221840 - 178.0T \text{ J/mol}. \end{aligned} \tag{3}$$

Reaction of Eq. (3) is called *combustion of coke*. Two molecules in carbon monoxide as reducing gas and 221.1 kJ in thermal energy is generated by reaction of Eq. (3) with one molecule in O_2. Temperature near the tuyere of blast furnace is 2570 K that is similar to adiabatic flame temperature of reaction of Eq. (3).

2.1.2. *Iron oxide reduction*

Reducing reaction of iron ore (oxide) in blast furnace is classified into two kind of reaction. One is *indirect* reducing reaction with carbon monoxide. The other is *direct* reducing reaction with solid carbon. "*Direct*" or "*idirect*" is called whether solid coke is directly gasified. Indirect reaction occurs at the top or middle part of blast furnace and direct reaction progresses at the bottom part. Indirect reducing reaction is written in

$$FeO_m(s) + CO(g) \rightarrow FeO_{m-x} + CO_2(g)\,. \tag{4}$$

This successive reaction is a desirable reaction from the viewpoint of the thermal balance in blast furnace because the reaction is an exothermic except reducing reaction from magnetite to wustite. Direct reducing reaction, on the other hand, is written as follows:

$$FeO_m(s) + mC(s) \rightarrow Fe + mCO(g)\,. \tag{5}$$

Reaction of Eq. (5) progresses at the bottom part of blast furnace where combustion of coke occurs and is endothermic. It negatively affects the amount of energy consumption that reaction of Eq. (5) mainly occurs. To improve carbon use efficiency (thermal efficiency), it is important to enhance indirect reducing reaction because reducing ratio of iron ore by indirect reducing reaction should be lifted rather than that of direct reducing reaction.

2.1.3. *Coke solution-loss reaction*

Reaction of Eq. (2) can be also expressed as equilibrium reaction.

$$\begin{aligned} &C(s) + CO_2(g) = 2CO(g) \\ &\log K_p = -8969 / T + 9.14 \\ &K_p = p_{CO}^2 / p_{CO_2}, \end{aligned} \tag{6}$$

where K_p is equilibrium constant. Equilibrium of Eq. (6) is called *Boudouard equilibrium*. The composition of this equilibrium relates with reaction of Eqs. (1)-(5), and dominates state of C-O-Fe system (e.g. composition of Fe_2O_3, Fe_3O_4, FeO, Fe, CO_2, CO, C and so on) in blast furnace. The reaction toward right hand of Eq. (6) is endothermic and is promoted with high temperature. At the bottom part of blast furnace where the temperature indicates 2570 K, ratio of $CO/(CO+CO_2)$ is almost 1.0. At the middle part of blast furnace, the ratio is about 0.9. This reaction at the middle part of blast furnace is, in particular, called *carbon solution-loss reaction*.

2.2. The effect of coke solution-loss reactivity on carbon use efficiency

2.2.1. Thermal reserve zone

Figure 2 shows conceptual diagram of temperature distribution along the height direction in blast furnace. TRZ (Thermal Reserve Zone) is where temperature slightly changes over the cohesive zone. In TRZ, indirect reducing reaction actively progress rather than direct one. The degree of progress of indirect reducing is affected by TRZ temperature because the temperature governs state of C-O-Fe equilibrium system in TRZ. Coke solution-loss reaction, also, occurs in TRZ and its reactivity strongly affects TRZ temperature. TRZ temperature decreases when solution-loss reactivity of charged coke is enhanced due to endothermic reaction. Therefore, coke with high solution-loss reactivity is used in blast furnace, and TRZ temperature decreases and the equilibrium point moves. However, Final conversion of coke gasification (ratio of weight loss based on carbon) is constantly 20 mass% regardless of gasification reactivity.

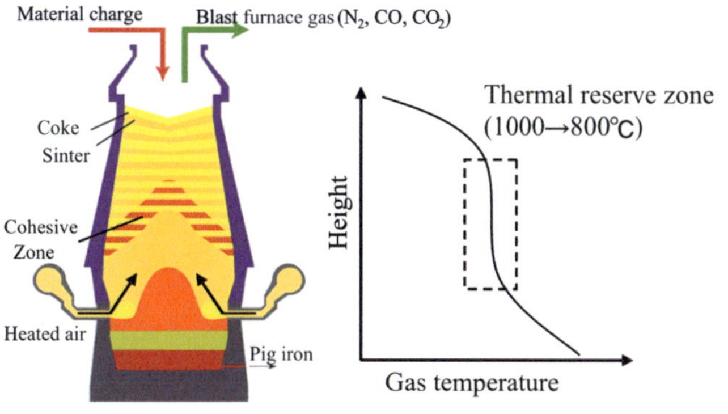

Figure 2. A conceptual diagram of temperature distribution along height direction of blast furnace

2.2.2. Rist diagram

Rist et al. proposed the model that represents a state of blast furnace operation based on thermal and chemical equilibrium state (Rist & Bonnivard, 1962; Rist & Meyerson, 1967). The both equilibrium states at temperature of TRZ and tuyere are plotted (Rist diagram), and the state of the operation (e.g. carbon use efficiency C/Fe that is amount of carbon use per unit reduced iron) can be estimated. Rist diagram is convenient tool to describe the effect of coke solution-loss reactivity on carbon use efficiency. Figure 3 shows an example of Rist diagram with the operation line. Horizontal and vertical axes show O/C and O/Fe, respectively. State of C-O-Fe equilibrium system can be understood in blast furnace. Gas equilibrium in C-O system at tuyere is shown in the lower left part of cut Rist diagram into quarters, state of C-O-Fe equilibrium system in TRZ is represented in the upper right part of the cut diagram, and state of iron oxide in O-Fe system is indicated in the lower right part of the cut diagram. Gradient of the line in Rist diagram means carbon use efficiency C/Fe = (O/C)/(O/Fe) . The "*W*" point shows ideal operation state that is in equilibrium state in TRZ.

If TRZ temperature decreases with enhancement of coke reactivity, gas equilibrium of C-O-Fe system is shifted to *oxidation state* (a decrease in ratio of CO/CO_2 in equilibrium). Then, the *new* "W' " point is plotted in Rist diagram. Indirect reaction of Eq. (4) is promoted because carbon dioxide generation is promoted due to the new equilibrium that indicates oxidation atmosphere. As a result, the carbon consumption for thermal conservation at the bottom part of blast furnace is curtailed because the amount of reducing iron caused by direct reducing that adsorbs the heat decreases (Ariyama et al., 2005; Ariyama, 2009; Naito et al., 2001). It means that carbon use efficiency C/Fe can be improved. There is some actual proof of this improvement using BIS (blast furnace inner-reaction simulator) (Naito et al., 2001) and commercial blast furnace (Nomura et al., 2005).

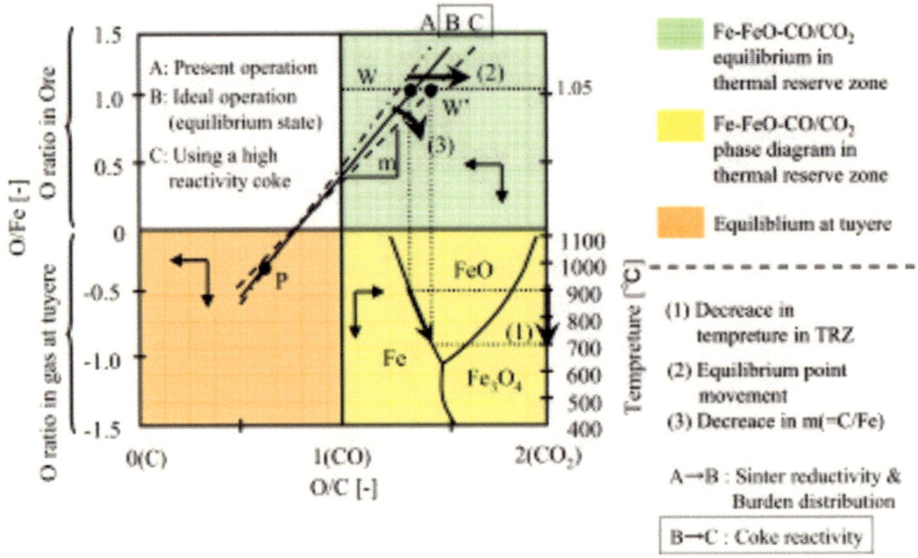

Figure 3. An example of Rist diagram with the operation line

3. The effect of catalyst on solution-loss reactivity

Reaction gas of coke solution-loss reaction (carbon dioxide) adsorbs and/or chemisorbs on the site of graphite structure. It is expected to be able to enhance the reactivity by a change of the site state. It is, however, difficult to change the site state by a change of only coke making process. Another idea is addition of metallic catalyst. This section will discuss about the element that indicates the catalyst activity, the effect of element and state of catalyst on solution-loss reactivity, and additional method.

The catalyst activity in each element have been organized and is shown in Fig. 4 (Lahaye & Ehrburger, 1991). Many kinds of alkali metal (Lahaye & Ehrburger, 1991; Tomita et al., 1983; Walker, 1968; Miura et al, 1989; Takarada et al., 1992; Jaran & Rao, 1978), alkaline earth metal (Tomita et al., 1983; Miura et al, 1989; Sears et al., 1980; Carzorla-Amoros et al., 1992; Yamada & Homma, 1979) and transition metal (Ohtsuka et al., 1986; Kashiwaya et al., 1991;

Tomita et al., 1983; Kamishita et al., 1980) shows good catalyst activity. Many kinds of inexpensive metal show the activity such as potassium, calcium and iron, respectively.

In metallurgical coke, on the other hand, there is mineral that is derived from coal ash and has catalyst activity. In addition, coke-matrix is not perfect graphite structure and amorphous structure coexists with crystalline structure. Therefore, it is necessary to consider the effect of the catalyst on solution-loss reactivity of *metallurgical coke*. Nomura et al. have compared catalyst activity among different element and compound (Nomura et al., 2005). Strontium, calcium and magnesium were selected. Figure 5 shows the relationship between the addition ratio of the alkaline earth metal and the JIS coke reactivity index. Rank of catalytic activity is Sr > Ca >> Mg. However, in this examination, they selected calcium as catalyst from these candidate materials because of cost performance of each material. In addition, he has reported that there is little difference between oxidation metal and carbonate. Grigore et al. have reported that not all iron, calcium, potassium and sodium crystalline mineral phases present in the coke catalyse the gasification reaction (Grigore et al., 2006). Coke reactivity increased with increasing total amount of catalysts in the crystalline phases. They concluded that the most likely materials responsible for the variation in coke reactivity are metallic iron and iron sulfides with a possible contribution by iron oxides and calcium sulfide. Kashiwaya et al. have particularly examined the reaction mechanism of solution-loss reaction with iron catalyst (Kashiwaya et al., 1991). Metallurgical coke, graphite and these materials with iron catalyst were gasified. The effect of iron addition to graphite is stronger than that to coke because of original mineral in raw coal. There are two ad- and/or chemisorption site on coke. First one is adsorbed by carbon dioxide and second one is adsorbed by carbon monoxide. If latter one increases with addition of iron catalyst, carbon dioxide adsorption is competitively inhibited and gasification rate decreases rapidly. It seems that adding iron to coke, latter site decreases and gasification is catalyzed.

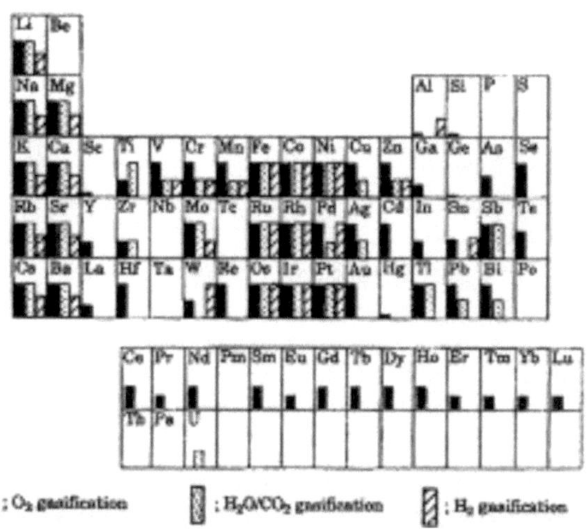

Figure 4. The catalyst activities in each element (Lahaye & Ehrburger, 1991)

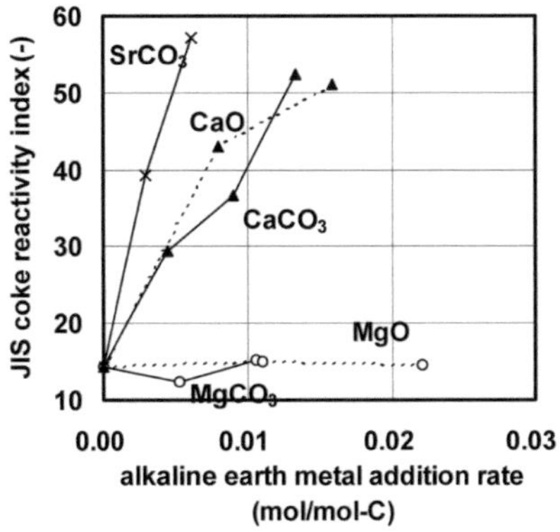

Figure 5. The relationship between the addition ratio of the alkaline earth metal and the JIS coke reactivity index (Nomura et al., 2005)

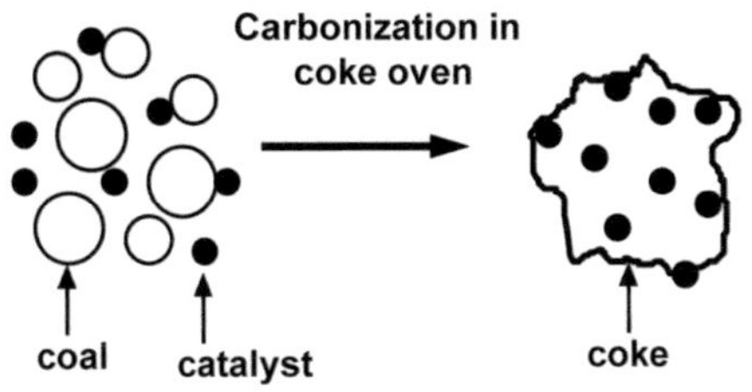

(a) 'Post-addition of catalyst to coke' method

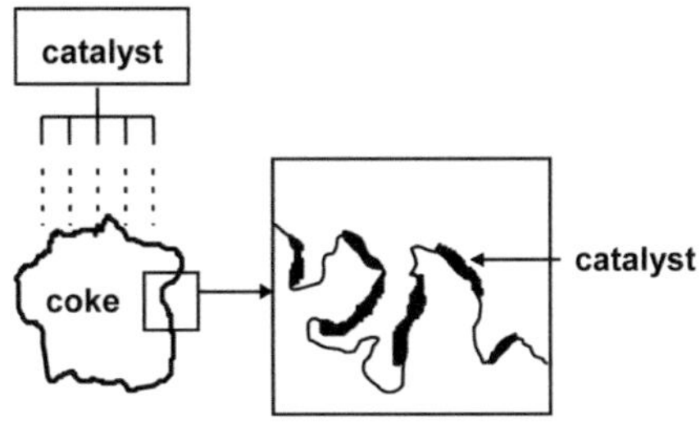

(b) 'Pre-addition of catalyst to coke' method

Figure 6. Conceptual diagrams of two methods of catalyst addition (Nomura et al., 2005)

As above, it seems that iron is the most useful source of catalyst because of cost, catalyst activity and source of pig-iron. Useful source of catalyst second to iron seems calcium, also, is very economical material but melt in by-product as slag.

Catalyst addition method can be classified into pre- and post-addition (Nomura et al., 2005). Pre- or post- mean before or after coke making. Figure 6 (Nomura et al., 2005) shows conceptual diagrams of both methods of catalyst addition. Post-addition is easier to enhance solution-loss reactivity (Kitaguchi et al., 2007) and to control the type of reaction and reaction surface than pre-addition (Nomura et al., 2007). On the other hand, process of pre-addition is simpler than that of post-addition, and is employed in extensive examination

(Nomura et al., 2005; Nomura et al., 2006; Nomura et al., 2007; Nomura et al., 2009; Fujimoto & Sato, 2010; Yamamoto et al., 2011; Yamazaki et al., 2010; Yamazaki et al., 2011). Particularly, if Fe addition is used, iron ore as iron/steel making material is useful as the source of catalyst (Nomura et al., 2009; Fujimoto & Sato, 2010; Yamazaki et al., 2010). Iron ore in briquetted material before carbonization (mixed-coal/ore = 70/30) is almost reduced during carbonization (reducing ratio is ca. 95%, Fujimoto & Sato, 2010).

4. Reaction type of metallurgical coke lump during solution-loss reaction with and without catalyst – The phenomena of coke-matrix state and controlling process in the lump

4.1. Introduction

4.1.1. The strength after gasification reaction and the spatial distribution of local porosity in coke lump

For the improvement of the carbon use efficiency, the important factors of highly reactive coke are not only the gasification reactivity of metallurgical coke but also the strength of one because metallurgical coke supports gas and liquid permeability in blast furnace. The issues on the strength of highly reactive coke are principally caused by catalyst addition. The issues can be divided into two main classes.

As the first one, at the time of *before changing into blast furnace*, coke strength changes (or mostly decreases) with catalyst addition regardless of adding method. Studies of coke strength degradation with catalyst addition have been performed (Nomura et al., 2005; Nomura et al., 2009; Fujimoto & Sato, 2010; Yamazaki et al., 2010). There are studies and knowledge in order to clarify mechanism of strength development or strength degradation as well as reports for practical and commercial making method. The cause of the strength degradation is the inhibition coal particle swelling and adhesion each other. A certain level of knowledge and technology is developed, and the coke that has appropriate strength *before charging into blast furnace* can be made now.

As another one, *after charging into blast furnace* and *after gasification reaction*, coke strength of highly reactive coke changes from that of non-reactivity-promoted coke. It is considered that coke pulverization and coke breeze generation are promoted after the gasification reaction because coke-matrix is more vanished and embrittled when the gasification reaction is catalyzed. Porosity or local porosity of porous media (rather than matrix strength of porous media strongly) affects its strength; hence a change of porosity or local porosity due to the coke-matrix vanishing. As practical knowledge, spatial distribution of *local porosity of coke lump* after gasification reaction strongly affects the strength of gasified coke (Kamijo et al., 1987). Nishi et al. have reported that coke after gasification has high pulverization resistance when there is unreacted-core observed as spatial distribution of local porosity of coke lump (Nishi et al., 1984; Nishi et al., 1987). Watakabe et al. have reported that the coke whose spatial gradient of gasification ratio (local porosity) near the outer region of coke lump is sharp has high pulverization and fracture resistance (Watakabe et al., 2001).

There are few fundamental (i.e. phenomenon analysing based) studies of *a change of coke strength after gasification reaction* (e.g. causal correlation between *gasification reaction* and *the strength* from the viewpoint of transport phenomena and reaction mechanism) although there are some reports in practical test. Meanwhile, it is certain that the coke has high pulverization resistance, if there is "unreacted-core" in spatial distribution of local porosity from empirical fact. It is because that the coke is planed from outer region with marked embrittlement, but fracture hardly occurs due to strength-reserved core. Hence, the fundamental studies should be used to develop the "unreacted-core" in spatial distribution of local porosity.

4.1.2. Reaction type of coke lump (resistance of reaction gas consumption in the lump and resistance of reaction gas diffusion into coke lump of as reaction-controlling process of coke lump) – a factor of the spatial distribution–

The spatial distribution of local porosity of coke lump is as a result of (1) reaction gas diffusion into coke lump, (2) reaction gas diffusion in coke lump, and (3) gasification reaction of carbon(coke)-matrix. In other words, the resistance of (1), (2), and (3) dominate the spatial distribution. Resistance of (2) and (3) govern *resistance of reaction gas consumption in the lump* and resistance of (1) governs *resistance of reaction gas diffusion into coke lump*. Reaction type of coke lump, which is represented as homogeneous reaction model or unreacted-core model discussed in reaction engineering, seems to be a result of balance of both the resistances (*reaction gas consumption in the lump* and *reaction gas diffusion into coke lump*). If diffusivity of reaction gas into inner region of the lump is more dominant than gasification reaction of carbon material, homogeneous reaction may be observed (resistance of gas diffusion into the lump >> resistance of gas consumption in the lump). Meanwhile, if the gasification is more dominant than the diffusivity, unreacted-core remains and embrittlement may be selectively observed from outer region of coke lump (resistance of gas consumption in the lump >> resistance of gas diffusion into the lump).

4.1.3. Coke-matrix state – another factor of the spatial distribution

However, metallurgical coke is porous media that contains pore of 50% in porosity. The size of metallurgical coke lump is from 25 mm to 50 mm. Hence, the important factors that dominate the spatial distribution are not only the *reaction type of coke lump* but also *coke-matrix state* as a result of above processes (2) and (3). Although resistance of (1) overcomes other resistances in whole process, the each rate of the processes (2) and (3) is finite after reaction gas diffusion into coke lump.

As a result of these phenomena, coke-matrix state, after gasification reaction, changes of coke microstructure in mm-scale have been observed as follows: (i) Coke-matrix (solid) is visually vanished (Watakabe & Takeda, 2001; Hayashizaki et al., 2009) and is as change of local porosity, and (ii) Elastic modulus of coke-matrix decreases (Hayashizaki et al., 2009). In former phenomenon (i), carbon dioxide diffuses into coke-matrix insufficiently, and coke-matrix on the surface reacts. In latter phenomenon (ii), a decrease of elastic modulus of coke-matrix is, on the other hand, correlated with nm-order micro pore volume. Hayashizaki et al reported the relationship between a decrease in the elastic modulus and an increase in nm-order pore volume during chemical reaction-controlling condition in which gasification rate of coke lump is not affected by reaction gas diffusion around the coke lump (Fig. 7, Hayashizaki et al., 2009). It has been known that volume of nm-order micro pore inside coke-matrix increases with progress of gasification (Kawakami et al., 2004) because carbon dioxide diffuses well into carbon-matrix.

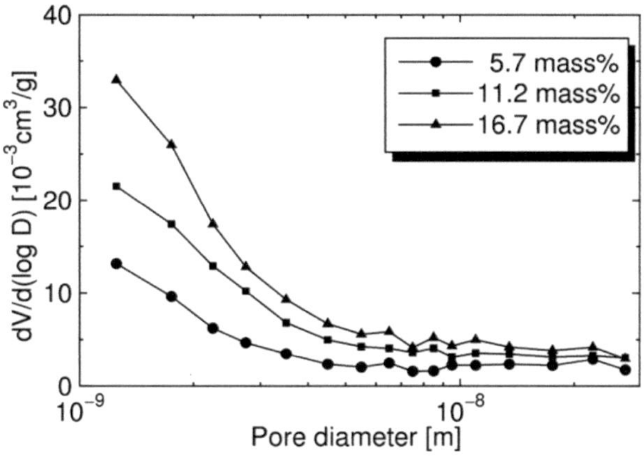

Figure 7. Change in micro pore size distribution of metallurgical coke with conversion correlated with elastic modulus by gas adsorption (Hayashizaki et al., 2009)

4.1.4. Purpose

If highly reactive coke reaction-promoted by catalyst is gasified, resistance of (3) decreases when both reaction temperature of highly reactive coke and ordinary coke is the same; hence the spatial distribution should become "unreacted-core". Actually, reaction temperature (i.e. TRZ temperature), however, decreases with use of highly reactive coke described as section 2.2.2. Additionally, Gasification reaction may be promoted on surface of the catalyst particle and coke-matrix. By existence of catalyst particle in coke lump, highly reactive coke will show different *reaction type of coke lump* and *coke-matrix state* from non-reactivity-promoted coke.

For this, as fundamental study, we have investigated the reaction mechanism and phenomena of coke before and after the gasification reaction. In section 4, we detail and discuss about these. The section is based on the research about these (Yamazaki et al., 2010; Yamazaki et al., 2011). Figure 8 shows position of the study in this section and whole picture of causal correlation between "coke gasification reaction" and "strength after gasification" with condition, phenomena and mechanism. At first, actual spatial distribution of local porosity of highly reactive coke in the TRZ condition when highly reactive coke is used in blast furnace is examined. Next, the factors that govern the spatial distribution (*reaction type of coke lump* and *coke-matrix state*) are estimated. Reaction mechanism estimation method is used to estimate the controlling process. Nano indentation method is used to measure the elastic modulus of coke-matrix. The elastic modulus is correlated with nm-order micro pore volume (Hayashizaki et al., 2009) that increases with progress of gasification (Kawakami et al., 2004).

Then, we propose a proper method to make highly reactive coke catalyzed by metals.

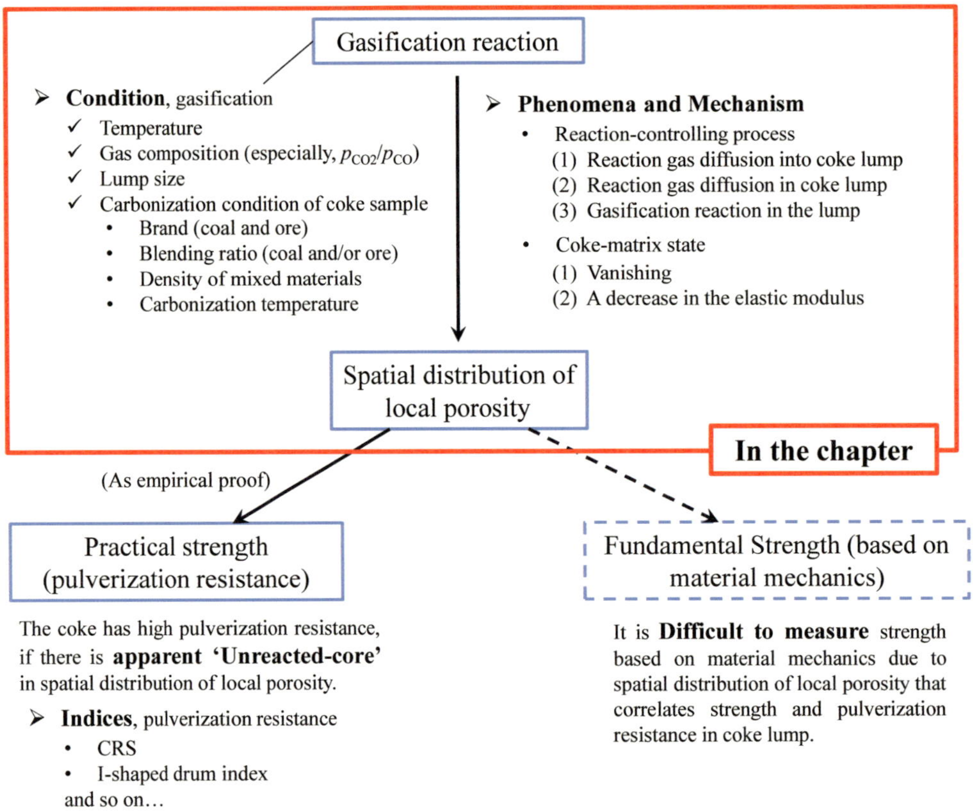

Figure 8. Position of the study in section 4 and whole picture of causal correlation between "coke gasification reaction" and "strength after gasification" with condition, phenomena and mechanism

4.2. Sample

4.2.1. Making

Coke lumps with and without iron-particles were made. Both cokes are called *ferrous coke* and *formed coke*, respectively. Slightly-caking coal and non-caking coal were used. Table 1 shows proximate and ultimate analysis of coals. Iron ore is, also, used as the source of iron catalyst. Table 2 shows major component of iron ore. Blending ratio of slightly-caking and non-caking coals whose diameter was less than 3 mm was 70/30 based on mass. Both coals were mixed well. In making of ferrous coke, mixed coal was also mixed with 30 mass % of iron ore whose diameter was under 250 μm. Mixed material was pressed into 6 mL of briquette at 296 MPa and was carbonized at 1273 K for 6h. After carbonization, the blending iron ore was completely reduced to metallic iron and was distributed uniformly. A representative photograph of sample after carbonization is shown in Fig. 13. In appearance, there is little difference for formed and ferrous coke. Form of sample is briquette whose size is 29 mm x 24 mm x 21 mm. Above mixing, pressing and carbonization process is the same as previous studies (Fujimoto & Sato, 2010; Yamazaki

Brand	Ash [db. %]	VM [db. %]	Fixed C [db. %]	Ultimate analysis [daf. %]				
				C	H	N	S	O
Slightly-caking coal	8.4	36.1	55.5	82.3	5.8	1.9	0.88	9.1
Non-caking coal	8.6	11.2	80.2	80.4	3.5	1.5	0.40	2.8

Table 1. Proximate and ultimate analysis of coals

T-Fe	FeO	SiO_2	CaO	Al_2O_3	MgO	P
67.5	0.21	1.31	0.01	0.73	0.01	0.033
S	Na	K	TiO_2	Mn	Zn	
0.010	0.01	0.01	0.07	0.11	0.003	

Table 2. Major component of iron ore [mass %]

Iron ore mixing ratio [mass%]	0 and 30
Reaction temperature [K]	1173
Reaction gas composition [vol.%]	CO_2/CO = 100/0 and 50/50
Gas flow rate [NL/min]	5
Gas velocity in reaction tube [m/s]	0.133
Conversion (Carbon basis) [mass%]	20

Table 3. Experimental conditions of CO_2 gasification reaction

4.2.2. Determination of loading mass% iron ore as Fe catalyst source

There are two purposes of Fe addition to metallurgical coke in iron/steel making process. First one is to decrease the amount of iron ore which must be reduced in blast furnace. Iron ore in briquetted material before carbonization (mixed-coal/ore = 70/30) is almost reduced during carbonization (reducing ratio is ca. 95%, Fujimoto & Sato, 2010). This fact shows that the required reducing gas that corresponds to carbon consumption to reduce iron ore can be decreased with the higher blending ratio of iron ore. Second one is to improve carbon use efficiency as described in section 2.2. Figure 9 shows that the initiation temperature of the gasification reaction decreases with an increase in the blending ratio of iron ore. The initiation temperature strongly correlates with TRZ temperature described in section 2.2. The initiation temperature saturated at 30 mass% in the blending ratio. From these viewpoints, the higher blending ratio of iron ore is better. In fact, TRZ temperature satisfactorily decreases by using ferrous coke that includes 30 mass% in the blending ratio shown in Fig. 10 (No. 1 and No. 6 shows traditional metallurgical coke and ferrous coke, respectively, Nomura et al., 2009).

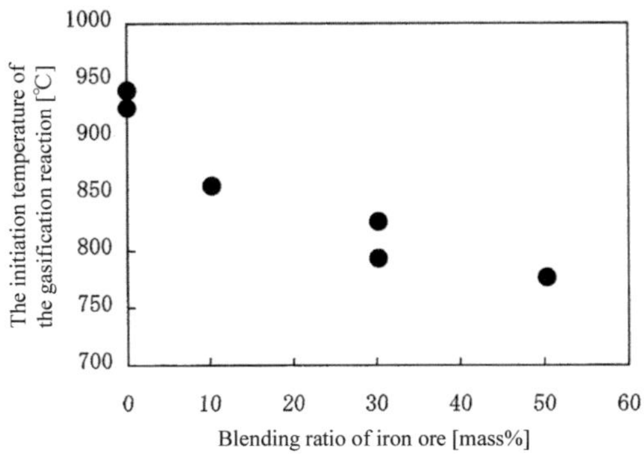

Figure 9. The effect of blending ratio of iron ore on the initiation temperature of the gasification reaction (Fujimoto & Sato, 2010)

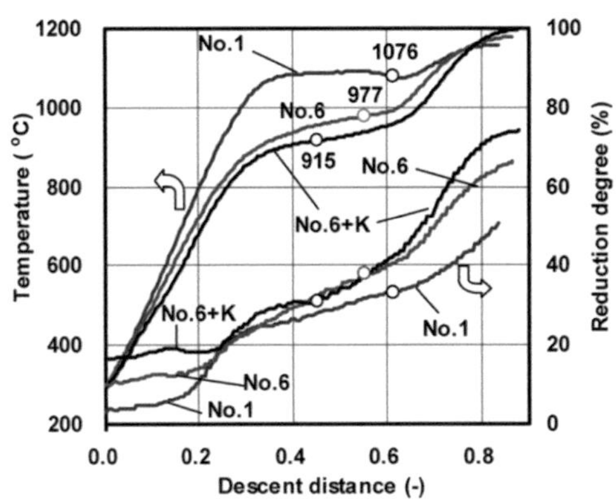

Figure 10. Temperature and reduction degree as a function of BIS descent distance (in the figure, No. 1 and No. 6 represent coke made by conventional coke and ferrous coke, respectively) (Nomura et al., 2009)

From the viewpoint of coke strength, the blending ratio of iron ore should be, on the other hand, limited. Figure 11 shows the relationship between the blending ratio of iron ore and tensile strength. Figure 12, also, shows relationship between the blending ratio of iron ore and agglomerated coal strength (the I-shaped drum index) that indicates pulverization resistance in blast furnace. By 30 mass% in the blending ratio, tensile strength and I-shaped drum index (ID600/10) are reserved, respectively. Both results suggest the same conclusion that the blending ratio of iron ore should be limited by 30 mass%. In Fig. 9, the effect of iron ore addition on TRZ temperature is satisfied at 30 mass% in the blending ratio. From the both viewpoint of the carbon use improvement and the strength (Figs. 11 and 12), 30 mass% in the blending ratio is proper in practical use.

Adding 30 mass% of iron ore is, hence, proper in practical use.

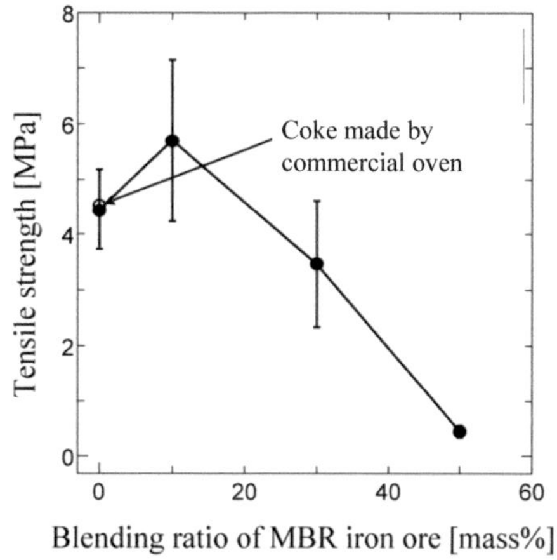

Figure 11. Relationship between tensile strength and blending ratio of iron ore (Yamazaki et al., 2010)

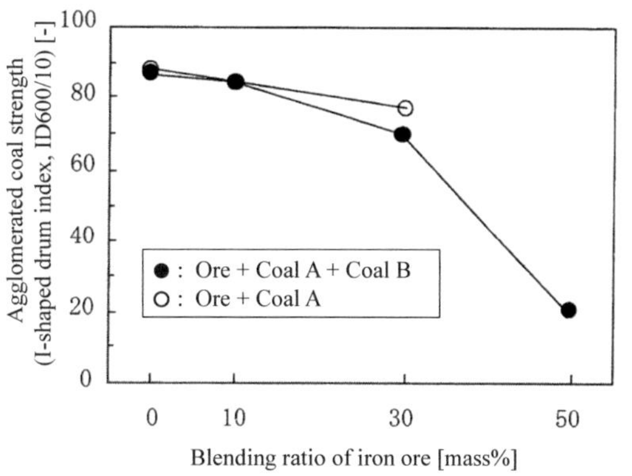

Figure 12. The effect of blending ratio of iron ore on agglomerated coal strength (I-shaped drum index, ID600/10) (Fujimoto & Sato, 2010)

4.3. Gasification (solution-loss reaction)

4.3.1. Experiment

Ferrous coke lump and formed coke lump were gasified by carbon dioxide – carbon monoxide mixture gas. Schematic diagram of experimental apparatus is shown in Fig. 14. A coke sample was hanged from the weighing scale to alumina reaction tube filled with alumina ball for heat transfer to gas. Reaction tube was heated by electric furnace. Reaction gases were led into the reaction tube from the bottom. The gases after gasification were cooled by water-cooling tube, and were then ejected outside. Weight loss with the gasification reaction was measured by weighing scale. Experimental conditions are shown in Table 3. Reaction temperature was set at 1173 K. Reaction gas compositions were set at 100/0 (Yamazaki et al., 2010) and 50/50 (Yamazaki et al., 2011) in ratio of CO_2/CO. Final conversion (ratio of weight loss based on carbon) x_B was 20 mass%.

Figure 13. A photograph of a sample after carbonization (in appearance, there is no difference for formed and ferrous coke) (Yamazaki et al., 2011)

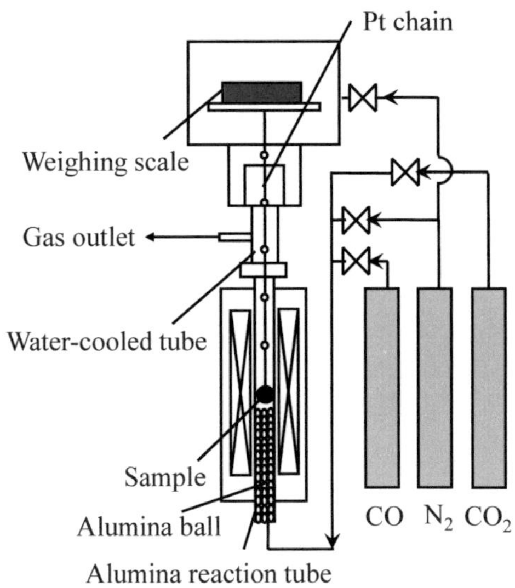

Figure 14. Schematic diagram of experimental apparatus (Yamazaki et al., 2011)

4.3.2. Determination of reaction gas composition in this study compared with actual gas composition at TRZ

Main component of the actual gas at TRZ is N_2 besides CO and CO_2. The actual gas composition $N_2/CO/CO_2$ is ca. 60/20/20. N_2 is from air origin and is, however, inactive for the gasification reaction. In case of gasification reaction by the mixture CO/CO_2, the reaction rate is governed by p_{CO}/p_{CO2} when the reaction gas contains above 10% of CO, and the ratio of N_2 does not affect the gasification rate although the actual gas contains a massive amount of nitrogen. The reasons are as follows:

In the gasification reaction of solid-carbon by the mixture CO/CO_2, the reactions in series are analogous to the resistance in series. The reaction rate is controlled by the reaction step which exerts most of the resistance to the overall reaction. There are two rate-controlling mechanisms during the gasification reaction in series: (1) dissociation of CO_2 on the surface of carbon, and (2) formation of CO on the surface of carbon. Carbon monoxide has a two-fold poisoning effect: (a) covering of the surface site due to strong adsorption, and (b) increasing the activity coefficient of the activated complex for the dissociation of CO_2; hence CO changes the rate-controlling mechanism. In the gasification of carbon material (e.g. charcoal, graphite and metallurgical coke), at CO contents above 10%, restance of (1) >> resistance of (2), and at low CO contents, resistance of (2) >> resistance of (1) (Turkdogan and Vinters, 1970).

In CO/CO_2 = 50/50, resistance (1) overcomes resistance (2). In this situation, the gasification rate is proportional to the difference of the partial pressure of actual gas CO_2 and the one governed by $2CO = C + CO_2$ equilibrium. The equilibrium is determined as p_{CO}/p_{CO_2}; thus the reaction rate is governed by actual gas p_{CO}/p_{CO_2} and equilibrium one. The actual gas composition indicated in the ratio of CO/CO_2 at TRZ is 50/50 (Nomura et al., 2006 and Nomura et at., 2009). So, the composition in the section, CO/CO_2 = 50/50, can represents the actual blast furnace condition, especially so-called at TRZ. Therefore, the results for the actual blast furnace gas composition like CO_2 18-20% and N_2 60% is expected to be similar to the result in this section.

In CO/CO_2 = 0/100, the gas composition is not actual. However, we investigate the behavior and phenomena as the model case in pure CO_2 condition due to gasification agent.

4.4. Spatial distribution of porosity

Cross-sectional digital images were taken by optical microscope (LV-100-POL, Nikon) . Spatial distribution of porosity after gasification reaction was measured by image analysis (Winroof 5.01, Mitani Corporation). Conceptual diagram of the taking procedure of digital image is shown in Fig. 15. Coke samples were buried into resin, cut and polished. From end to end of coke samples, digital images were three times taken in each sample. Taking area of digital images (3.14 mm x 2.35 mm, 2.45 μm/pixel) was slid aside in half length of image size.

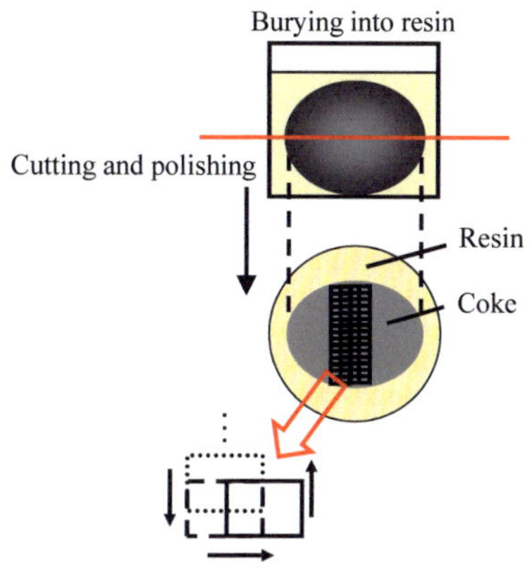

Figure 15. A conceptual diagram of the taking procedure of digital image (Yamazaki et al., 2011)

4.5. Reaction mechanism estimation of coke lump

In case of CO_2/CO = 100/0, unreacted-core model was used to estimate the gasification reaction mechanism of coke lump. Using time change of conversion x_B, dominant reaction mechanism of coke lump gasification can be estimated. Relationships between dimensionless reaction time t/t^* and conversion x_B when reaction controlling process is diffusion in boundary film, diffusion in product layer or reaction on the lump surface are expressed in Eqs. (7)-(9), respectively.

Diffusion in boundary film:

$$t/t^* = x_B \equiv f(x_B)\,. \tag{7}$$

Diffusion in product layer:

$$t/t^* = 1 - 3(1-x_B)^{2/3} + 2(1-x_B) \equiv f(x_B)\,. \tag{8}$$

Reaction on or in the lump surface:

$$t/t^* = 1-(1-x_B)^{1/3} \equiv f(x_B)\,. \tag{9}$$

Relationship between t/t^* and $f(x_B)$ of the dominant mechanism shows linear plot.

In case of CO_2/CO = 50/50, homogeneous reaction model was used. Time change of weight loss can be represented by this model when chemical reaction progress uniformly in whole lump. Mass balance is expressed as equation connected with chemical reaction rate and time derivative of mass. If reaction gas concentration is constant while reaction of lump progresses, chemical reaction rate is proportional to ratio of residual solid. Mass balance is written as

$$dx_B / dt = k(1-x_B)\,. \tag{10}$$

Integrated with initial conditions $t = t_0$ and $x_B = x_{B0}$, this can be written as

$$x_B = 1-(1-x_{B0})\exp\{-k(t-t_0)\} \tag{11}$$

If initial conditions t_0 and x_{B0} are equal to zero, the curve of Eq. (11) is through the origin. Weight loss curve is equal to Eq. (11) when lump reaction is controlled by chemical reaction.

4.6. Elastic modulus of coke-matrix

Elastic modulus of coke-matrix was measured by nano-indentation method. Load cycle indentation using sub-micron (or nano) indentation instruments is now a means of determining the deformation properties such as hardness and elastic modulus. A diamond tipped indenter with a precise geometry is pressed into a specimen with an increasing load up to a predetermined limit, and is then removed. The deformation properties can be determined using the load and displacement data obtained during the loading-unloading sequence. In this study, calculating method of elastic modulus was based on the method proposed by Oliver et al. When Berkovich triangular indenter which has 115-degree in angle is used, elastic modulus E_{eff} can be calculated by following formula:

$$h_s = 0.75\frac{P_{max}}{S} \tag{12}$$

$$h_c = h_{max} - h_s \tag{13}$$

$$A \approx 23.97h_c^2 \tag{14}$$

$$S = \frac{dP}{dh} = \frac{2}{\sqrt{\pi}}E_{eff}\sqrt{A} \tag{15}$$

$$\frac{1}{E_{eff}} = \frac{1-\nu^2}{E} + \frac{1-\nu_i^2}{E_i} \tag{16}$$

Resin-mounted specimens which are the same as ones mentioned *in 4.2.4* were used again. Measurement parts of test specimen were outer region (vicinity of surface) and inner region (vicinity of center). Measurement conditions are shown in Table 4. The number of measuring points was 50 by each sample and gas composition.

Indenter		Berkovich triangular pyramid
Loading/unloading velocity	[mN/s]	3
Maximum load	[mN]	100
Holding time in maximum load	[s]	2
The number of measurements		50

Table 4. Measurement condition of nano-indentation method

4.7. Results and discussion

4.7.1. CO_2/CO = 100/0

Spatial distribution of porosity after the gasification reaction

Figure 16 shows spatial distributions of porosity before and after gasification. Plots are denoted as average value. In formed coke, porosity was distributed uniformly along the radial direction. In ferrous coke, in outer region, porosity was significantly large. Relationships between porosity of each part and conversion based on carbon mass of ferrous coke lump are shown in Fig. 17. Plots and error bars are denoted as average value and standard deviation, respectively. In outer region, porosity increased with an increase in progress of gasification. On the other hand, in inner region, porosity hardly changed. Figures 16 (b) and 17(a) show that there is "unreacted-core" in local porosity distribution in ferrous coke after gasification reaction (CO_2/CO =100/0). It is suggested that chemical reactivity of gasification is advanced by the presence of iron-particles, and gasification in outer the coke lump is selectively progressed.

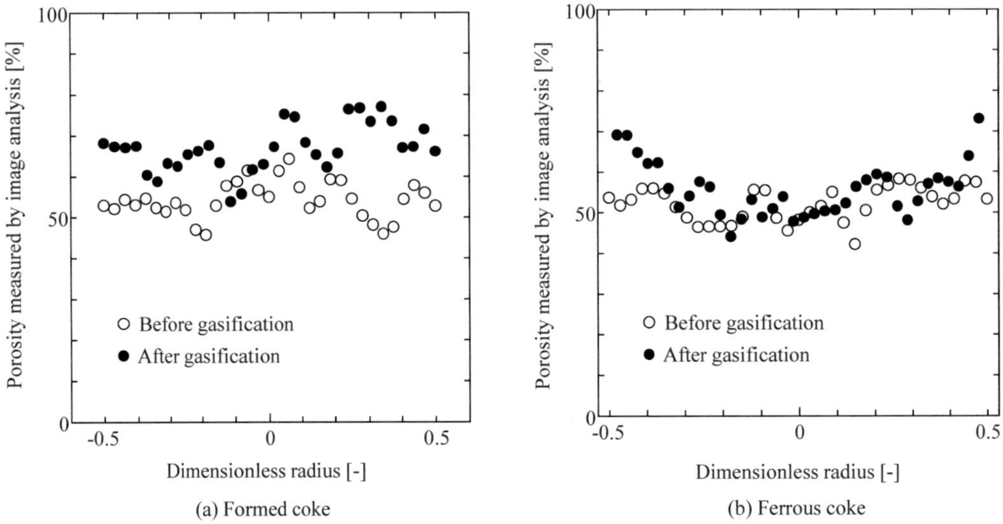

Figure 16. Spatial distributions of porosity before and after gasification (CO_2/CO = 100/0) of (a) formed coke and (b) ferrous coke (Yamazaki et al., 2010)

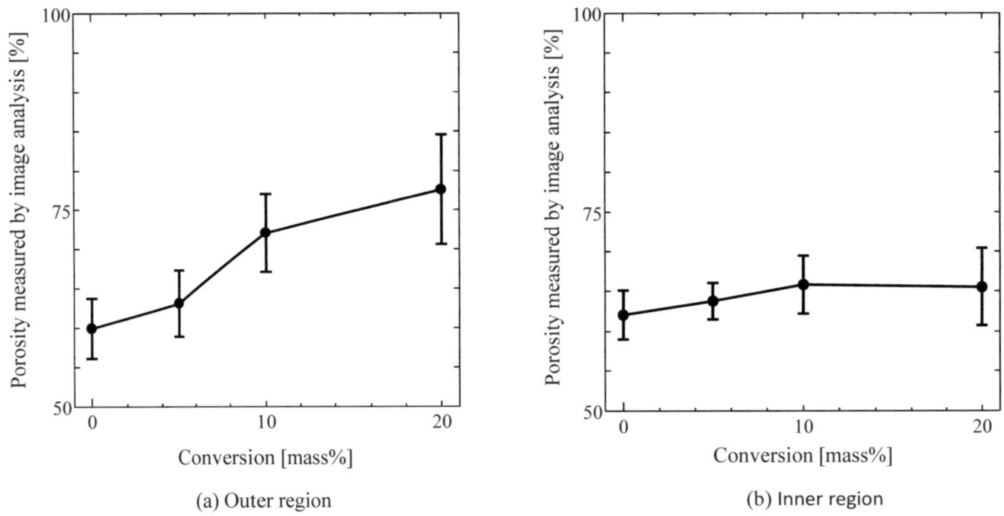

Figure 17. Relationships between porosity of (a) outer region and (b) inner region and conversion based on carbon mass of coke lump in ferrous coke lump at CO_2/CO = 100/0 (Yamazaki et al., 2011)

Cross-sectional images of formed coke and ferrous coke at 0.1, 0.2, 0.3, 0.4 and 0.5 in dimensionless radius before and after reaction under CO_2/CO = 100/0 gas atmosphere is shown in Fig. 18. In *formed coke after gasification* ((b) in Fig. 18), coke-matrix between the two pores (pore-wall) is thinning compared with *before gasification* reaction ((a) in Fig. 18). Microstructure at any dimensionless radius has the similar trend. An increase in local porosity shown in Fig. 16 should be due to the pore-wall thinning. There results show that whole coke lump uniformly embrittles. It is a matter of odds at where fracture origin is generated. Lump size may dramatically decrease due to lump fracture as split in fragments depends on the part of fracture origin. From the viewpoint of securing gas and liquid permeability in blast furnace, the increasing local porosity should be, therefore, avoided. In *ferrous coke after gasification* ((d) in Fig. 18), pore-wall in the region of 0.5 in dimensionless radius is dramatically thinning. Additionally, although most coke-matrix is continuing in other region, coke-matrix continuity is broken down in the region of 0.5 in dimensionless radius. These results indicate that the embrittlement occurs in the outer region of coke lump selectively; hence strength of whole coke lump can be maintained.

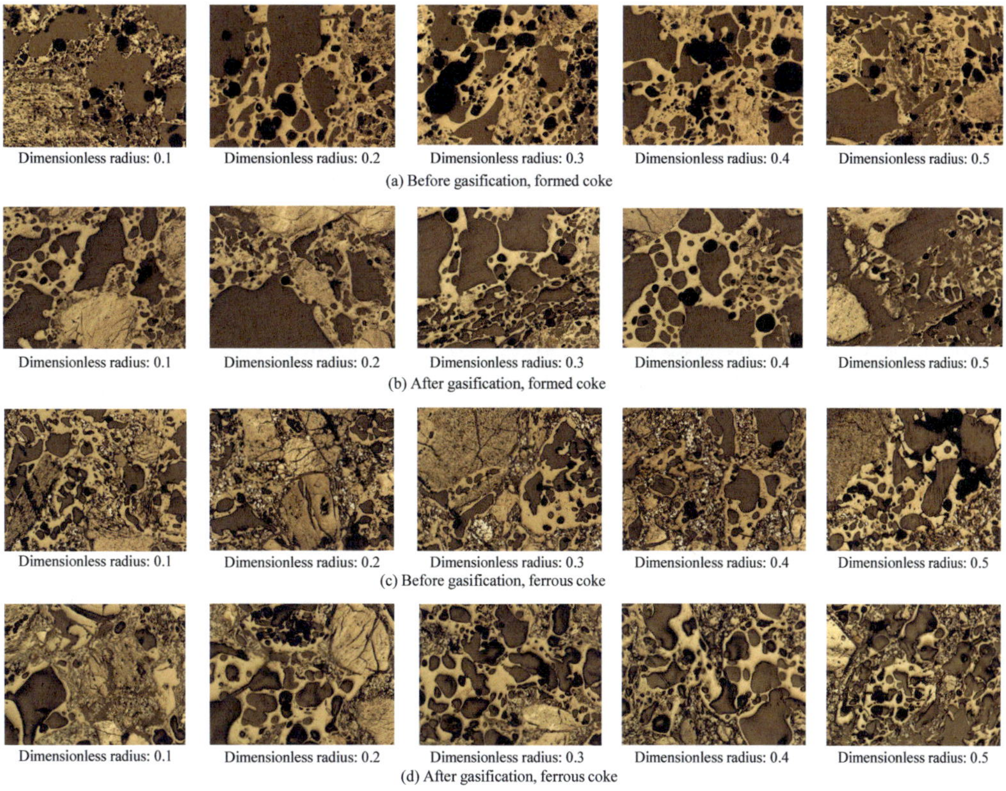

Figure 18. Cross-sectional images of (a, b) formed coke and (c, d) ferrous coke at 0.1, 0.2, 0.3, 0.4 and 0.5 in dimensionless radius, respectively, before and after reaction under CO_2/CO = 100/0 gas atmosphere; Taking area of each picture is 3.14 mm x 2.35 mm.

Reaction-controlling process of ferrous coke lump and formed coke lump are estimated. Figure 19 shows relationship between reaction time and conversion. In formed coke, weight loss behavior is not homogeneous reaction behavior despite uniform porosity distribution. In ferrous coke, at start of gasification, lump weight apparently increased due to oxidation of iron-particles by CO_2. Then lump weight decreased. Gasification reaction was terminated at 0.2 minus minimum value of conversion. Reaction time from minimum conversion to termination conversion was similar to reaction time of formed coke. Figure 20 shows results of the reaction controlling process estimation for formed and ferrous coke. Equations (7)-(9) are plotted, respectively. The lines in Fig. 20 are regression line using least squares method. The largest R^2 (correlation coefficient) is focused since the dominant mechanism shows linear plot. Both dominant mechanism of ferrous coke and formed coke are *diffusion in boundary film-controlling*. Despite not the same spatial distribution of porosity (Fig. 16 (a) and (b)), each reaction controlling process is the same.

Reaction gas diffuses into inner region of coke lump after passing through the boundary film around coke lump. The fact shown in Fig. 16 suggests that there are different behaviors in the inner region after the reaction gas diffusion through boundary film although the each dominant mechanism is the same. Figures 16 and 17 suggest that the rate-controlling process of whole lump reaction is "diffusion in boundary film around the lump" in both formed and ferrous coke in pure CO_2 condition. In the whole gasification process, resistance of "diffusion in boundary film" overcomes resistance of other process; hence the rate of other process can be assumed infinite. In local process after diffusion into the coke lump, on the other hand, the both processes of formed and ferrous coke (i.e. chemical reactivity and its topology) should be compared as the process which has finite rate to tell the difference of not the same spatial distribution of porosity. In ferrous coke, the chemical reactivity is catalyzed by metal iron catalyst. It is supposed that reaction gas gasifies coke-matrix and consumes rapidly soon after diffusing into the lump due to iron catalysis. Hence, gasification in outer region of lump progresses selectively. On the other hand, in formed coke, chemical reactivity is not catalyzed; hence chemical reaction rate (not whole weight loss rate since reaction–controlling process is gas diffusion through the boundary film) should be slower than ferrous coke. In other words, formed coke shows the chemical reaction on or in coke-matrix slower than diffusion in the lump. The gas is, in addition, easier to diffuse into inner coke lump due to 50 vol. % in porosity. As a result, there is no the "unreacted-core" in spatial distribution of porosity.

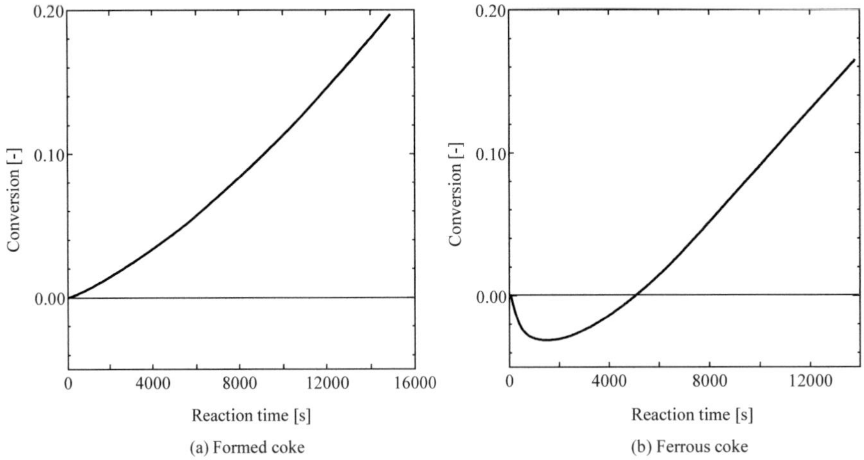

Figure 19. Relationships between reaction time and conversion of (a)formed coke and (b) ferrous coke lump in CO_2/CO = 100/0 (Yamazaki et al., 2010)

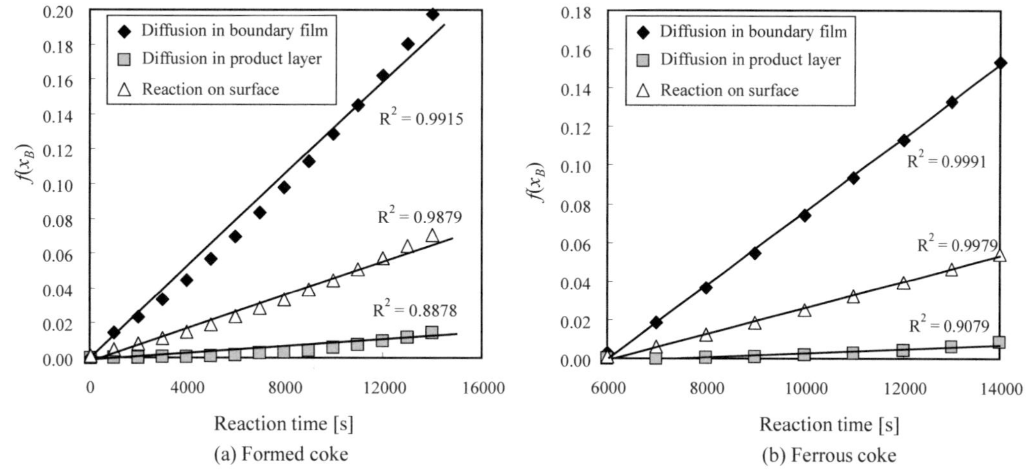

Figure 20. Results of the reaction controlling process estimation for (a) formed coke and (b) ferrous coke lump in CO_2/CO = 100/0 (Yamazaki et al., 2010)

Elastic modulus of coke-matrix

Elastic moduli of formed and ferrous coke-matrix before and after gasification are shown in Fig. 21 ((a) Outer region, formed coke, (b) Inner region, formed coke, (c) Outer region, ferrous coke and (d) Inner region, ferrous coke).

In inner region, both cokes, elastic moduli were not significantly changed. In formed coke, despite spatial distribution of porosity after gasification reaction was uniform, elastic moduli between before and after reaction is not significantly different. Coke-matrix vanishing occurred at surface between coke-matrix and mm-order pore. Meanwhile, for a decrease in elastic modulus, gas must diffuse into nm-order pore. Therefore, it seems that there is the difference between gasification rate of the vanishing and a decrease in elastic modulus. In ferrous coke, the inner region is unreacted-core.

In outer region, significant difference of elastic moduli is shown between formed and ferrous coke. In formed coke, elastic modulus of coke-matrix significantly decreased. It is suggested that the gas sufficiently diffuses into nm-order pore in outer region, and nm-order pore increased. However, in ferrous coke, the elastic modulus did not decrease with gasification reaction. In outer region, also, it is suggested that coke-matrix vanishing is more rapid than the gas diffusion into the nm-scale pores. In other words, it is suggested that weight loss of whole ferrous coke lump is caused not by an increase in nm-order pore but by the coke-matrix vanishing. Microstructures of ferrous coke before and after gasification in outer region are shown in Fig. 22. Before gasification, iron-particles were completely surrounded by coke-matrix. After gasification, coke matrix surrounding iron-particles did not exist. Iron particle contacts with coke-matrix. Therefore, only coke-matrix vanishing may be promoted.

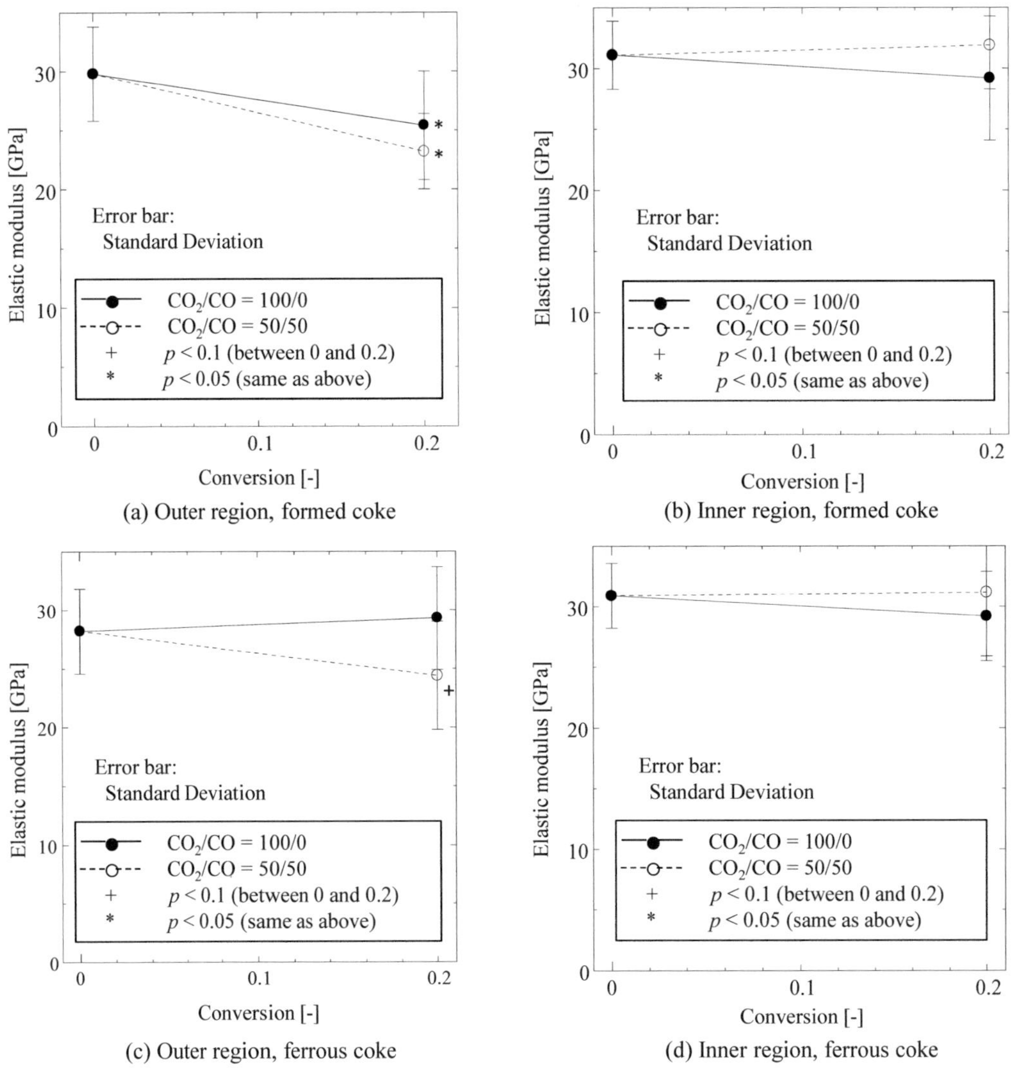

Figure 21. The difference of elastic modulus of coke matrix with gasification reaction. (a) Outer region, formed coke, (b) Inner region, formed coke, (c) Outer region, ferrous coke and (d) Inner region, ferrous coke.

4.7.2. $CO_2/CO = 50/50$

Spatial distribution of porosity after the gasification reaction

Figure 23 shows spatial distributions of porosity before and after gasification. In both ferrous coke and formed coke, porosity was distributed uniformly along the radial direction. In ferrous coke, also, there was no unreacted-core. Figure 24 shows relationships between reaction time and conversion and results of the reaction controlling process estimation using Eq. (8). In ferrous coke, lump weight apparently increased due to oxidation of iron-particles by CO_2 the same as Fig. 19 (b). In order to estimate only the controlling process of gasification reaction, the results after weight increase are plotted. Both ferrous coke and formed coke, weight loss behavior of gasification was similar. In addition, the conversion curves are very closely followed by the estimation results. Lump reaction mechanisms are described by homogeneous reaction model. Solution-loss reaction is inhibited when CO is contained with reaction gas since CO adsorbs competitively with CO_2 to active site on coke-matrix. Thus, the chemical reactivity in CO_2/CO = 50/50 is smaller than that in CO_2/CO = 100/0. So there is no unreacted-core since reaction gas CO_2 can diffuse enough into inner region of lump.

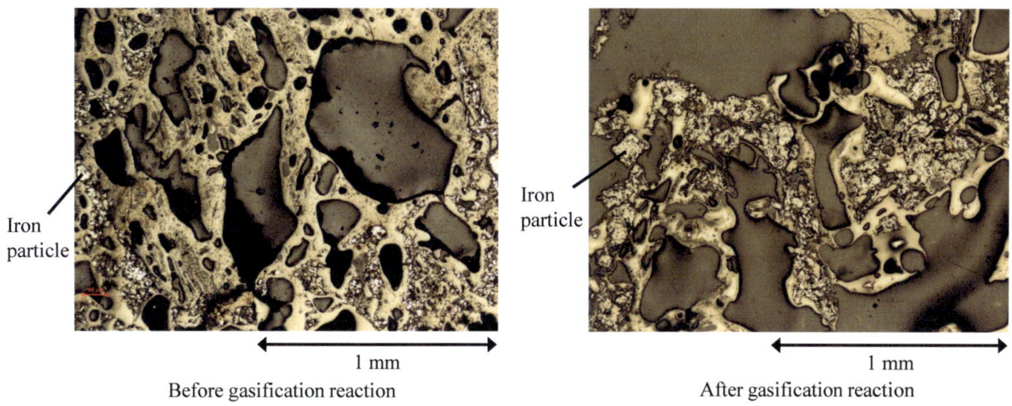

(a) Before gasification (b) After gasification

Figure 22. Microstructure of ferrous coke (a) before and (b) after gasification in outer region of coke lump in CO_2/CO = 100/0 (Yamazaki et al., 2010)

Cross-sectional images of formed coke and ferrous coke at 0.1, 0.2, 0.3, 0.4 and 0.5 in dimensionless radius before and after reaction under CO_2/CO = 50/50 gas atmosphere is shown in Fig. 25.*After gasification* ((b) in Fig. 25), both cokes have thinner pore-wall than cokes *before gasification* ((a) in Fig. 25). In ferrous coke, coke-matrix around iron particle, however, vanishes selectively, and continuity of coke-matrix is broken down; hence the strength degradation of ferrous coke seems to be more significant than that of formed coke.

Elastic modulus of coke-matrix

Elastic moduli of formed and ferrous coke-matrix before and after gasification are shown in Fig. 21 ((a) Outer region, formed coke, (b) Inner region, formed coke, (c) Outer region, ferrous coke and (d) Inner region, ferrous coke). In both cokes, elastic modulus in outer region decreases with gasification reaction. It is suggested that nm-order pore increases with gasification. However, a decrease in elastic modulus of ferrous coke-matrix (3.8 GPa) is smaller than that of formed coke (6.6 GPa). It seems that nm-order pore increment in volume of ferrous coke is smaller than that of formed coke.

Microstructures of ferrous coke before and after gasification are shown in Fig. 26. After gasification, coke matrix surrounding iron-particles did also not exist. Iron particles promote coke-matrix vanishing, but inhibit a decrease in elastic modulus. Coke-matrix is vanished selectively around iron particles. This phenomenon indicates that gasification reaction part in coke lump can be controlled by addition of iron particle.

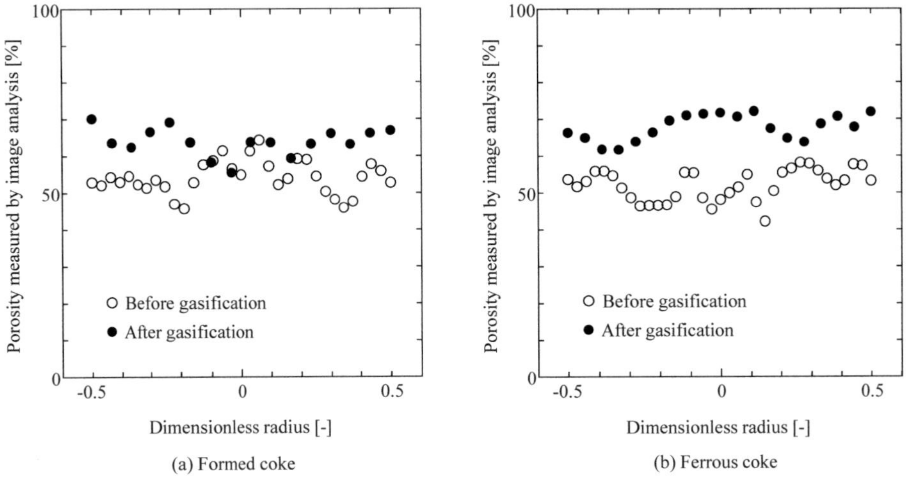

Figure 23. Spatial distributions of porosity before and after gasification ($CO_2/CO = 50/50$) of (a) formed coke, (b) ferrous coke (Yamazaki et al., 2011)

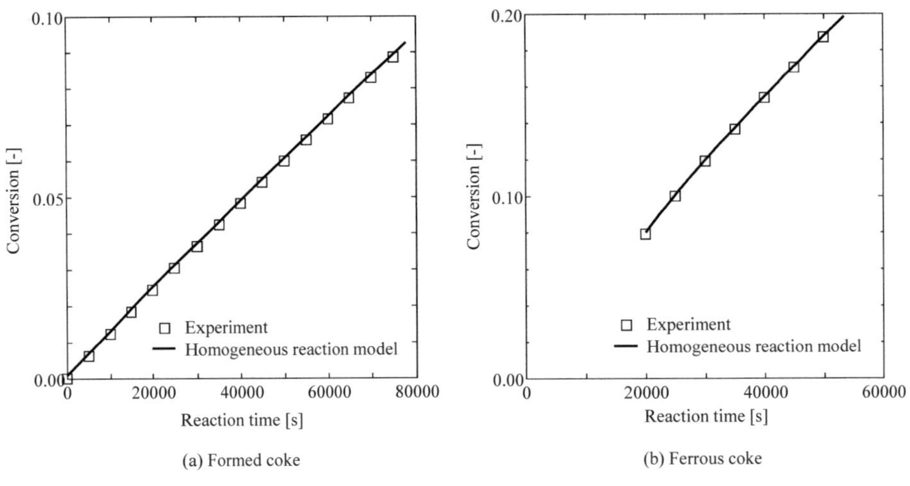

Figure 24. Results of the reaction controlling process estimation and relationship between reaction time and conversion based on carbon mass, (a) formed coke, (b) ferrous coke (Yamazaki et al., 2011)

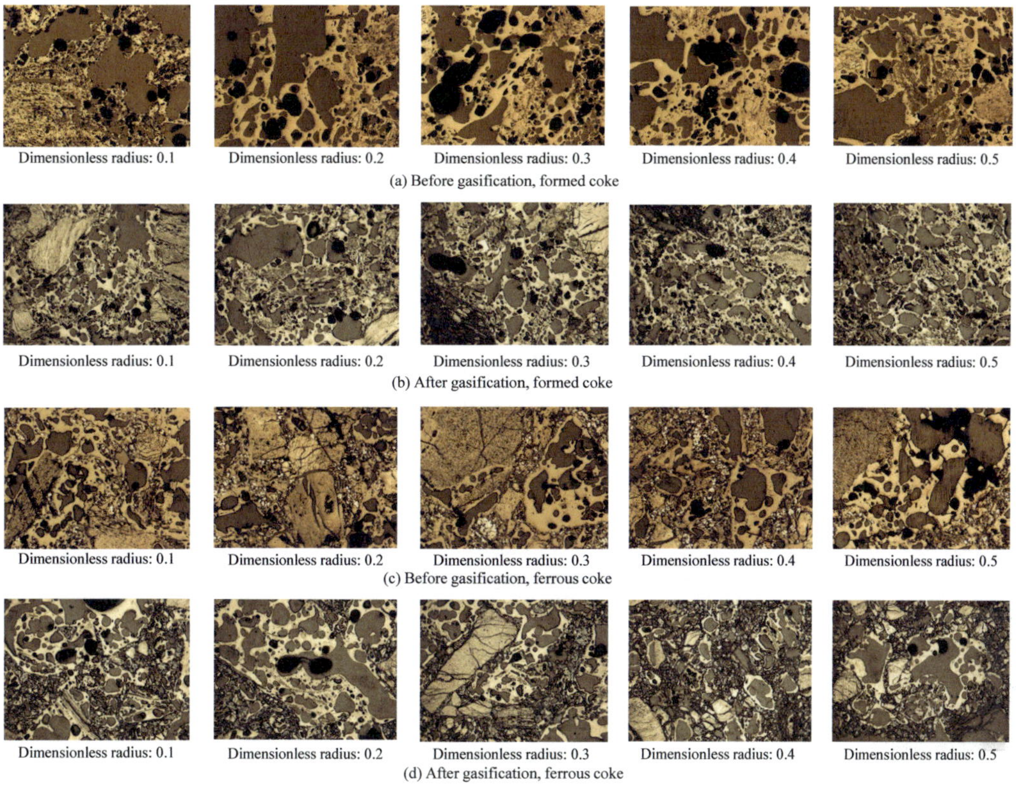

Figure 25. Cross-sectional images of (a, b) formed coke and (c, d) ferrous coke at 0.1, 0.2, 0.3, 0.4 and 0.5 in dimensionless radius, respectively, before and after reaction under CO_2/CO = 50/50 gas atmosphere; Taking area of each picture is 3.14 mm x 2.35 mm.

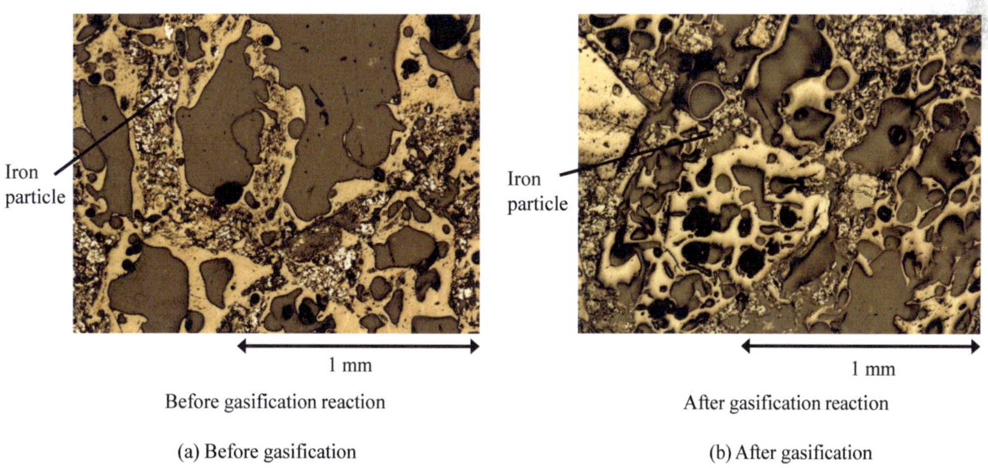

Figure 26. Microstructure of ferrous coke (a) before and (b) after gasification in outer region of the lump in CO_2/CO = 50/50 (Yamazaki et al., 2011)

4.7.3. Discussion about the difference between CO_2/CO = 100/0 and 50/50 from the viewpoint of the reaction type of coke lump (reaction-controlling process) and coke-matrix state

Reaction-controlling mechanism of coke lump

The rate constant of graphite or metallurgical coke in the case of CO_2/CO = 50/50 is 10 times smaller than that in the case of CO_2/CO = 100/0 (Miyasaka & Kondo, 1968; Turkdogan & Vinters, 1970). The reaction-controlling mechanism of coke lump changes from *diffusion in boundary film* to *reaction on or in the lump* surface due to the difference of chemical reactivity between each case of reaction gas composition.

In the case of CO_2/CO = 100/0, both of formed coke and ferrous coke show similar weight loss curve due to the same reaction-controlling mechanism. The same rate of whole weight loss is, hence, shown by the both cokes (Fig. 19). However, in the case of CO_2/CO = 50/50, both formed coke and ferrous coke show much different rate of weight loss each other due to chemical reaction-limited process of coke lump gasification. Ferrous coke shows the gasification rate about five times larger than the rate of formed coke. In comparison between the case of CO_2/CO = 50/50 and 100/0, ferrous coke and formed coke show the rate difference of weight loss about three and seven times, respectively. Weight loss rate can be assumed to be similarly equal to chemical reaction rate in this condition. Thus, the difference of weight loss rate should be considered as the difference of chemical reactivity. Due to these differences, there are the changes of the state of coke-matrix in coke lump after gasification reaction.

Coke-matrix vanishing

Formed coke:

As shown in Figs. 16(a) and 23(a), local porosity uniformly increases due to gasification reaction in both gas compositions CO_2/CO = 50/50 and 100/0. The porosity changes from about 50% to about 60 - 65% and each change shows a similar increase. Although there is the difference of elastic modulus between both gas compositions as shown in Fig. 21(a), the uniform change value of local porosity corresponds to 20% in the amount of weight loss based on the carbon weight before gasification reaction. To summarize above facts, both gas compositions CO_2/CO = 50/50 and 100/0 have some gasification behaviors of coke lump in common as follows: (1) After gas diffusion into coke lump, reaction gas diffuses over a range of whole lump through the mm-order (macroscopic) pore, (2) In outer region of coke lump, a decrease in the elastic modulus is observed due to an increase in nm-order pore in coke-matrix; however, amount of a decrease in the elastic modulus slightly affects amount of coke-matrix vanishing, (3) In TRZ temperature during using highly reactive coke, not-catalyzed coke shows *uniform* spatial distribution of local porosity *regardless of reaction gas composition*.

Ferrous coke:

In gas composition *CO_2/CO = 100/0*, coke-matrix in outer region of coke-lump is significantly vanished as shown in Figs. 16(b) and 17. Coke-matrix around the iron particle is more vanished than the matrix not around the iron particle. In inner region, on the other hand, there is scarcely any change of coke-matrix. In addition, there is scarcely any change of the elastic modulus as shown in Fig. 21; hence coke gasification reaction in inner region hardly occurred. Reaction gas gasifies coke-matrix around the iron particle soon after diffusion into coke-lump from the outside and consumes. Hence, as a result, reaction gas hardly diffuses into inner region.

In gas composition *$CO_2/CO = 50/50$*, the behavior of coke-matrix vanishing is different from the case of $CO_2/CO = 100/0$. The vanishing occurs uniformly in whole ferrous coke lump as shown in Fig. 23(b). Although the iron particle vanishes coke-matrix surrounding it as the same for the condition of $CO_2/CO = 100/0$, spatial distribution of local porosity is uniform over a range of whole lump after gasification reaction. The uniform change value of local porosity of formed coke is the same as that of formed coke.

Formation of unreacted-core in local porosity:

Due to addition of iron catalyst, the unreacted-core is formed in the condition (1173K, $CO_2/CO = 100/0$) in which formed coke forms no unreacted-core. On the other hand, in the condition (1173K, $CO_2/CO = 50/50$), there is no unreacted-core after reaction. However, the behavior is observed that the coke-matrix around the iron particle is preferentially vanished. This fact suggests that it is possible to form arbitrary spatial distribution of local porosity by location of iron particle in coke lump due to the effect of preferential vanishing of coke-matrix.

Elastic modulus of coke-matrix

Inner region:

Both gas compositions $CO_2/CO = 50/50$ and 100/0 show no difference of elastic modulus before and after gasification reaction whether or not formed coke or ferrous coke is gasified. Reaction gas diffusion seems to be difficult to occur into the bulk of coke-matrix because coke-matrix vanishing at surface of the matrix and mm-order pore is easier to occur than the diffusion into the bulk.

Outer region:

A decrease in elastic modulus occurs only in the outer region in both formed coke and ferrous coke. Each coke shows the different behavior of the decrease.

In *formed coke*, a decrease in elastic modulus significantly ($p < 0.05$) occurs before and after gasification reaction in both gas compositions $CO_2/CO = 50/50$ and 100/0. Outer region of coke lump seems the part where reaction gas concentration is higher and reaction gas-exposed duration is also longer than inner region of the lump; hence reaction gas diffuses into bulk of coke-matrix, and nm-order pore volume increases.

In *ferrous coke*, no significant decrease in elastic modulus is observed in the case of $CO_2/CO = 100/0$. Coke-matrix around the iron particle is vanished in outer region, and the conversion of whole lump weight loss reaches 20mass% before occurring of coke-matrix embrittlement. On the other hand, in the case of $CO_2/CO = 50/50$, elastic modulus significantly ($p < 0.10$)

decreases. As discussed previously, in comparison between the case of CO_2/CO = 50/50 and 100/0, ferrous coke shows the rate difference of weight loss about three times. The longer time causes reaction gas diffusion into bulk of coke-matrix, and nm-order pore volume increases.

4.7.4. Concluding remarks and proposal

The results in section 4 are concluded as follows:

- Gasification of metallurgical *coke lump* was principally observed as microscopic vanishing of coke-matrix in mm scale.
- A decrease in elastic modulus involved with an increase in volume of nm-order pore in bulk of coke-matrix was hardly observed.
- Iron particle in coke lump plays the role of the catalyst of not so much the degradation but the vanishing of coke-matrix
- Regardless of the reaction-controlling process difference of coke lump gasification, the chemical reactivity of coke-matrix which includes the effect of iron catalyst affects the spatial distribution of local porosity after gasification reaction. Therefore, controlling of iron particle alignment permits to control spatial distribution of local porosity and to form unreacted-core.

The study in this section shows relationship between gasification condition (temperature, gas composition and so on) and the formed spatial distribution of local porosity that affects the coke strength after gasification reaction. Particularly, the distinctive characteristic of coke-lump gasification which differs from gasification of other carbon material (i.e. electrode graphite and charcoal) is observed. Generally, a phenomenon of carbon material gasification in *fine granule* is observed as *opened microscopic- or mesoscopic-pore increasing* in carbon material, but a phenomenon of gasification of *metallurgical coke lump* is mostly observed as *coke-matrix vanishing*. Although coke-matrix degradation observed as a decrease in elastic modulus is also shown, the effect of the degradation on the relationship between coke-matrix vanishing and weight loss of whole lump is limited.

In previous study, it is noted that mm-pore structure in coke-lump affects the rate of weight loss of whole lump or pore structure after gasification in *TRZ condition* (temperature and reaction gas composition). In other words, gasification rate of whole lump is affected by reaction gas diffusion into inner region of lump rather than the gas diffusion into bulk of coke-matrix. Reaction gas diffuses into inner region of lump through the mm-pore of coke lump, and the gasification reaction mainly progresses on the surface between mm-pore and coke-matrix. The above mean that coke-matrix vanishing preferentially occurs than an increase in nm-pore volume and a decrease in elastic modulus and correspond to the results in this section.

In ferrous coke, the iron particle promotes vanishing of coke-matrix surrounding it rather than a decrease in elastic modulus. As discussed previously, degradation of coke-matrix is only limited, and the part where local porosity does not change seems to be never damaged. These facts suggest that spatial distribution of local porosity (in other words, formation of unreacted-core) can be controlled by alignment of iron particle.

The pre-addition method which is used in this section disperses catalyst iron particle in whole coke lump. Hence, coke-matrix in inner region of coke-lump is catalyzed as well as that in outer region. On the other hand, the post-addition method can disperse the catalyst particle locally. In addition, if catalyst particle is supported only in outer region of the lump, catalyst particle can remain in coke lump after gasification reaction (Yamamoto et al, 2010). To form unreacted-core that is not damaged, catalyst addition to only outer region seems to be useful. For the future, the method to control the spatial distribution of local pore or to form unreacted-core should be investigated as well as pre-addition method.

5. Conclusion

To realize the blast furnace operation in high carbon use efficiency, making of the coke which is satisfying *high solution-loss reactivity* and *high strength* is required. In order to make the coke which satisfies both the high solution-loss reactivity and the high strength, spatial distribution of local porosity of coke lump should be controlled as well as gasification reactivity of whole lump. In section 2, at first, the background of this issue was discussed. In section 3, the method to enhance coke solution-loss reactivity was briefly discussed. In section 4, the fundamental investigation and the proposal to support *the reactivity* and *the strength* at the same time was carried out. The essences discussed in these sections are summarized as follows:

Section 2:

- There are two iron reducing reactions (*direct* and *indirect* reaction).
- The *direct* reaction consumes solid carbon which is mainly metallurgical coke and is endothermic. In addition, the *direct* one occurs in the bottom part of blast furnace and affects the thermal balance in blast furnace and gas and liquid permeability, respectively; hence the ratio of *direct* reaction compared with that of *indirect* one should be decreased.
- The *indirect* reaction can be enhanced by change of C-O-Fe equilibrium state. The useful method for the change of equilibrium state is a decrease in TRZ temperature. The decrease can be achieved using highly-reactive coke.

Section 3:

- There are many elements which have catalyst activity of carbon gasification reacted by carbon dioxide (solution-loss reaction).
- Although there are the many kinds of catalyst that have good activity, calcium and iron are better element due to cost performance. Particularly, if Fe addition is used, iron ore as iron/steel making material is useful as the source of catalyst. In addition, iron ore in briquetted material before carbonization (mixed-coal/ore = 70/30) is almost reduced during carbonization (reducing ratio is ca. 95%). Hence, iron ore is available to use as catalyst source of highly-reactive coke. Therefore, iron seems the best material for highly-reactive coke in iron making process.
- There are two catalyst addition methods (post- and pre-addition).

Section 4:

- To maintain the strength after gasification, controlling of spatial distribution of coke lump is required.
- Gasification of metallurgical *coke lump* was principally observed as microscopic vanishing of coke-matrix in mm scale.
- A decrease in elastic modulus involved with an increase in volume of nm-order pore in bulk of coke-matrix was hardly observed.
- Iron particle in coke lump plays the role of the catalyst of not so much the degradation but the vanishing of coke-matrix
- Regardless of the reaction-controlling process difference of coke lump gasification, the chemical reactivity of coke-matrix which includes the effect of iron catalyst affects the spatial distribution of local porosity after gasification reaction. Therefore, controlling of iron particle alignment permits to control spatial distribution of local porosity and to form unreacted-core.

To obtain useful knowledge, all we only need to know is the relationship between the quality of practical use and the history during falling to the bottom part of blast furnace or the coke making condition (procedure and material). However, to propose design and making guideline of metallurgical coke considering condition surrounding coke, investing the phenomena of gasification of *whole coke lump* is very important. For understanding the phenomena, both of macroscopic (reaction type of coke lump or coke-matrix state discussed in section 4) and microscopic understanding and relationship between both are needed (cf. Fig. 8).

Microscopic investigations for metallurgical coke gasification have been performed based on chemical approach (e.g. reaction mechanism analysis (Turkdogan & Vinters, 1970), gaseous adsorption property (Turkdogan et al, 1970; Kashiwaya et al, 2003; Kawakami et al, 2004), crystal structure analysis (Kashiwaya & Ishii, 1990) and so on) since the middle in 20th century. These investigations are fundamental and test specimen finely crushed, but coke is used as lump. The knowledge from the investigations is very important but is indirectly linked to the quality of practical use (e.g. strength before and after gasification and whole lump reactivity). In this chapter, the macroscopic phenomena were discussed.

In the future, it is expected that the model which combines microscopic phenomena and transport phenomena in coke lump during gasification and derives macroscopic phenomena will be developed. This helps to understand the gasification phenomena inclusively. In addition, it will be able to propose the proper guideline of design and making aggressively.

Author details

Yoshiaki Yamazaki
Department of Chemical Engineering, Graduate School of Engineering, Tohoku University, Sendai, Japan

Acknowledgement

The author gratefully acknowledge the key contributions of principal researchers Takatoshi Miura, Hideyuki Aoki, Yohsuke Matsushita, Masakazu Shoji, Yasuhiro Saito, Seiji Nomura, Takashi Arima and Hidekazu Fujimoto and the valued contributions of numerous student researchers, including Kenta Ueoka, Hideyuki Hayashizaki and Tetsuya Kanai. In addition, *Section 4* in this chapter partly includes findings in the study carried as a part of the research activities "Fundamental Studies on Next Innovative Iron Making Process" programmed for the project "Strategic Development of Energy Conservation Technology Project". The financial support from New Energy and Industrial Technology Development Organization (NEDO) is gratefully acknowledged.

6. References

Ariyama, T. (2009). *Ferrum (bulletin of ISIJ)*, Vol. 14: p. 781

Ariyama, T., Murai, R., Ishii, J. & Sato, M. (2005). *ISIJ International*, Vol. 45: p.1371

Bannya, S. (2000). *Ferrous Process Metallurgy*, The Japan Institute of Metals, ISBN 4-88903-013-1, Sendai, Japan

Carzorla-Amoros, D., Linares-Solano, A. & Salinas-Martinez de Lecea, C. (1992). *Carbon*, Vol. 30: p. 995

Fujimoto, H. & Sato, M. (2010). *J. Jpn. Inst. Energ.*, Vol. 89: p. 21

Grigore, M., Sakurovs, R., French, D & Sahajwalla, V. (2006). *ISIJ International*, Vol. 46: p. 503

Hayashizaki, H., Ueoka, K., Kajiyama, M., Yamazaki, Y., Hiraki, K., Matsushita, Y., Aoki, H., Miurata, T., Fukuda, K. & Matsudaira, K. (2009). *Tetsu-to-Hagané*, Vol. 95: p. 593

Hayashizaki, H., Ueoka, K., Ogata, T., Yamazaki, Y., Matsushita, Y., Aoki, H., Miura, T., Fukuda, K. & Matsudaira, K. (2009). *Tetsu-to-Hagané*, Vol. 95: p.460

Jaran, B. P. & Rao, Y. K. (1978). *Carbon*, Vol. 16: p. 175

Kamijo, T., Iwakiri, H., Kiguchi, J., Yabata, T., Tanaka, H. & Kitamura, M., (1987). *Tetsu-to-Hagané*, Vol. 73: p.2012

Kamishita, M., Tsukashima, Y., Saga, M., Miyagawa, T. & Tanihara, H. (1980). *Nenryo Kyokaishi*, Vol. 59: p. 757

Kashiwaya, Y. & Ishii, K. (1991). *Tetsu-to-Hagané*, Vol. 76: p. 1254

Kashiwaya, Y., Nakaya, S. & Ishii, K. (1991). *Tetsu-to-Hagané*, Vol. 77: p. 759

Kawakami, M., Taga, H., Takenaka, T. & Yokoyama, S. (2004). *ISIJ International*, Vol. 44: p. 2018

Kitaguchi, H., Nomura, S. & Naito M. (2007). *Kagaku-Kogaku-Ronbunshu*, Vol. 33: p. 339

Lahaye, J. & Ehrburger, P. (1991). *Fundamental Issues in Control of Carbon Gasification Reactivity*, Kluwer Academic Publishers, ISBN-13:9780792310808, Norwell, USA

Miyasaka, N. & Kondo, S., (1968). *Tetsu-to-Hagané*, Vol. 54: p. 1427

Miura, K., Hashimoto, K. & Silveston, P. L. (1989). *Fuel*, Vol. 68: p. 1461

Naito, M., Okamoto, A., Yamaguchi, K, Yamaguchi, T. & Inoue, Y. (2001). *Tetsu-to-Hagané*, Vol. 87: p. 357

Nishi, T., Haraguchi, H. & Miura, Y. (1984). *Tetsu-to-Hagané*, Vol. 70: p. 43

Nishi, T., Haraguchi, H. & Okuhara, T. (1987). *Tetsu-to-Hagané*, Vol. 73: p. 1869

Nomura, S., Ayukawa, H., Kitaguchi, H., Tahara, T., Matsuzaki, S., Naito, M., Koizumi, S., Ogata, Y., Nakayama, T. & Abe, T. (2005). *ISIJ International*, Vol. 45: p. 316

Nomura, S., Higuchi, K., Kunitomo, K. & Naito, M. (2009). *Tetsu-to-Hagané*, Vol. 95: p. 813

Nomura, S., Kitaguchi, H., Yamaguchi, K. & Naito, M. (2007). *ISIJ International*, Vol. 47: p. 245

Nomura, S., Terashima, H., Sato, E. & Naito, M. (2007). *ISIJ International*, Vol. 47: p. 823

Nomura, S., Terashima, T., Sato, E. & Naito, M. (2006). *Tetsu-to-Hagané*, Vol. 92: p. 849

Nomura, S., Higuchi, K., Kunitomo, K. and Naito, M. (2009). *Tetsu-to-Hagané*, Vol. 95: p. 813

Ohtsuka, Y., Kuroda, Y., Tamai, Y. & Tomita, A. (1986). *Fuel*, Vol. 65: p. 1476

Rist, A. & Bonnivard, G. (1962). *Rev. Metall., Cah. Inf. Tech.*, Vol. 59: p. 401

Rist, A. & Meyerson, N. (1967). *J. Met*, Vol. 19: p. 50

Sears, J. T. Muralidhara, H. S. & Wen, C. Y. (1980). *Ind. Eng. Chem. Process Des. Dev.*, Vol. 19: p. 358

Takarada, T., Ichinose, S. & Kato, K (1992). *Fuel*, Vol. 71: p. 883

Tomita, A., Ohtsuka, Y. & Tamai, Y. (1983). *Fuel*, Vol. 62: p. 150

Tomita, A., Takarada, T. & Tamai, Y. (1983). *Fuel*, Vol. 62: p. 62

Turkdogan, E., Vinters, J., (1970). *Carbon*, Vol. 8: p. 39

Turkdogan, E., Olsson, R. & Vinters, J., (1970). *Carbon*, Vol. 8: p. 545

Walker, P. L. Jr. (1968). *Chem. Phys. Carbon*, Vol. 4: p. 287

Watakabe, S. & Takeda, K. (2001). *Tetsu-to-Hagané*, Vol. 87: p. 467

Yamada, T. & Homma, T. (1979). *Nenryo Kyokaishi*, Vol. 58: p. 11

Yamamoto, Y., Kashiwaya, Y., Miura, S., Nishimura, M., Kato, K., Nomura, S., Kubota, Y., Kunitomo, K. & Naito, M. (2010). *Tetsu-to-Hagané*, Vol. 96: p. 297

Yamamoto, Y., Kashiwaya, Y., Miura, S., Nishimura, M., Kato, K., Nomura, S., Kubota, Y., Kunitomo, K. & Naito, M. (2010). *Tetsu-to-Hagané*, Vol. 96: p. 288

Yamamoto, T., Sato, T., Fujimoto, H., Anyashiki, T., Fukada, K., Sato, M., Takeda, K. & Ariyama, T. (2011). *Tetsu-to-Hagané*, Vol. 97: p. 501

Yamazaki, Y., Hayashizaki, H., Ueoka, K., Hiraki, K., Matsushita, Y., Aoki, H. & Miura, T. (2010). *Tetsu-to-Hagané*, Vol. 96: p. 536

Yamazaki, Y., Hiraki, K., Kanai T., Zhang X., Shoji, M., Aoki H. & Miura T, An Experimental Study on the Effect of Metallic Iron Particles on Strength Factors of Coke after CO_2 Gasification Reaction, *The 27th Annual International Pittsburgh Coal Conference*, 48-3, Istanbul, Turkey, October 11-14, 2010

Yamazaki, Y., Hiraki, K., Kanai, T., Zhang, X., Matsushita, Y., Shoji, M., Aoki, H. & Miura, T. (2011). *J. Therm. Sci. Tech.*, Vol. 2: p. 278

Section 2

Biomass Gasification

Chapter 4

Gasification of Wood Bio-Oil

Younes Chhiti and Sylvain Salvador

Additional information is available at the end of the chapter

http://dx.doi.org/10.5772/48553

1. Introduction

Energy and environmental issues are two common concerns of modern society. Energy is a central part of every human being's daily life. In all its forms, such as chemical energy (food), thermal energy (heat), or electricity, energy has the ability to transform the daily lives of humans across the world by easing workloads, boosting economies and generally increasing the comfort of our lives. Worldwide energy consumption has been increasing rapidly. This has been accelerated by the improvement of the quality of life that almost directly relates to the amount of energy consumed. At present, fossil fuels based energy resources, such as coal, gas, and oil supply the majority of the total world energy requirement.

The global warming owing to the emissions of greenhouse gas is the most drastic consequence of the use of fossil fuels. According to experts in the field, global warming can disturb the natural equilibrium of the Earth's ecosystem. If CO_2 emissions are not regulated, global warming can have severe consequences for environment. These consequences, although some of them are not fully corroborated, are increasing sea and ocean levels, ocean acidification, change in rainfall patterns, hurricanes, volcanic eruptions, earthquakes and plant or animal extinctions, among others.

The development of non-conventional sources like wind, sunlight, water, biomass, etc., is inevitable. Among the renewable sources of energy, substantial focus of research is currently on the use of biomass. Besides being a renewable source of energy, there are many other advantages associated to the use of biomass. It is available abundantly in the world. Its use does not increase the net amount of CO_2 in the atmosphere and can reduce the emissions of SO_2 and NO_x remarkably.

Biomass gasification is a promising technology, which can contribute to develop future energy systems which are efficient and environmentally friendly in order to increase the share of renewable energy for heating, electricity, transport fuels and higher applications.

The gasification of carbon-containing materials to produce combustible gas is an established technology. Biomass gasification is a thermochemical process that produces relatively clean and combustible gas through pyrolytic and reforming reactions. The syngas generated can be an important resource suitable for direct combustion, application in prime movers such as engines and turbines, or for the production of synthetic natural gas (SNG) and transportation fuels e.g. Fischer-Tropsch diesel.

For energy production, the major concerns about syngas are its heating value, composition, and possible contamination. The proportion of the combustible gas hydrogen (H_2), methane (CH_4), carbon monoxide (CO), and moisture determines the heating value of the gas. The composition of syngas depends on the biomass properties and gasifier operating conditions. For a specific gasification system, operating conditions play a vital role in all aspects of biomass gasification. These include carbon conversion, syngas composition, tars and soot formation and oxidation (Devi et al., 2003).

The main hurdles for large-scale implementation of energy production from solid biomass are the nature of biomass - non uniform, low-energy density, sometimes large ash content - together with the usual inconsistency between the local availability of biomass and the demand for biomass related products: heat, electricity, fuels and chemicals. Usually, import/transport of fossil fuels is cheaper. Pyrolysis may be a process to overcome these hurdles: biomass is transformed into a versatile liquid called bio-oil, easy to handle and to transport. This bio-oil would then be transported to centralized air/steam gasification units. Bio-oil is an intermediate product which is produced from relatively dry biomass via fast pyrolysis process. It is a liquid with similar elemental composition to its original feedstock and with high bulk and energy density. The high bulk and energy density of bio-oil can reduce transportation costs to large scale centralized gasification plants; these costs have been a detrimental factor in large scale use of solid biomass resource. Bio-oil can be produced where the biomass is available and then be transported over long distances to central processing units of similar scales as the current petrochemical industry. Besides technical and logistic advantages, this conversion chain may also give incentives for economic development and job creation especially in rural areas.

The essential features to obtain high yields of bio-oil (up to 75 wt% on dry basis) are a moderate pyrolysis temperature (500°C), high heating rates (10^3-10^5°C/s), short vapour residence times (<2 s) and rapid quenching of the pyrolysis vapours.

The combination of fast pyrolysis of biomass followed by transportation in large units for steam reforming has attracted considerable attention of the research community, as one of the most promising viable methods for hydrogen production. For the high temperature applications such as gasification, steam reforming or even combustion, it is of particular interest to understand the behavior of bio-oils during the very first step of its decomposition under pyrolysis conditions. However, only few works can be found on the understanding of processes occurring during thermal conversion of bio-oils.

The earliest combustion tests of bio-oil droplets were conducted in Sandia National Laboratory (Wornat et al., 1994). Streams of monodispersed droplets were injected into a laminar flow reactor. The experimental conditions were as follows: droplet diameter of about 300 μm, reactor temperature of 1600 K and O_2 concentrations of 14–33%. In-situ video imaging of burning droplets reveals that biomass oil droplets undergo several distinct stages of combustion. Initially biomass oil droplets burn quiescently in a blue flame. The broad range of component volatilities and inefficient mass transfer within the viscous biomass oils bring about an abrupt termination of the quiescent stage, however, causing rapid droplet swelling and distortion, followed by a microexplosion.

Thermogravimetric analysis (TGA) is widely used to characterize the evaporation, thermal decomposition and combustion properties of bio-oils. The weight loss process of bio-oils in inert atmospheres can be divided into two stages: the evaporation of light volatiles (<150-200°C) and the subsequent thermal decomposition of unstable heavier components (<350-400°C). In the case of TGA tests performed in the presence of air, the weight loss of bio-oils can be divided into three stages. The first two stages are similar to those in inert atmospheres and the third stage is the combustion of chars formed in the first two stages (>400°C) (Ba et al., 2004a, 2004b).

Branca et al. (Branca, 2005a) studied the devolatilization and heterogeneous combustion of wood fast bio-oil. Weight loss curves of wood fast bio-oil in air have been measured, under controlled thermal conditions, carrying out two separate sets of experiments. The first, which has a final temperature of 600 K, concerns evaporation/cracking of the oil and secondary char formation. A heating rate of 0.08 °C/s was applied. The yield of secondary char varies from about 25% to 39% (on a total oil basis). After collection and milling, in the second set of experiments, heterogeneous combustion of the secondary char is carried out to temperatures of 873 K. In another study, Branca et al. (Branca et al., 2005b) found that thermogravimetric curves of bio-oil in air show two main reaction stages. The first (temperatures ≤ 600 K) concerns evaporation, formation and release of gases and formation of secondary char (coke). Then, at higher temperatures, heterogeneous combustion of secondary char takes place. They found that the pyrolysis temperature does not affect significantly weight loss dynamics and amount of secondary char (approximately equal to 20% of the liquid on a dry basis).

Hallet et al. (Hallett et al., 2006) established a numerical model for the evaporation and pyrolysis of a single droplet of bio-oil derived from biomass. The model is compared with the results of suspended droplet experiments, and is shown to give good predictions of the times of the major events in the lifetime of a droplet: initial heating, evaporation of volatile species, and pyrolysis of pyrolytic lignin to char.

Guus van Rossum et al. (Van Rossum et al., 2010) studied the evaporation of bio-oil and product distribution at varying heating rates (~1.5×10^{-2}–1.5×10^{4}°C/s) with surrounding temperatures up to 850°C. A total product distribution (gas, vapor, and char) was measured using two atomizers with different droplet sizes. A big difference is seen in char production between the two atomizers where the ultrasonic atomizer gives much less char compared to the needle atomizer, ~8 and 22% (on carbon basis), respectively. Small droplets (88-117μm generated by ultrasonic atomizer, undergoing high heating rate) are much quicker evaporated than larger droplets (~ 1.9 mm, generated by needle atomizer, undergoing low heating rate).

Calabria et al. carried out lots of studies on the combustion behaviors of fibre-suspended single bio-oil droplets. The droplet size varied between 300 and 1100 μm and the furnace temperature changed in the range of 400–1200°C. The droplets were observed to undergo initial heating, swelling and microexplosion before ignition. During this stage, the temperature–time curves showed two zones with constant temperatures (100 and 450°C), which corresponded to the evaporation of light volatiles and the thermal cracking of unstable components, respectively. The droplets were ignited at around 600°C. The combustion of the droplets started with an enveloping blue flame. Then, the flame developed a yellow tail with its size increasing, which indicated the formation of soot. After that, the flame shrank and extinguished, and the remaining solid carbonaceous residues burned leading to the formation of ash (Calabria et al., 2007).

In air/steam gasification process the essential steps are pyrolysis, partial oxidation, cracking of tar, solid carbon residue gasification, reforming (steam and/or dry), and water gas shift to yield syngas, water, carbon dioxide, and unwanted products like tars, methane and carbon (Levenspiel et al., 2005). As a summary, a schematic representation of air/steam gasification of single droplet of bio-oil is proposed in Figure 1.

The steam reforming of the bio-oil can be simplified as the steam reforming of an oxygenated organic compound ($C_nH_mO_k$) following:

$$C_nH_mO_k + (n - k)H_2O \leftrightarrow nCO + (n + m/2 - k)H_2 \quad (1)$$

During the last decade, catalytic steam reforming of bio-oil components has been widely studied, focusing on acetic acid as one of the most representative compounds.

Production of hydrogen from catalytic steam reforming of bio-oil was extensively investigated by NREL (Wang et al., 1997, 1998). Czernik et al. obtained hydrogen in a fluidized bed reactor from the carbohydrate derived fraction of wood bio-oil with a yield of about 80% of theoretical maximum (Czernik et al., 2002). The catalytic steam reforming of the bio-oil or the model oxygenates (e.g., ethanol, acetic acid) has been widely explored via various catalysts, e.g., Ni-based catalysts (Sakaguchi et al., 2010), Mg-doped catalysts (Garcia et al., 2000) and noble metal-loaded catalysts (Goula et al., 2004; Rioche et al., 2005; Trimm et al., 1997). Noble metals (Pt, Ru, Rh) are more effective than the Ni-based catalysts and less carbon depositing. Such catalysts are not common in real applications because of their high cost. Catalytic steam reforming of bio-oil is a costly process and presents several disadvantages such as carbon deposit and the deactivation of catalysts due to coke or oligomer deposition even in the presence of an excess of steam ($S/C > 5$) (Trimm et al., 1997; Rostrup-Nielsen et al., 1997). For these reasons, there is an interest in developing non catalytic gasification of bio-oil.

Only very few works can be found on the non catalytic reforming of whole bio-oil. Bimbela et al. studied catalytic and non catalytic steam reforming of acetol (bio-oil model compound)

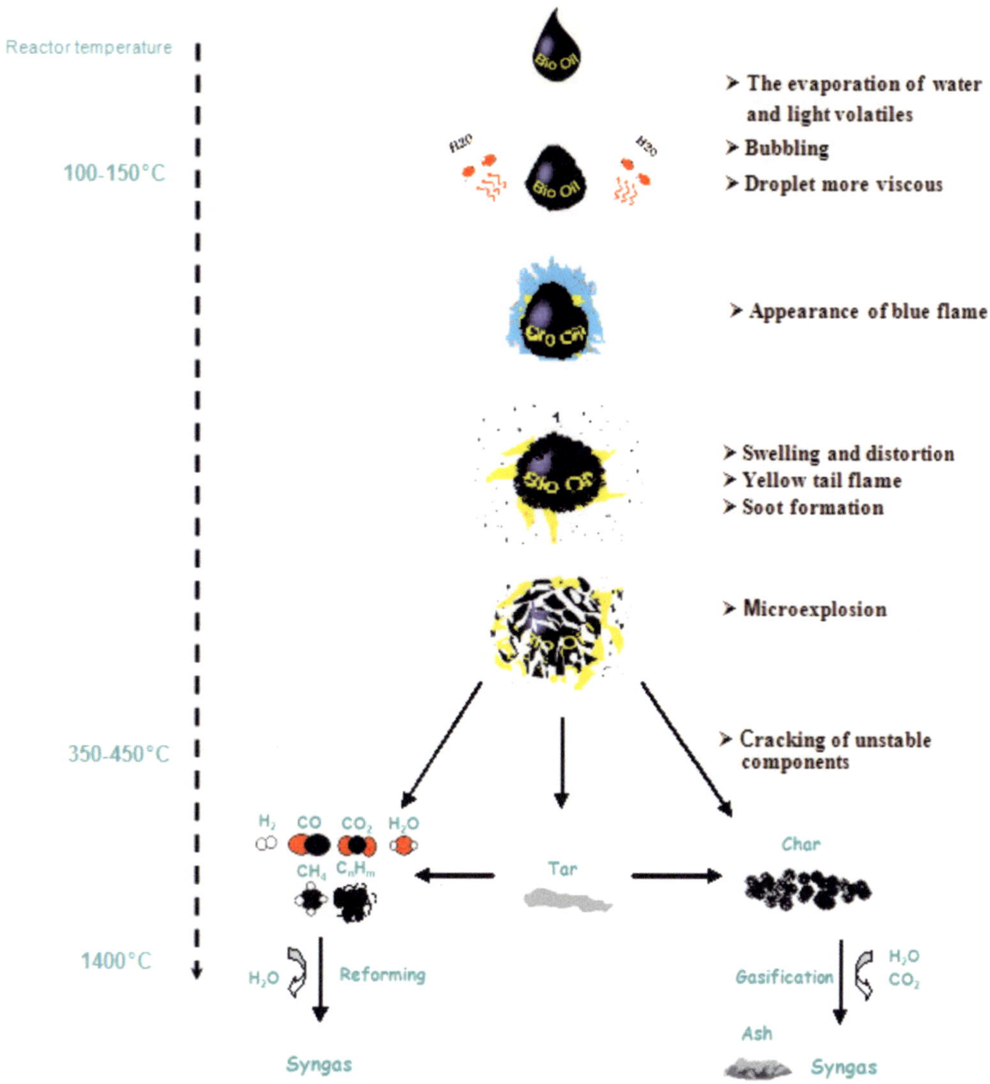

Figure 1. Schematic representation of air steam gasification of bio-oil droplet

in fixed bed at low temperature (550-750°C) in order to highlight the specific role of the catalyst in this process (Bimbela et al., 2009). The same study is carried out by Guus van Rossum et al. concerning catalytic and non catalytic gasification of bio-oil in a fluidized bed over a wide temperature range (523-914°C) (van Rossum et al., 2007). Marda et al. has developed a system for the volatilization and conversion of a bio-oil mixed with methanol to syngas via non-catalytic partial oxidation (NPOX) using an ultrasonic nozzle to feed the mixture. The effects of both temperature (from 625 to 850°C) and added oxygen (effective O/C ratio from 0.7 to 1.6) on the yields of CO and H_2 have been explored. They obtained hydrogen yield of about 75% of theoretical maximum (Marda et al., 2009). Panigrahi et al. gasified biomass-derived oil (BDO) to syngas and gaseous fuels at 800°C. They obtained syngas (H_2 + CO) yield ranging from 75 to 80 mol % (Panigrahi et al., 2003). Henrich et al. gasified lignocellulosic biomass. The first process step is a fast pyrolysis at atmospheric pressure, which produces large condensate that was mixed to slurries. The slurries are pumped into a slagging entrained flow gasifier and are atomized and converted to syngas at high operating temperatures and pressures (Henrich et al., 2004).

The objective of this work is to bridge the lack of knowledge concerning the physicochemical transformation of bio-oil into syngas using non catalytic steam gasification in entrained flow reactors. This complex process involves vaporization, thermal cracking reactions with formation of gas, tars and char that considered as undesirable product. This is followed by steam reforming of gas and tars, together with char conversion. To better understand the process, the first step of gasification (pyrolysis) and thereafter the whole process (pyrolysis + gasification) were studied. The pyrolysis study focused on the influence of the heating rate and the final pyrolysis temperature, for this aim, two complementary devices namely: a Horizontal Tubular Reactor (HTR) and a High Temperature - Entrained Flow Reactor (HT-EFR) were used to study on the one hand a wide range of heating rates, in the range from 2 to 2000°C/s and on the other hand final temperature ranging from 550 to 1000°C. Concerning gasification, the effect of temperature on syngas yield and composition was studied over a wide range from 1000°C to 1400°C, for this aim HT-EFR was used.

2. Materials and methods

2.1. Description of the laboratory device and of the procedure

Two complementary devices, namely: a Horizontal Tubular Reactor (HTR) and a High Temperature - Entrained Flow Reactor (HT-EFR), were used to study a wide range of heating rates, in the range from 2 to 2000°C/s and final temperature from 550 to 1000°C.

2.1.1. Horizontal Tubular Reactor HTR

The experiments of fast pyrolysis were carried out in a HTR (Fig. 2). This device allowed carrying out experiments in conditions of fast pyrolysis which is not possible in a thermobalance. The reactor consisted of a double-walled quartz pipe. The length and inside diameters were 850 mm and 55 mm respectively for the inner tube, and 1290 mm and 70 mm respectively for the outer tube. The reactor temperature can reach 1100°C.

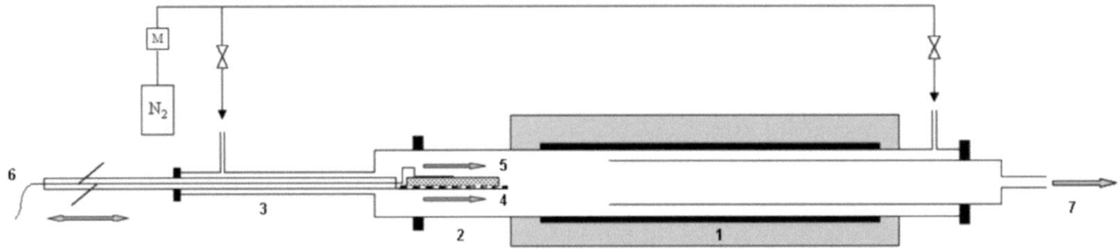

1- Furnace; **2-** Quartz reactor; **3-** Movable sample boat; **4-** Metal grid; **5-** Refractory ceramic wool soaked with 1g of bio-oil; **6-**Thermocouple; **7-** Gas outlet; **M-** Mass flow meters and controllers

Figure 2. Horizontal Tubular Reactor (HTR) ready for sample introduction

The procedure carried out for an experiment was the following. First, the furnace was heated and the gas flowrate (nitrogen) was adjusted using a mass flow meter controller. When the temperature was stabilized, the sample was placed on the metal grid at the unheated section of the reactor. This section was swept by half of the total cold nitrogen flow injected, in order to maintain it cold and under inert atmosphere, and therefore avoid its degradation. Meanwhile the second half of the nitrogen flow was preheated through the double-walled annular section of the reactor as shown in Fig. 1. The sample consisted in 1g of bio-oil was placed inside a crucible of 25 mm diameter and 40 mm height for studying the effect of temperature and of ash content. In order to achieve higher heating rates, some runs were performed with 1g of bio-oil previously soaked in a refractory ceramic wool sample of 100x20 mm length and width and 3 mm thickness. The choice of this sample holder allowed increasing the exchange surface and subsequently obtaining larger heating rates. We proved that this wool has no catalytic effect on bio-oil pyrolysis. Indeed, previous bio-oil pyrolysis experiments were carried out first with a single crucible, and secondly in the refractory ceramic wool deposited in the crucible. The wool didn't induce any change on the products yield.

The reactor outlet was first connected to an O_2 gas analyser to ensure that there is no oxygen in the reactor. Afterwards, a manual insertion enabled to move the sample in the furnace at different velocities, ranging between 0.06 and 30 cm/s. The sample temperature evolution was measured using a thermocouple placed in the middle of the sample in order to determine a heating rate for each experiment. Variation of the heating rate was obtained by varying the sample introduction through the tubular reactor. Four different durations have been used: 16, 8, 4 and finally 0.03 min resulting in four different heating rates. The sample then remained in the middle of the reactor for a definite time and is brought back out of the furnace; the solid residue was weighed after cooling. Even after several experiments, no char deposit was observed inside the reactor. Only tar deposits were observed in the cold outlet of the reactor. The reactor outlet was connected to a sampling bag at t = 0 just before sample introduction. The gases formed by pyrolysis were collected in the bag. The duration of all experiments was 10 min with a 2 NL/min N_2 flowrate which enabled to know accurately the volume of N_2 sampled in the bag. In HTR reactor, the volume of formed gas never exceeded 1% of the volume of N_2 sampled in the bag. After the experiment the bag was disconnected from HTR, and connected to the micro-chromatograph analyser (μGC). From the total volume of gas in the bag and measure of the gas concentration, the quantity of each gas formed by 1g of bio-oil can be precisely calculated.

2.1.2. *Entrained Flow Reactor EFR*

A laboratory scale high temperature entrained flow reactor HT-EFR was used in this work. It consisted in a vertical tubular reactor electrically heated by a total of 18 kW three-zones electrical furnace, and was able to reach 1600°C in a 1m long isothermal reaction zone, as illustrated in Fig. 3.

The atmosphere gas is generated by feeding the controlled flow of nitrogen in a 2 kW electrical steam generator. This atmosphere gas is then preheated to 900°C using a 2.5 kW electrical battery of heating elements before reaching the isothermal reaction zone. The HT-EFR was initially set up to achieve high heating-rate gasification of solid biomass, and was equipped for the present work with a specially designed bio-oil pulverization feeder, in order to obtain a very constant mass flowrate spray.

The feeder consists of a 1 m long and 14 mm o.d. probe cooled with water at 30°C. At its end a commercial stainless steel nozzle is integrated. This allows uniform distribution with fine atomization. Nozzle type (DELAVAN WDB) is a solid cone, with orifice diameter of 0.46mm and a spray angle of 60°.

The oil is fed with a syringe which is automatically pushed. The expected mass flowrate of 0.3 g/min was too low for direct pulverization. Therefore, a 3.5 NL/min N_2 flowrate was used to entrain oil in the feeding probe and to ensure a thin spray of the oil. The spray of droplets is dispersed on the section of a 75 mm i.d. alumina reactor swept by 15 $NL.min^{-1}$ of atmosphere gas. The steam gasification experiments were carried out in HT-EFR with steam to fuel mass ratio (fuel includes inherent water in bio-oil) of S/F=4.5.

Reactions take place along the reactor during a controlled gas residence time, which was of about 3-4s. The residence time of droplets or solid residue after reaction is assumed to be similar to that of the gas because of the very small particle size. The gas residence time was calculated as the ratio of the reaction zone length to the average gas velocity in the reactor. At 1650 mm downstream of the injection point, gases and solid residue were sampled by a hot-oil cooled probe at 150°C. Gas and solid residue were separated using a settling box and a filter, both heated to avoid water condensation. The water and potential remaining tars were first condensed in a heat exchanger, and non-condensable gases were forwarded to a micro-chromatograph analyser (μGC) to quantify H_2, CO, CO_2, CH_4, C_2H_2, C_2H_4, C_2H_6, C_3H_8 and C_6H_6. The μGC offers excellent resolutions of all analyze species at higher concentrations with repeatability of ± 2 percent relative standard deviation, the system offers also a minimum detectable quantity of about 10 ppm for most gases species.

Gases were also analyzed by other analyzers that allowed checking the absence of O_2, to confirm the analysis and to control continuously gas production: a Fourier Transform InfraRed (FTIR) analyser, a Non-Dispersive InfraRed (NDIR) analyser coupled with a paramagnetic analyser for O_2 and a Thermal Conductivity Detector (TCD) to quantify H_2.

2.2. Feedstock

The feedstock used for all experiments was a bio-oil produced by fast pyrolysis of softwood on an industrial-scale fluidized bed unit (Dynamotive, West Lorne, Ontario) and provided by CIRAD, France. Its physico-chemical properties have been measured (see Table 1). The water content of the bio-oil measured by Karl Fischer method (ASTM E203) is around 26 wt % which is in agreement with the average values reported in the literature. It can be noticed that the solid particles content is rather high (2.3 wt.%) while the ash content remains low (around 0.06 wt.%). This confirms that the solid particles mainly consist of high-carbon content char particles. These particles were entrained during bio-oil production by the gas stream to the bio-oils condensers. Ultimate analysis and LHV of the bio-oil are very

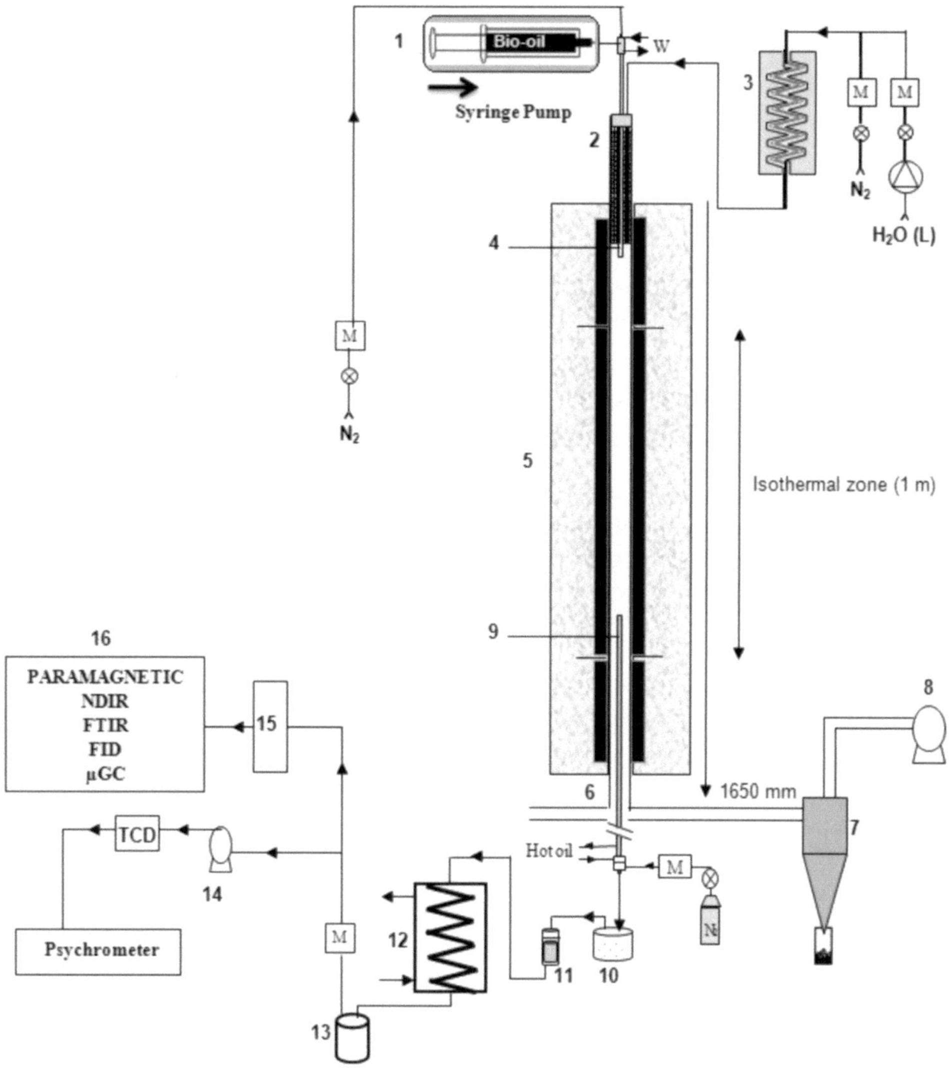

1- Injection system; **2-** Electrical preheater; **3-** Steam generator; **4-** Water cooled feeding probe; **5-** Three zones electrical furnace; **6-** 75 mm i.d. alumina reactor; **7-** Cyclone collector; **8-** Exhaust fan; **9-** Oil cooled sampling probe; **10-** Hot settling box; **11-** Hot particle collector (filter); **12-** Water cooler; **13-** Condensate collector; **14-** Sampling pump; **15-** Gas dryer; **16-** Gas analyser; **M-** Mass flow meters and controllers; **N_2-** Nitrogen; **W-** Water (probes cooling)

Figure 3. Entrained flow reactor

similar to those of wood. From the ultimate analysis, the chemical formula of the bio-oil can be established as $CH_{1.18}O_{0.48}.0.4H_2O$.

Ultimate analysis (wt.%)					H_2O	Ash	Solids	LHV	Kinematic viscosity
C	H	O	S	N	(wt.%)	(wt.%)	(wt.%)	($MJ.kg^{-1}$)	at 20 °C ($mm^2.s^{-1}$)
42.9	7.1	50.58	< 0.10	< 0.10	26.0	0.057	2.34	14.5	103

Table 1. Ultimate analysis and several characteristics of bio-oil derived from hardwood fast pyrolysis

3. Results and discussions

3.1. Thermal decomposition of bio-oil: focus on the products yields under different pyrolysis conditions

3.1.1. Preliminary runs of bio-oil pyrolysis at two final reactor temperatures (The experiments were carried out in a HTR)

Two reactor temperatures were tested in order to evaluate the effect of the final pyrolysis temperature on devolatilization process affecting the yield of gas, condensate and residual solid:

- Moderate temperatures at 550°C;
- High temperature 1000°C to approach the severe conditions of gasification.

The yields of final products are listed in Figure 4. With temperature increasing from 550 to 1000°C, the total gas yield sharply increases from 12.2 to 43.0 wt.%, while condensate (tar + water) decreases from 73.2 to 47.5 wt.%. Varying temperature shows a great influence on the gas composition as well.

Figure 5 shows that the main gas products are H_2, CO, CO_2, CH_4 and some C_2 hydrocarbons (C_2H_2, C_2H_4 and C_2H_6). Among them, the H_2 and CO content increased significantly from 0.056 wt.% to 1.65 wt.% and from 5.9 to 23.9 wt.% respectively as temperature increased from 550 to 1000°C. Yields of CH_4 also increased from 1.2 to 5.0 wt.% whilst that of CO_2 increased from 4.2 to 10.8 wt.%. The yields of C_2H_2, C_2H_4 and C_2H_6 are relatively small. The specie C_2H_6 only appears at 550°C while C_2H_2 only appears at 1000°C. The thermal cracking of gas-phase hydrocarbons at high temperature might explain the variation of gas product composition observed.

Finally, with increasing temperature from 550°C to 1000°C, the char yield decreased significantly from 14.5 to 9.4 wt.%. However changing the reactor temperature implies a change of both the heat flux density imposed to bio-oil (and hence its heating rate) but also the final temperature reached by the char produced. Therefore the later trend observed might be due to two reasons:

- The char formed at 550°C contains residual volatile matters which are released when the emperature increases to 1000°C;
- Increasing the heating rate results in the decrease of the char yield. This is actually in good agreement with what is usually observed in the literature from pyrolysis of biomass (Ayllón et al., 2006; Haykiri-Acma et al., 2006; Mani et al.,2010).

To check the first assumption, a char first prepared at 550°C was submitted to a second heating step at 1000°C. During this second step, the mass of char did not change, which excluded the first assumption, and highlighted actually the effect of heating rate. In order to confirm this trend, additional experiments were carried out to separate the effect of these two parameters. This is studied in details in the following section.

3.1.2. *Effect of heating rate and final temperature on the product yields*

The temperature profiles obtained in the HTR are illustrated in Figure 6. A calculation of the highest heating rate is then made taking into account only the linear part of curves. Details of the calculated heating rates and products yields obtained from experiments are given in Table 2.

The temperature profiles curves show that the heating rate ranges from 2 to 14°C/s at the final pyrolysis temperature of 550°C, and from 2 to 100°C/s at the final pyrolysis temperature of 1000°C. The response time of temperature measurement system was characterized by placing the thermocouple alone and the thermocouple placed in the refractory ceramic wool without bio-oil sample together inside the reactor in 0.03 min. The results are also plotted in Figure 6. At 1000°C we can notice that the response of the thermocouple and refractory ceramic wool does not exceed 100°C/s. But, it appears that the actual heating rate for the sample introduced in 0.03min may be still higher than 100°C/s. This is further illustrated on Figure 6.

	t (min)[a]	Heating rate °C/s	Solid %wt	Total gas %wt	condensate %wt
Pyrolysis at 550 °C	16	2	14.4	14.1	71.4
	8	5	12.4	13.7	73.8
	4	10	11.4	13.3	75.2
	0.03	14	10.5	13.3	76.0
	flash	>2000	1.2	13.6	85.1
Pyrolysis at 1000 °C	16	2	11.5	41.6	46.8
	8	5	10.4	41.7	47.8
	4	14	8.6	40.9	52.2
	0.03	100	3.8	43.8	53.4
	flash	>2000	0.9	40.3	58.7

[a] Duration of sample introduction in the reactor

Table 2. Product yield of bio-oil pyrolysis at different temperatures and heating rates

In order to highlight the effect of heating rate and final temperature on the yields of char, they were plotted in figure 7, with the heating rate as the x scale, using a log scale. The low heating rate experiments gave higher yields of char. Char yield then decreased significantly: from 14.4 wt.% down to 10.5 wt.% when heating rate increased from 2 to 14°C/s at the final temperature of 550°C, and from 11.5 to 3.8 wt.% when heating rate was increased from 2 to 100°C/s at the final temperature of 1000°C.

In order to increase still the heating rate and reach the flash pyrolysis conditions, we have performed additional experiments in the HT-EFR. This process allows achieving very high heating rate.

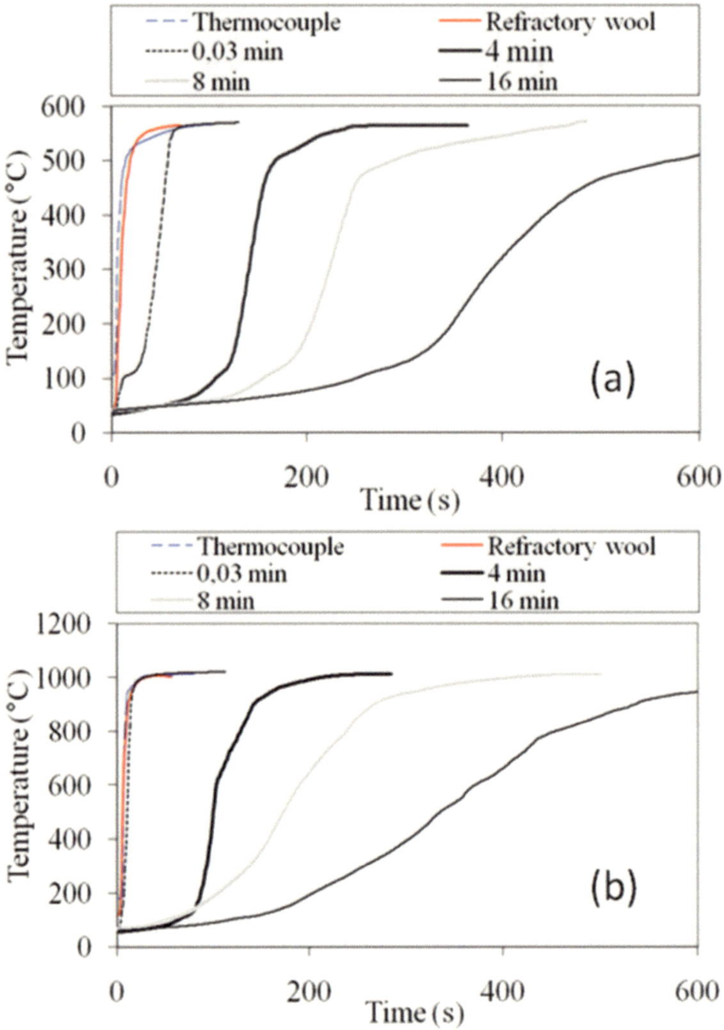

Figure 4. Temperature evolution of the sample during bio-oil pyrolysis in HTR at different heating rates and two final temperatures. **a:** 550°C, **b:** 1000°C

Indeed it is shown that when a particle or droplet is transported by a cold spraying gas, its heating rate is controlled by mixing of the cold gas with the hot gas in the reactor. CFD modeling was used and derived this order of magnitude. Heating rate was estimated at 2000°C/s (Van de Steene et al., 2000). Under these conditions, the char yield measured is very low: < 1 wt.%. As can be seen in Figure 7, the char yield obtained with HT-EFR is in rather good agreement with the values obtained in HTR and extrapolated to high heating rates. This result is in agreement with the work carried out by Guus van Rossum et al. (Van Rossum et al., 2010). They found that small droplets (undergoing high heating rate) are much quicker evaporated and give fewer char compared to larger droplets (undergoing low heating rate pyrolysis).

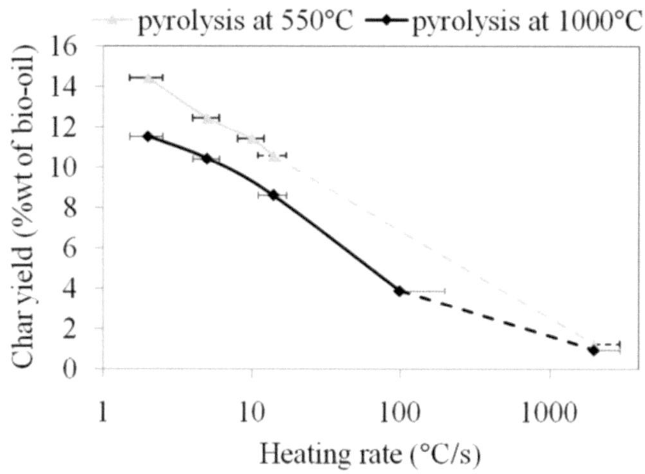

Figure 5. Char yield obtained from pyrolysis of bio-oil at two final temperatures: 550°C and 1000°C - effect of heating rate

Globally from all the data collected, the char yield depends very much on the heating rate, and less on the final temperature, confirming the observation from section 3-1. These results give important information for understanding the pathways occurring during gasification of bio-oil in reactors such as EFR: the amount of char formed by pyrolysis and submitted to subsequent steam-gasification reactions will be very low whereas the main reactions will occur in the gas phase (reforming, partial oxidation…). Considering that solid gasification is rate-limiting, this might be an advantage of using bio-oil instead of biomass as feedstock for EFR gasification.

Figure 8 shows the effect of heating rate on the product yields at two final pyrolysis temperatures. There is no apparent impact of the heating rate but a drastic influence of the temperature on the total gas yield which remains of about 13–14 wt.% and 40–43 wt.% at 550°C and 1000°C, respectively.

On the other hand, we can notice that the total condensate yield increased when the heating rate increased and when the final temperature decreased. A value of 76 wt.% is obtained at 14°C/s and a final temperature of 550°C, which is about 5 wt.% higher than that obtained at 2°C/s. In the same manner, at 1000°C the total condensate yield increased with the heating rate, up to 53.4 wt.% at 100°C/s. This value was about 6 wt.% higher than that of 2°C/s.

All these trends can be summarized and explained as follows.

i. Pyrolysis inside the sample

The volatile matters yield increases with the heating rate of bio-oil, to the detriment of the char yield as reported earlier. The primary volatiles may undergo secondary reactions through two competitive pathways (Zaror et al., 1985; Seebauer et al., 1997):

- re-polymerizing to form char;
- cracking to form lighter volatiles which implies less tar repolymerisation.

The re-polymerization pathway is probably favored by lower heating rates. Indeed, low heating rates lead to longer volatiles residence times inside the sample, and favor secondary reactions of re-polymerization to form solid residue. These conditions are known to favor the formation of secondary char from biomass pyrolysis experiments (Zaror et al., 1985) and apparently, this could be extended to the case of bio-oil pyrolysis.

ii. Gas phase reactions outside the sample

Once the volatiles have escaped from the sample, they can undergo additional secondary gas-phase cracking reactions as previously presented. The conversion rate of this reaction highly increases with the gas temperature, leading to higher gas yields to the detriment of condensates. This result is in agreement with number of pyrolysis works carried out on biomass (Seebauer et al., 1997).

Let's notice that due to the procedure described, higher heating rate leads to lower residence time of tars in the hot zone because the bio-oil sample is introduced more rapidly to the centre of the heated zone. The estimate of the gas residence time in the HTR was calculated, from their release at the sample position (which varies with time according to the duration of sample introduction) to the exit of the reactor. It varies from 8 to 16s at 550°C and from 5 to 10s at 1000°C.

3.2. Gasification of wood bio-oil (The experiments were carried out in a HT-EFR)

3.2.1. Effect of temperature

The first objective was to study the influence of temperature - over a wide range - on the syngas yield and composition.

Generally the gas mixture formed from catalytic reforming of bio-oil is composed of hydrogen, carbon monoxide and dioxide, methane, acetylene, unconverted steam, coke (carbon) and soot. Figure 9, presents the mole fraction of the gaseous products from this work (in dry basis and without N_2) as a function of temperature in the range 1000 to 1400°C. Error bars were established by repeating each test 2 or 3 times. The species C_2H_4, C_2H_6, C_3H_8 and C_6H_6 are not detected by chromatography. Whatever the operating temperature between 1000°C and 1300°C, bio-oil is mainly decomposed to H_2, CO, CO_2, CH_4 and C_2H_2. Above 1300°C C_2H_2 disappears, while CH_4 disappears above 1400°C. As the temperature rises, the fraction of H_2 increases monotonically at the expense of carbon monoxide, methane and acetylene. Above 1300°C the hydrogen content remains almost stable. At 1400°C hydrogen mole fraction reaches the maximum value of 64 mol% of the syngas.

The reactions that may explain the increase of hydrogen with temperature are:

$$\text{The steam reforming of } CH_4 \text{ and } C_2H_2 \text{ into } H_2 \text{ and CO} \quad (2)$$

$$\text{The water gas shift reaction } CO + H_2O \leftrightarrow CO_2 + H_2 \quad (3)$$

The water gas shift reaction can also explain the increase of carbon dioxide and the decrease of carbon monoxide between 1000 and 1200°C. Above 1200°C, carbon monoxide slightly

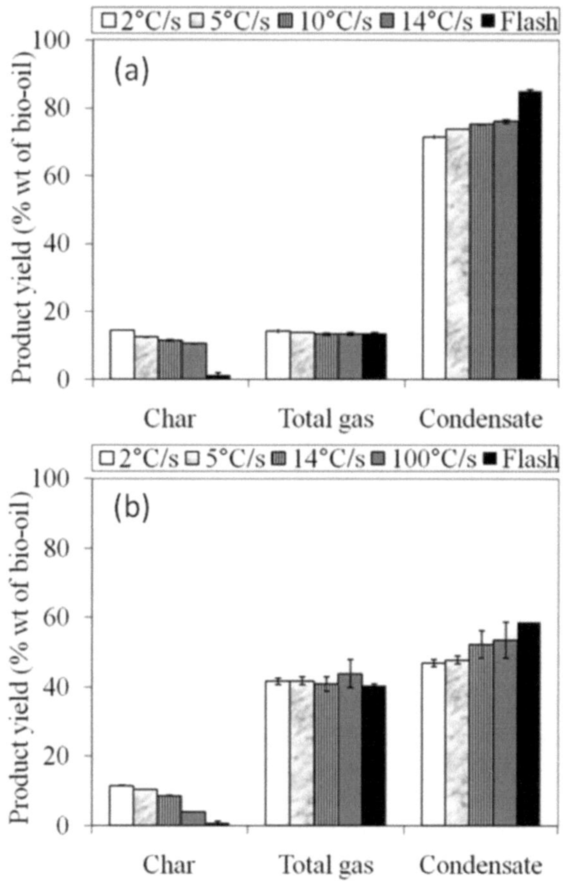

Figure 6. Product yield obtained from bio-oil pyrolysis at two final temperatures. **a:** 550°C, **b:** 1000°C- effect of heating rate

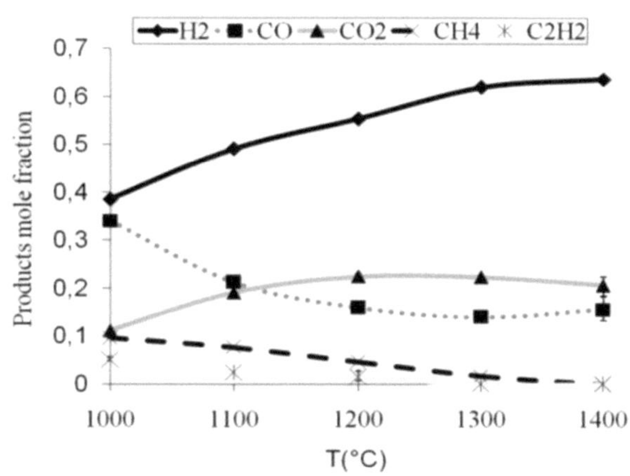

Figure 7. Composition of the produced syngas (dry basis and without N_2) - effect of temperature, at S/F=4.5

increases. This may be explained by steam gasification of the solid carbon residue (char and soot) resulting from the pyrolysis of oil droplets to yield carbon monoxide and hydrogen following the reaction:

$$C + H_2O \leftrightarrow CO + H_2 \quad (4)$$

and potentially following the Boudouard reaction which would explain the slight decrease of CO_2:

$$C + CO_2 \rightarrow 2CO \quad (5)$$

It was observed that as the temperature increases the amount of collected solid decreases significantly above 1000°C. The process allows achieving very high heating rate estimated at 2000°C/s (Van de Steene et al., 2000). Under these conditions, the char yield measured is very low: < 1 wt.%. At 1400°C more than the 99.9% the bio-oil is converted to gas.

3.2.2. Equilibrium calculation

The thermodynamic equilibrium calculation is independent of reactor and predicts the yield of final products, based on the minimization of the Gibbs free energy of the system. It was conducted here using FactSage software 5.4 to establish whether the syngas was close or not to equilibrium at the different temperatures. Operating temperature varied from 1000°C to 1400°C; pressure was fixed at 1 atm. The software is not presented in detail in this paper; details of the thermodynamic calculation could be found on FactSage web site.

The results of prediction are presented in Figure 10, expressed in g of gas produced per g of crude bio-oil injected. As the temperature increases from 1000 to 1400°C the calculated equilibrium yield of H_2 remains approximately constant at 0,11g/g, while the yield of CO increases from 0.3 at 1000°C to 0.45g/g at 1400°C. The CO_2 yield decrease from 1.1 g/g at 1000°C down to 0.9 at 1400°C.

At 1000°C the calculation yields are far away from the experimental results. The deviation from equilibrium at lower temperatures is also reported by Sakaguchi et al (Sakaguchi et al., 2010). At 1200°C the thermodynamic equilibrium begins to establish. The calculation nevertheless does not retrieve the presence of CH_4 and C_2H_2. At 1400°C the experimental yields are very close to the equilibrium calculation yields: 0.11 and 0.12 respectively for H_2, 0.45 and 0.45 respectively for CO, and 0.86 and 0.88 respectively for CO_2. It can be concluded that at this temperature the equilibrium is reached.

It is also interesting to compare the obtained experimental yields at 1400°C to the theoretical yields corresponding with complete gasification of oil that would follow:

$$CH_{1.18}O_{0.48}.0.4H_2O + 1.12\ H_2O \rightarrow CO_2 + 2.11\ H_2 \quad (6)$$

The maximum stoichiometric H_2 yield for this oil would be 0.150g per 1g crude bio-oil while a value of 0.126 g was obtained experimentally. This shows that under our experimental conditions and at 1400°C steam reforming of bio-oil lead to a production of H_2 with a yield of about 84% of theoretical yields corresponding with complete gasification of oil (reaction 6).

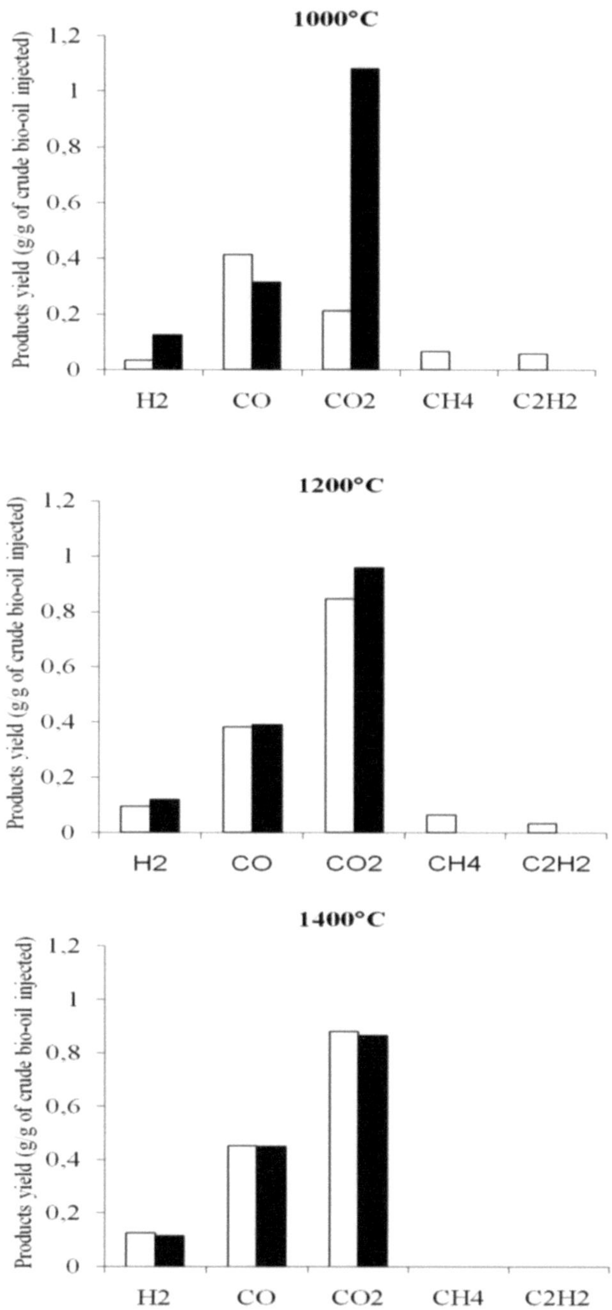

Figure 8. Gas yield from bio-oil reforming at 1000, 1200 and 1400°C, S/F=4.5. □ Experiments; ■ Equilibrium calculation

4. Conclusion

Gasification of biomass is one of the leading near-term options for renewable energy production. When large scale units are considered, bio-oil shows lots of advantages compared to solid biomass. The combination of decentralized fast pyrolysis of biomass followed by transportation and gasification of bio-oil in bio-refinery has attracted great attention.

The overall purpose of this research was to investigate the feasibility of a whole bio-oil non catalytic steam gasification process for the production of high quality syngas in entrained flow reactor.

From a chemical point of view, bio-oil gasification process is quite complex and consists of the following main stages: vaporization, thermal cracking reactions with formation of gas, tars and char that considered as undesirable products. This is followed by steam reforming of gas and tars, together with char oxidation. To better understand the process, the first step of gasification (pyrolysis) and thereafter, the whole process (pyrolysis+gasification) were separately studied.

In the pyrolysis step, a temperature increase from 550°C to 1000°C greatly enhanced the gas yield, whilst solid and liquid yields decreased significantly in agreement with the literature. The heating rate of bio-oil has little impact on the gas yield, but plays a major role on the char yield. Hence the char yield decreases from 11 wt.% with a heating rate of 2°C/s down to 1 wt.% for flash heating rate of 2000°C/s at a final temperature of 1000°C. At very high heating rate, the final temperature has little influence on the char yield. These results show that for gasification under industrial EFR conditions, the quantity of char is very small. Thus the gasification process mainly consists in gas/tar reforming. Nevertheless, the production of clean syngas will require either complete gasification of char or its removal from the gas produced by the gasifier.

In steam gasification process, whole bio-oil was successfully steam gasified in HT-EFR. An increase in the reaction temperature over a wide range from 1000°C to 1400°C implies higher hydrogen yield and higher solid carbon conversion. A thermodynamic equilibrium calculation showed that equilibrium was reached at 1400°C. At this temperature steam reforming of bio-oil leads to yield of equal 84% of theoretical yields corresponding with complete gasification of oil.

Author details

Younes Chhiti
Université de Pau/LaTEP-ENSGTI

Sylvain Salvador
Ecole des Mines d'Albi – Carmaux/RAPSODEE

Acknowledgement

The authors gratefully acknowledge the financial support from EnerBio Program of Fondation Tuck France, and express their gratitude to Mr. Bernard AUDUC technician in

Ecole des Mines d'Albi-Carmaux for his assistance and contribution to experimental device design and operation.

5. References

Ayllón, M.; Aznar, M.; Sànchez, JL.; Gea, G. & Arauzo, J. (2006). Influence of temperature and heating rate on the fixed bed pyrolysis of meat and bone meal. *Chemical Engineering Journal,* Vol.121, No.2-3, (June 2006), pp. 85–96, ISSN 13858947

Ba, T.; Chaala, A.; Pérez, MG.; Rodrigue, D. & Roy, C. (2004a).Colloidal properties of bio-oils obtained by vacuum pyrolysis of softwood bark. Characterization of water soluble and water-insoluble fractions. *Energy Fuel,* Vol.18, No.5, (August 2004), pp. 704–12, ISSN 0887-0624

Ba, T.; Chaala, A.; Pérez, MG. & Roy, C. (2004b). Colloidal properties of bio-oils obtained by vacuum pyrolysis of softwood bark. Storage stability. *Energy Fuel,* Vol.18, No.1, (December 2003), pp. 188–201, ISSN 08870624

Branca C.; Di Blasi C. & Elefante R. (2005a). Devolatilization and Heterogeneous Combustion of Wood Fast Pyrolysis Oils. *Ind. Eng. Chem. Res* Vol. 44, No.4, (January 2005), pp. 799-810, ISSN 08885885

Branca C.; Di Blasi C. & Russo C. (2005b). Devolatilization in the temperature range 300–600 K of liquids derived from wood pyrolysis and gasification. *Fuel,* Vol.84, No.1, (August 2004), pp. 37–45, ISSN 0016-2361

Bimbela, F.; Oliva M.; Ruiz J.; Garcia L. & Arauzo, J. (2009). Catalytic steam reforming of model compounds of biomass pyrolysis liquids in fixed bed: Acetol and n-butanol. *J. Anal. Appl. Pyrolysis,* Vol.85, No.1-2, (December 2008), pp. 204-213, ISSN 01652370

Calabria, R.; Chiariello, F. & Massoli, P. (2007). Combustion fundamentals of pyrolysis oil based fuels. *Exp Therm Fluid Sci,* Vol.31, No.5, (July 2006), pp. 413–20, ISSN 0894-1777

Czernik, S.; French, R. & Feik, C. (2002). Hydrogen by catalytic steam reforming of liquid byproducts from biomass thermoconversion processes. *Ind Eng Chem Res,* Vol.41, No.17, (July 2002), pp. 4209–15, ISSN 0888-5885

Garcia, L.; French, R.; Czernik, S. & Chornet, E. (2000). Catalytic steam reforming of bio-oils for the production of hydrogen: effects of catalyst composition. *Appl. Catal.,* Vol.201, No.2, (January 2000), pp. 225–239, ISSN 0926860X

Devi, L.; Ptasinski, KJ. & Janssen, FJ.JG. (2003). A Review of the Primary Measures for Tar Elimination in Biomass Gasification Processes. *Biomass and Bioenergy,* Vol.24, No.2, (September 2002), pp. 125-140, ISSN 09619534

Goula, MA.; Kontou, SK. & Tsiakaras PE. (2004). Hydrogen production by ethanol steam reforming over a commercial Pd/γ-Al2O3 catalyst. *Appl. Catal.,* Vol.49, No.2, (February 2004), pp. 135–144, ISSN 09263373

Hallett, WLH & Clark, NAA. (2006). Model for the evaporation of biomass pyrolysis oil droplets. *Fuel,* Vol.85, No.4, (September 2005), pp. 532–544, ISSN 0016-2361

Haykiri-Acma, H.; Yaman, S & Kucukbayrak, S. (2006). Effect of heating rate on the pyrolysis yields of rapeseed. *Renewable Energy,* Vol.31, No.6, (May 2005), pp. 803–810, ISSN 0960-1481

Henrich, E. & Weirich, F. (2004). Pressurized Entrained Flow Gasifiers for Biomass. *Environmental Engineering Science,* Vol.21, No.1, (July 2004), pp. 53-64, ISSN 1092-8758

Levenspiel, O. (2005). What will come after petroleum. *Ind. Eng. Chem. Res.,* Vol.44, No.14, (February 2005), pp. 5073-5073, ISSN 0888-5885

Mani, T.; Murugan, P.; Abedi, J & Mahinpey, N. (2010). Pyrolysis of wheat straw in a thermogravimetric analyzer: Effect of particle size and heating rate on devolatilization and estimation of global kinetics. *Chemical Engineering Research and Design,* Vol.88, No.8, (August 2010), pp. 952-958, ISSN 0263-8762

Marda, JR.; DiBenedetto, J.; McKibben, S.; Evans, RJ.; Czernik, S.; French, RJ. & Dean, AM. (2009). Non-catalytic partial oxidation of bio-oil to synthesis gas for distributed hydrogen production. *International journal o f hydrogen energy,* Vol.34, No.20, (October 2009), pp. 8519–8534, ISSN 0360-3199

Panigrahi, S.; Dalai, AK.; Chaudhari, ST. & Bakhshi, NN. (2003). Synthesis Gas Production from Steam Gasification of Biomass-Derived Oil. *Energy Fuels,* Vol.17, No.3, (April 26, 2003), pp. 637–642, ISSN 0887-0624

Rioche, C.; Kulkarni, S.; Meunier, FC.; Breen, JP. & Burch, R. (2005). Steam reforming of model compounds and fast pyrolysis bio-oil on supported noble metal catalysts. *Appl. Catal.,* Vol. 61, No.1-2, (October 2009), pp. 130–139, ISSN 09263373

Rostrup-Nielsen, JR. Industrial relevance of coking. (1997). *Catal. Today,* Vol.37, No.3, (August 1997), pp. 225–23, ISSN 0920-5861

Sakaguchi, M.; Paul Watkinson, A. & Naoko, E. (2010). Steam Gasification of Bio-Oil and Bio-Oil/Char Slurry in a Fluidized Bed Reactor. *Energy Fuels,* Vol.24, No.9, (August 2010), pp. 5181–5189, ISSN 0887-0624

Seebauer, V.; Petek, J & Staudinger, G. (1997). Effects of particle size, heating rate and pressure on measurement of pyrolysis kinetics by thermogravimetric analysis. *Fuel,* Vol.76, No.13, (October 1997), pp. 1277-1282, ISSN 0016-2361

Trimm, DL. (1997). Coke formation and minimization during steam reforming reactions. *Catal. Today,* Vol.37, No.3, (August 1997), pp. 233–238, ISSN 09205861

Van de Steene, L.; Salvador, S & Charnay, G. (2000). Controlling powdered fuel combustion at low temperature in a new entrained flow reactor. *Combustion Science and Technology,* Vol.159, No.1, (January 2000), pp. 255-279, ISSN 0010-2202

Van Rossum, G.; Kersten, SRA. & van Swaaij, WPM. (2007). Catalytic and Noncatalytic Gasification of Pyrolysis Oil. *Ind. Eng. Chem. Res.,* Vol.46, No.12, (March 2007), pp. 3959-3967, ISSN 0888-5885

Van Rossum, G.; Güell, BM.; Balegedde Ramachandran, RP.; Seshan, K.; Lefferts, L.; Van Swaaij, WPM & Kersten, SRA. (2010). Evaporation of pyrolysis oil: Product distribution and residue char analysis. *AIChE Journal,* Vol.56, No.8, (August 2010), pp. 2200–2210, ISSN 0001-1541

Wang, D.; Czernik, S. & Montane, D. (1997). Biomass to hydrogen via pyrolysis and catalytic steam reforming of the pyrolysis oil and its fractions. *Ind Eng Chem Res,* Vol.36, No.5, (May 1997), pp. 1507–18, ISSN 0888-5885

Wang, D.; Czermik, S. & Chornet, E. (1998). Production of hydrogen from biomass by catalytic steam reforming of fast pyrolytic oils. *Energy Fuels,* Vol.12, No.1, (January 1998), pp. 19–24, ISSN 08870624

Wornat, MJ.; Porter, BG & Yang, NYC. (1994). Single droplet combustion of biomass pyrolysis oils. *Energy Fuel,* Vol.8, No.5, (September 1994), pp. 1131–1142, ISSN 1520-5029

Zaror, CA.; Hutchings, IS.; Pyle, DL.; Stiles, HN & Kandiyoti, R. (1985). Secondary char formation in the catalytic pyrolysis of biomass. *Fuel,* Vol.64, No.7, (july 1985), pp. 990–994, ISSN 0016-2361

Chapter 5

Biomass Downdraft Gasifier Controller Using Intelligent Techniques

A. Sanjeevi Gandhi, T. Kannadasan and R. Suresh

Additional information is available at the end of the chapter

http://dx.doi.org/10.5772/48564

1. Introduction

In this world of deteriorating amount of non-renewable resources, the relevance of a biomass gasifier is immense. Biomass is the biological material from living, or recently living organisms. As an energy source, biomass can either be used directly, or converted into other energy products such as biofuel. Biomass is carbon, hydrogen and oxygen based. It is used as a good source of power generation. The gas composition in the producer gas, the final product of the gasification process is as follows: - CO: 15-20 %, H_2: 15-20 %, CH_4: 2-6 %, CO_2: 7-10 %, N_2: 40-50 %.

Power generation from biomass has become a complement to conventional sources of energy due to its contribution to the reduction of greenhouse effect. Biomass ranks fourth as an energy source and, in developing countries, it provides 35% of their energy. It must be noted that gasification is cheaper as well as having considerable efficiency compare with non renewable energy sources. Also, downdraft gasifiers with throat are known to produce the best quality gas for engines. There are mainly two types of biomass gasifiers, which are the fixed and the fluidized bed types. The fixed bed gasifiers have been the traditional setup used for gasification, operated at temperatures around 1000 ^{0}C. Among the fixed bed gasifiers, there are three major types and these are updraft, downdraft and cross-draft gasifiers. The updraft configuration is the simplest and oldest form of gasifier and is still used for coal gasification. In this, the biomass is introduced at the top of the reactor and a grate at the bottom of the reactor supports the reacting bed. The downdraft gasifier has the same mechanical configuration as the updraft gasifier except that the oxidant and product gases flow down the reactor, in the same direction as the biomass. Crossdraft gasifiers are used for charcoal gasification.

In the updraft gasifier, gas leaves the gasifier with high tar vapour which may seriously interfere the operation of internal combustion engine. This problem is minimized in downdraft gasifier. In this type, air is introduced into downward flowing packed bed or solid fuels and gas is drawn off at the bottom. A lower overall efficiency and difficulties in handling higher moisture and ash content are common problems in small downdraft gas producers.

The time (20-30 minutes) needed to ignite and bring plant to working temperature with good gas quality is shorter than updraft gas producer. It undergoes in four different zones namely combustion zone, reduction zone, pyrolysis zone and drying & heating zone. The full view of the process zones are shown in the Figure 1. [1, 2].

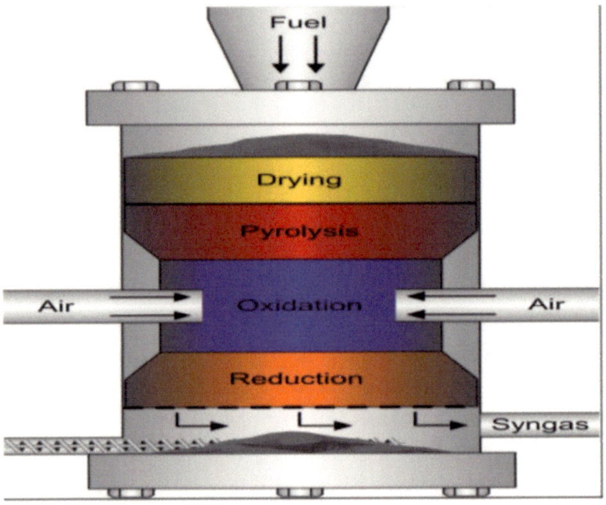

Figure 1. Downdraft gasifier producers

The various zones in the downdraft gasifier are as follows:

i. Combustion zone

In complete combustion of solid fuel composed of Carbon, Hydrogen and Oxygen, CO_2 is obtained from Carbon is fuel and water is obtained from Hydrogen usually as steam. This takes place in the combustion zone. Also, exothermic reaction takes place here. The main reactions are:

$$C + O_2 = CO_2$$

$$2H_2 + O_2 = 2H_2O$$

ii. Reduction zone

The products of partial combustion (water, CO_2 and combusted partially cracked pyrolysis products) now pass through a red-hot charcoal bed where the reduction reaction occurs as follows:

$$C + CO_2 = 2CO$$

$$C + H_2O = CO + H_2$$

These reactions being endothermic have the capability of reducing the gas temperature. The temperatures are normally 800-1000 ^{0}C.

iii. Pyrolysis zone

Wood pyrolysis is an intricate process. Products depend on temperature, pressure and heat losses. Up to 200 ^{0}C, only water is driven off. In 200-280 ^{0}C range, CO_2, acetic acid and water are given off. In 280-500 ^{0}C range, real pyrolysis occurs and produces large quantities of tar and gases containing CO_2. In the range of 500-700 ^{0}C, gas production is small and contains Hydrogen.

iv. Drying zone

Wood is being dried in the drying zone. Usually moisture content of wood is 10-30 per cent. Some organic acids come out during drying process which may cause corrosion of gasifiers.

2. Biomass gasification process

2.1. Wood gasification plant

The biomass downdraft gasifier is mostly used for power generation applications. It is basically a reactor into which fuel/feed stock is fed along with a limited supply of air. The heat that is required for gasification is generated through partial combustion of the feed material. This incomplete combustion leads to chemical breakdown of the fuel through internal reactions resulting in production of a combustible gas usually called Producer Gas. The calorific value of this gas varies between 4.0 and 6.0 MJ/Nm^3 or about 10 to 15 percent of the heating value of natural gas. Producer gas from different fuels produced in different gasifier types may considerably vary in composition. However, it consists always of a mixture of the combustible gases namely Hydrogen (H_2), Carbon Monoxide (CO), and Methane (CH_4) as well as incombustible gases such as Carbon Dioxide (CO_2) and Nitrogen (N_2). Because of the presence of CO, producer gas is toxic in nature. In its raw form, the gas tends to be extremely dirty, containing significant quantities of tars, soot, ash and water. In downdraft gasifier the fuel slowly moves down by gravity. During this downward movement, the fuel reacts with air, which is supplied by the suction of a blower or an engine and is converted into combustible producer gas in a complex series of oxidation, reduction, and pyrolysis reactions [3]. Ash is removed from the bottom of the reactor. The simplified diagram of this electric power plant (100 kW) is shown in Figure 2, where the following parts can be seen biomass and air feeding, ash removal, gas cleaning and conditioning.

The gasifier is a cylindrical reactor of 0.45 m inside diameter with a throat diameter of 0.36 m and 2 m of bed height. The moving bed of biomass rests on a perforated eccentric rotating grate which is at the bottom of the gasifier. The grate is driven by an electric motor, which operates at programmable time intervals. The frequency of motion could be modified to control the biomass residence time inside the reactor. The ashes fall through the perforated grate to be collected in a lower chamber. From this chamber the ashes are extracted by a screw conveyor. A roots blower supplies air into the gasifier through a circular pipe located in the reactor throat, which has three injectors with a radial distribution that enters 4cm inside the bed. Temperature is measured inside the reactor using four Type-K thermocouples located at different levels. An online gas analyzer allows continuous measurements of CO and CO_2 using infrared absorption.

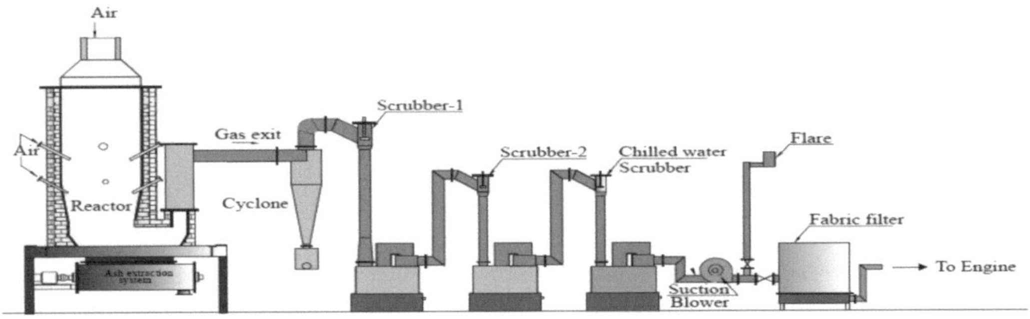

Figure 2. Biomass electric power plant

2.2. Parameters selection

The simple block diagram of the gasifier control system is shown in Figure. 3, the basic variables of a feedback controller are classified as process variables, which are important to maintain under control, and manipulated variables which are adjusted by the controller to obtain the desired behavior for the process variables. In addition, there exist disturbance variables which cannot be adjusted by the controller. The set points are the desired values for the process variables. Manipulated variables were used through the controller to obtain the desired effect on the process variables. The sampling time to measure the process variables, temperature and CO/CO_2 was 120s. The heating value of the produced gas should be higher than 4 MJ/Nm^3 in order to be used in a gas engine [4].

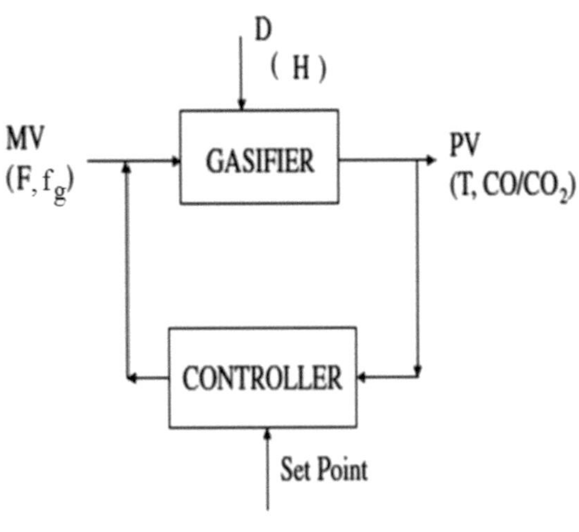

Figure 3. Basic block diagram of the process

The few of main parameters involved in the gasifier process are:

i. Manipulated variable (MV): Air flow rate (F_A), and frequency of grate (f_g)
ii. Process variables(PV) :Throat temperature (T), and CO/CO_2 ratio
iii. Disturbance(D): Moisture content (H_P)

The throat temperature is very closely related to the quality of the gas being produced. The CO/CO_2 ratio is also very crucial in the process. The heating value of the produced gas was calculated from the average gas composition during each run. Based on experimental study, it seems that CO/CO_2 ratio can be a useful measure of combustion performance for downdraft gasifier in service. Also investigations are in hand to make comparisons with measurements of CO alone. This is the reason for taking it as another process variable. In order to control the temperature (T) and CO/CO_2 ratio, air flow rate (F_A) and frequency of grate are manipulated respectively. The correct composition of producer gas found that during the temperature ranges from 650 to 700 ^{0}C, while CO/CO_2 ratio has an optimum value in the range of 1-1.5. The detailed block diagram of gasifier control system is shown in Figure 4. From the process, the measured value of temperature and CO/CO_2 ratio are fed back. Thus, error in temperature and CO/CO_2 ratio can be fed in to the controller in order to generate the necessary control signal to control the process variables.

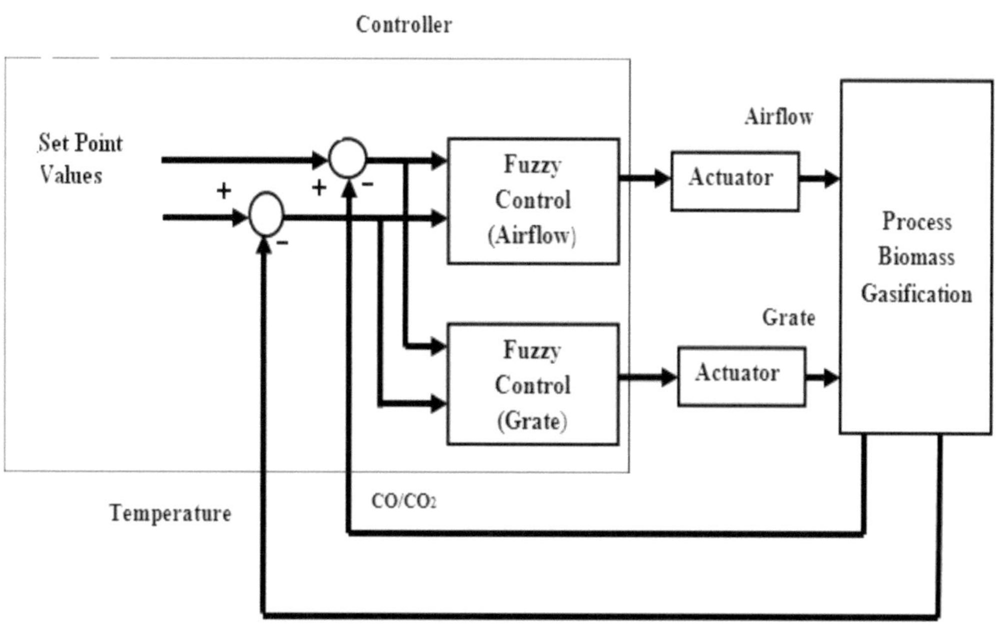

Figure 4. Block diagram of the gasifier control system

3. Transfer function model of the gasifier

Experiment have been carried out on the development of models of the downdraft gasifier to investigate the effects of varying biomass moisture, amount of fluidizing agent, gasification temperature and gas composition, viewing that gasification temperature has the highest influence on the efficiency. For studying the dynamic response of the gasifier, only temperature (T) is considered as a controlled variable and air flow rate (F_A) as a manipulated variable. Multi input and multi output system (MIMO) of the gasifier has been proposed later in this chapter for developing fuzzy controller as shown in figure 4. The Step response method is based on momentary response tests. Many industrial processes have step responses of the system in which the step response is varied after an initial time. A system with step response can be approximated by the transfer function. From the data of temperature values at different times obtained from the gasifier plant. A steady state response is plotted as shown in Figure 5. and from the response assumed that the process is a first order. The general transfer function of first order system is G(s) = K/ Ts+1. Where K is static gain and T is time constant.

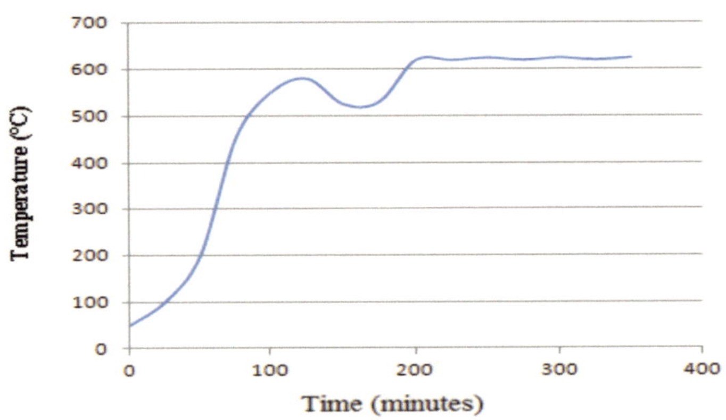

Figure 5. Temperature versus time curve

The calculations are as follows from the response:

i. (K) = (Final steady state value- Initial steady state value) / step change

= (620-50) / (115-15)

= 570/ 100

= 5.7

ii. Time constant (T) = time for the response to reach temperature T1

T1 = 63.2 % of (change in process variable) + offset

= 63.2 % of (620-50) + 50

= 410.24 °C

Time constant = 80 minutes = 4800 seconds

Then substitute the values of K and T in the Transfer function G(s) = 5.7/ (80 s + 1). The step change from 15-115 is optimum region for controlling the particular gasifier.

4. PID controller

Proportional-Integral-Derivative (PID) algorithm is the most common control algorithm used in industry presently. Often, people use PID to control processes that include heating and cooling systems, fluid level monitoring, flow control and pressure control. PID controller is not an adaptive controller, hence the controller has to be tuned frequently and whenever load changes. Auto- tuning of these controllers becomes difficult for complex systems [5, 6]. In order to prove the drawbacks of conventional controller in downdraft gasifier a little attempt is made to design a PID control which is designed to ensure the specifying desired nominal operating point for temperature control of gasifier and regulating it, so that it stays closer to the nominal operating point in the case of sudden disturbances, set point variations, and noise. The proportional gain (Kp), integral time constant (Ti), and derivative time constant (Td) of the PID control settings are designed using Zeigler- Nichols tuning method. The simulink model of PID control is shown in Figure 6. The results of PID controller for temperature set point 750°C shown in Figure 7 and it is observed that the performance of the gasifier system with PID controller is almost oscillating and takes more time to settle with reference temperature. The conventional controller has not suitable for this type of highly non-linear and slow process. In order to improve the gasifier control process the intelligent control techniques are proposed further in this paper.

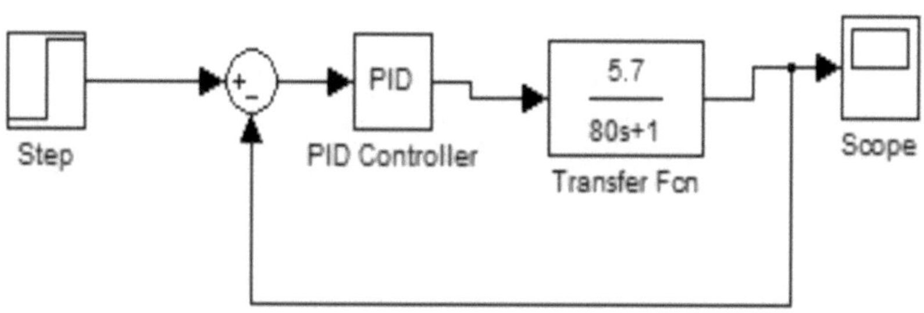

Figure 6. PID controller for downdraft gasifier

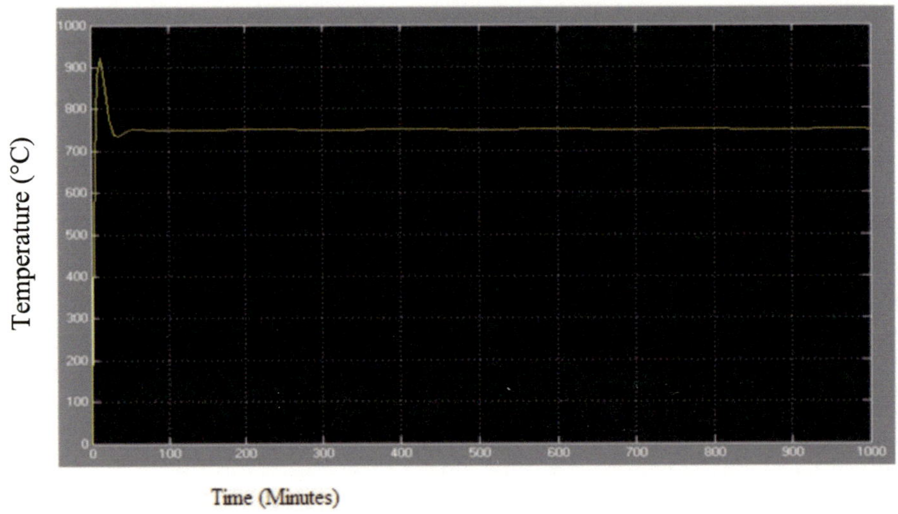

Figure 7. PID controller response

5. Static model of gasifier

The process of gasification is a highly non-linear and slow process, and hence the development of an accurate model is very difficult. The model must be representing the non-linear dynamic characteristics of the process. A plant model for biomass gasification process of woody wastes is proposed for control purpose, based on the plant data in a typical biomass gasification process in the biomass gasifier [7]. In this paper, a steady state model is developed with the collected plant data. In order to fit the collected plant data to the steady state model of the plant, certain simple mathematical equations were developed by adjusting the mathematical relations between the variables with reference to the recorded data. Four sub-systems were thus developed using the MATLAB. The recorded plant data are as shown in the Table 1. One of the objectives of the control model developed here is to tune the controller.

EX No	F_A (m^3/h)	f_g (s/s)	H_p (%)	F_{hb} (kg/h)	ER (%)	T -^{0}C	CO/CO_2 Ratio (%)
1	3	0.006	7.15	1.69	0.3348	720	0.17
2	3.9	0.006	7.15	1.65	0.4356	760	0.31
3	4.8	0.006	7.15	1.69	0.5316	628.2	0.35
4	5.7	0.006	7.15	1.73	0.6232	629.2	0.5
5	6.3	0.006	7.15	1.77	0.7107	633.3	0.5
6	6.9	0.050	7.15	1.81	0.7943	594.4	0.81
7	12	0.050	7.15	1.85	0.8742	810	0.78
8	18	0.080	7.15	1.79	0.1154	860	1.6
9	21	0.080	7.15	1.81	0.1250	900	1.9
10	22	0.085	7.15	1.84	0.1343	910	2.1
11	24	0.090	7.15	1.87	0.1432	920	2.2
12	25	0.095	7.15	1.90	0.1520	924	2.4
13	3.0	0.006	20	1.97	0.1604	655	0.28
14	3.9	0.050	20	1.98	0.1686	640	0.29

Table 1. Data collected from the plant

5.1. Biomass consumption (Fhb)

It is the amount of biomass consumed for the gasification process being considered for monitoring. It depends upon the flow rate (F_A), frequency of rotation of the grate (f_g) and moisture content (H_p). Figure 8 shows the simulink model for biomass consumption. The mathematical expression for biomass consumption is equation (1).

$$F_{hb} = [0.077\ F / H^{1/3}] + 1.5 + f_g \quad (1)$$

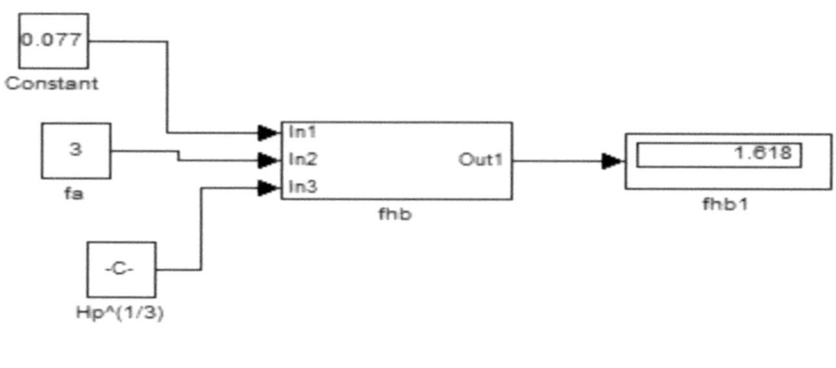

Figure 8. Subsystem for biomass consumption

5.2. Equivalence ratio (ER)

It depends on as F_{hb}, F_A, H_p and the type of material expressed as a function of a material factor m_b which represents the amount of of air needed to obtain combustion of 1 kg of dry biomass. Figure 9 shows the simulink model of the equivalence ratio. The expression for ER is as follows equation (2).

$$ER= [F / (254.7\ F_h\ (H-1)m_b)\ ]\ 100 \quad (2)$$

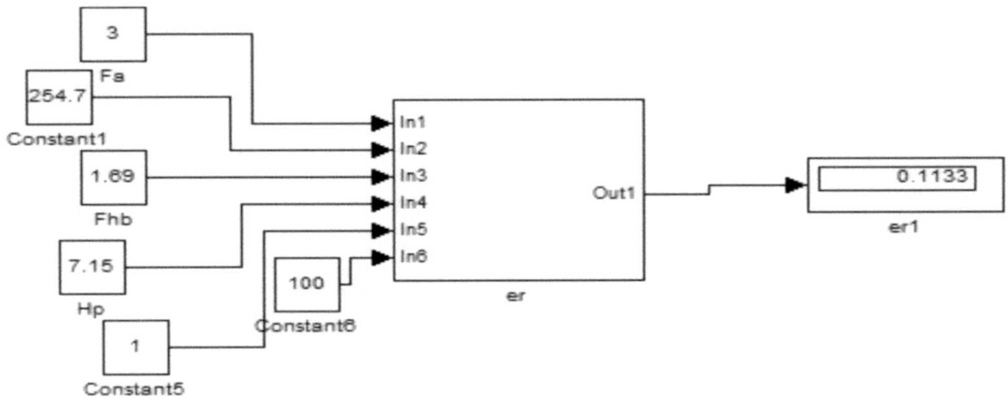

Figure 9. Subsystem for equivalence ratio

5.3. CO/CO2 ratio

It depends on ER and H_p. When H_p is low, the ratio increases with ER to reach a maximum. Figure 10 shows the simulink model of CO/CO_2 ratio. The expression derived for the ratio is eqution (3).

$$CO/CO_2\ ratio = (0.3\ H + 0.5)\ ER - 0.2761 \quad (3)$$

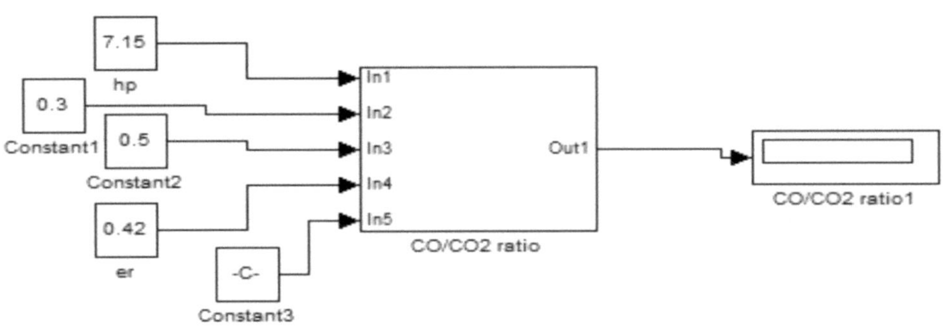

Figure 10. Subsytem for CO/CO_2 ratio

5.4. Temperature (T)

Temperature is related to the quality of produced gas very closely. It depends on ER, H_P and F_A. The value of T first decreases with ER, then increases and finally, again decreases. Figure 11 shows the simulink model for the temperature.

The expression for T is equation (4).

$$T = m_c ER + 33/H + m_a + 220 + 100\ ER + 100\ f_g \tag{4}$$

The equations 1-4 some constant values are assumed in order to fit with experimental data.

Using the simulink models of four subsystems, the complete steady state model of the biomass gasifier was developed. The gasifier model for control is shown in figure.12. This model can be used to validate the rules and membership functions of the fuzzy model.

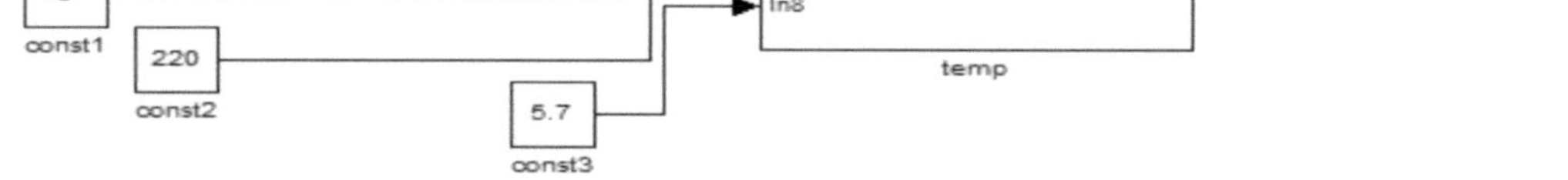

Figure 11. Subsystem for temperature

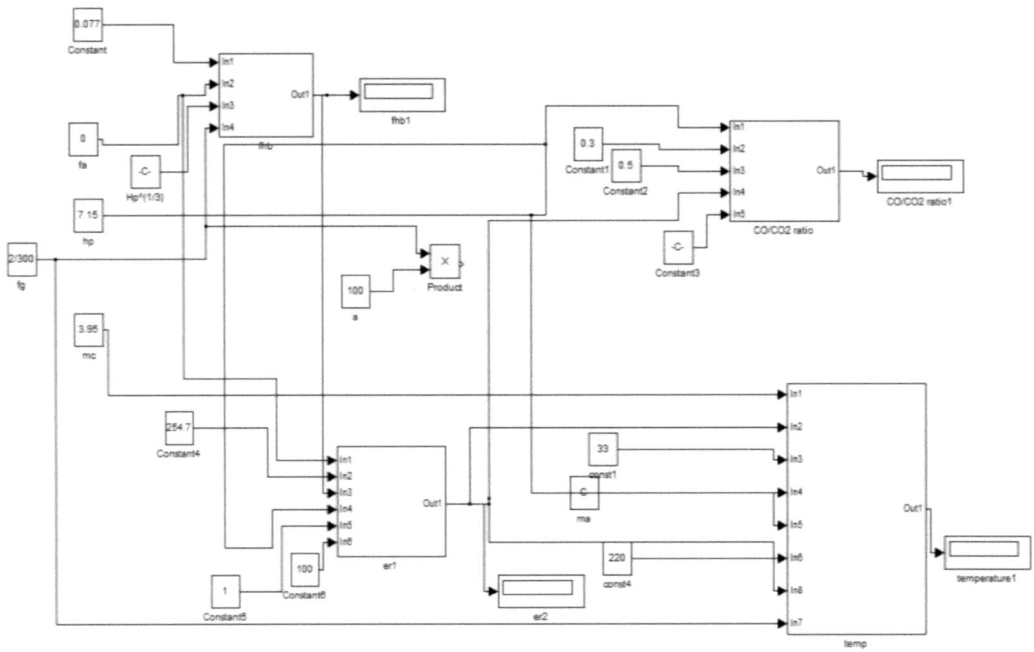

Figure 12. MATLAB simulink model of the gasifier

6. Fuzzy logic controller

A fuzzy system is a static nonlinear mapping between its inputs and outputs. Fuzzy system provides a formal methodology for representing, manipulating and implementing a human heuristic knowledge about how to control a system [8, 9-10, 11].

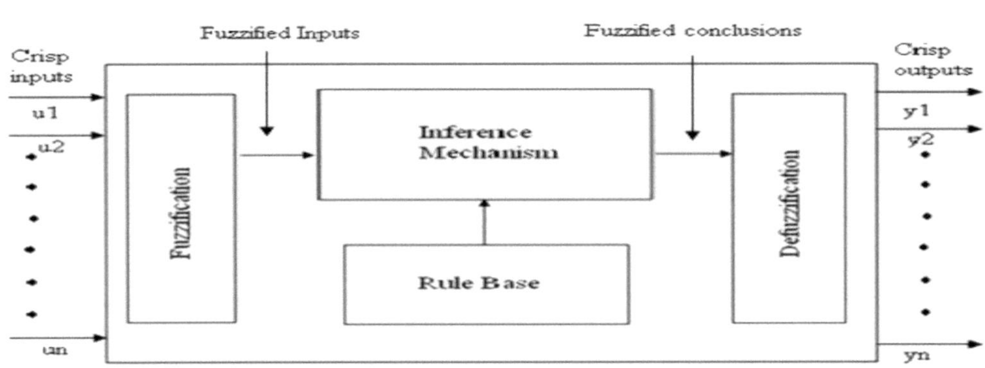

Figure 13. Fuzzy system

The block diagram of a Fuzzy system as shown in figure 13 includes fuzzification, inference mechanism, rule base and defuzzification.

Rule Base

It contains the fuzzy logic quantification of the expert's linguistic description in the form of a set of rules of how to achieve good control.

Inference Mechanism

The Inference mechanism also called an "inference engine" or "fuzzy inference module" emulates the expert's decision making in interpreting and applying knowledge about how best to control the plant. The inference mechanism evaluates which control rules are relevant at the current time and then decides what the input to the plant should be.

Fuzzification

A Fuzzification interface converts the inputs into information that the inference mechanism can easily use to activate and apply rules in the rule base.

Defuzzification

The defuzzification interface converts the conclusions reached by the inference mechanism into the actual inputs to the plant. Hence this is a process of converting decisions into actions.

6.1. Fuzzy system design

The design procedure of a Fuzzy system involves the following steps.

1. Choosing the fuzzy system inputs and outputs
2. Putting control knowledge into rule base.
3. Fuzzy quantification of knowledge.
4. Matching/Determining which rules to use.
5. Inference Step of determining conclusions.
6. Converting decisions into actions.

6.1.1. Choosing inputs & outputs

The fuzzy system is to be designed to automate how a human expert who is successful at the task would control the system. Essentially the objective is to make sure that the fuzzy system will have the proper information available to be able to make good decisions and have proper inputs and outputs needed to achieve high performance operation.

6.1.2. Putting control knowledge into rule base

Once the inputs and the outputs to the fuzzy system are decided, we have to load into the Fuzzy system, the linguistic description of the control knowledge of how best to control the plant as suggested by a human expert. This process consists of specifying the Linguistic descriptions, Rules and Rule- bases.

6.1.2.1. Linguistic descriptions

The Linguistic Description provided by the expert can generally be broken into two major parts as Linguistic variables and Linguistic values.

The **linguistic variables** describe each of the time-varying inputs and outputs of the Fuzzy system such as error, error change, force etc.

The **Linguistic values** are those characteristics that the Linguistic variables take on over time change dynamically. Examples of Linguistic values are negative large, negative small, zero, positive small, positive large etc.

6.1.2.2. Rules

The mapping of the inputs to the outputs for a fuzzy system is in part characterized by a set of condition → action rules or in If-Then form.

If premise **Then** consequent.

Usually the inputs to the fuzzy system are associated with the premise and the outputs with the consequent. These if-then rules can be represented in many forms. Two standard forms are MIMO (Multiple Input Multiple Output) and MISO (Multiple Input Single Output).

MIMO

If premise_1 **and** premise_2 and…….premise_n **then** consequent_1 **and** consequent_2.

MISO

If premise_1 **and** premise_2 and…….premise_n **then** consequent.

6.1.2.3. Rule base

This contains the rules for all possible combinations of the inputs and the outputs. For example, a fuzzy system with three inputs and five linguistic values, there can be at the most $5^3 = 125$ rules possible (all possible combinations of premise linguistic values for the three inputs)

6.1.3. Fuzzy quantification of knowledge

In a fuzzy system, the membership functions quantify, in a continuous manner, the values of a Linguistic variable into Fuzzy sets. Some of the membership function choices are Triangle, Trapezoid, and Gaussian etc.

6.1.4. Inference process

The inference process of matching/determining which rules to use involves two steps:

1. The premises of all rules are compared to the controller inputs to determine which rules apply to the current situation. This matching process involves the determining of the uncertainty that each rule applies, strongly taking into account, the recommendations of rules that we apply at the current situation.
2. The conclusions (what controls actions to take) are determined using the rules that have been determined to apply at the current time. The conclusions are characterized with a fuzzy set (or sets) that represent(s) the certainty that the input to the plant should take on various values.

6.1.4.1. Defuzzification

The final component of a fuzzy system is the defuzzification block. It operates on the implied fuzzy sets produced by the inference mechanism and combines their effects to provide the "most certain" controller output which is the input to the plant. Defuzzification can also be thought of as the process of decoding the fuzzy set information produced by the inference process (i.e. implied fuzzy sets) into numeric fuzzy controller outputs. Two commonly used defuzzification methods are Center of Gravity (COG) Method and Center - Average Method

7. Design of FLC for downdraft gasifier

Since the process of gasification is a highly non-linear and slow process [12], the formation of an accurate model is very difficult .The model must be representing the non-linear dynamic characteristics of the process. The knowledge of experts about biomass gasification can be determined in fuzzy if/then rules in such a way that the fuzzy systems can deal with the indistinguishable and inaccurate biomass condition. Here, fuzzy system is modeled with three inputs and two outputs. The three inputs are ErrorT (error in temperature), Error CO/ CO2 and Hp (moisture content) and the output is airflow and frequency of grate [12, 13-14, 15].

Fuzzy rules are formulated based on error temperature ,error CO/ CO2 ratio and Hp (moisture content) which are converted to non-fuzzy values by defuzzification,. These values are fed to the final control element for control action is shown in Figure 14.

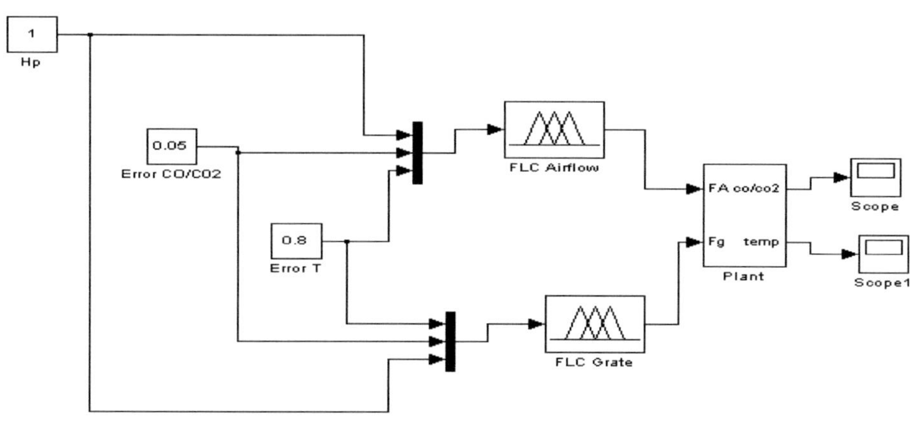

Figure 14. Simulation of control system

For the Error CO/ CO2 variable, five linguistic values were defined: very low, low, zero, high, very high, and for the Error T variable, five linguistic values, very low, low, zero, high, very high,were also defined. For the output variable Airflow five fuzzy values, extreamly low, very low, base, very high, extremely high and for the Grate variable, the values f1, f2, f3, f4, f5, were defined. The membership functions are shown in Figures 15.

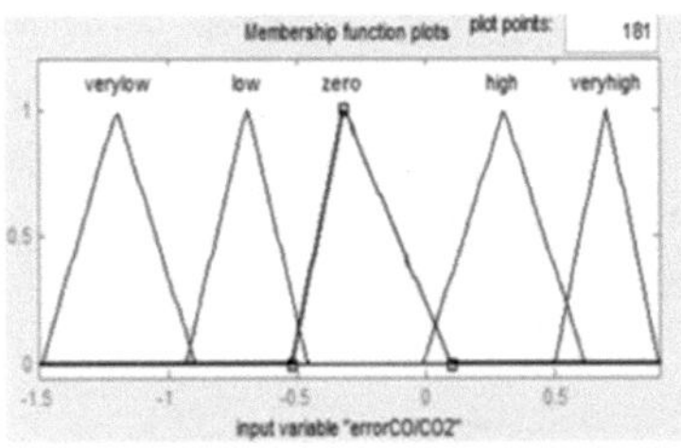

(a) Error CO/ CO2 Ratio

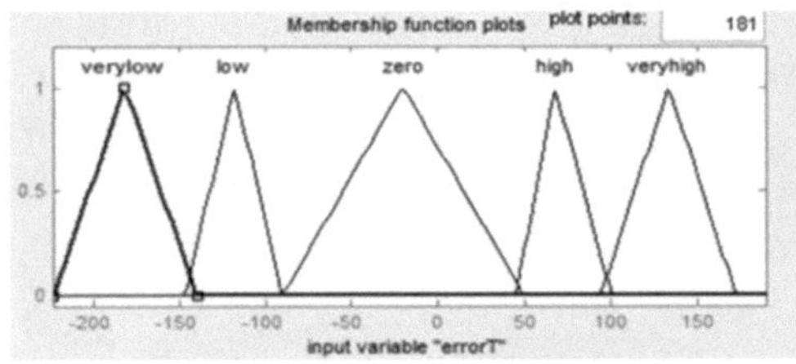

(b) Error temperature

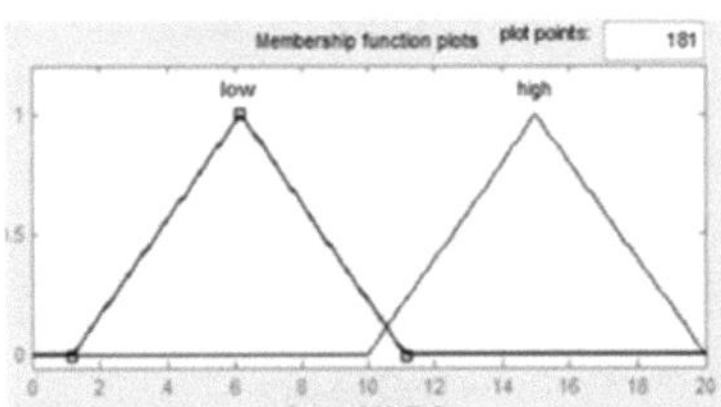

(c) hp

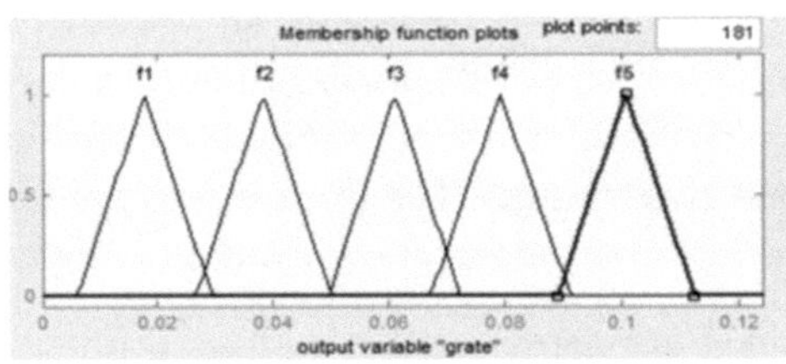

(e) Frequency of Grate

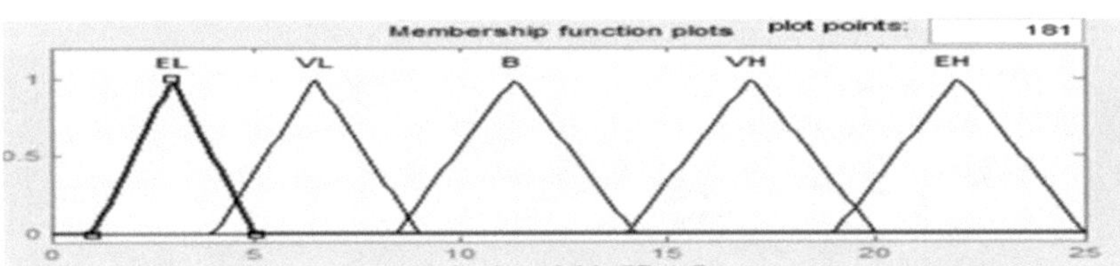

(d) Airflow

Figure 15. Figure 15. Membership function of (a)Error CO/ CO_2 Ratio , (b) Error temperature (c) hp, (d) Airflow and (e) Frequency of Grate

The rules that have been framed for the controller are as shown in Table 2 and 3.

Sl.No	**If** hp	**and** errorT	**and** errorCO/CO_2	**Then** grate
1.	low	verylow	verylow	f4
2.	low	verylow	low	f3
3.	low	verylow	zero	f1
4.	low	verylow	high	f3
5.	low	verylow	veryhigh	f1
6.	low	low	verylow	f3
7.	low	low	low	f2
8.	low	low	zero	f2
9.	low	low	high	f2
10.	low	low	veryhigh	f1
11.	low	zero	verylow	f3

Table 2. Rules for adjusting frequency of grate

Sl .No	**If** hp	**and** errorT	**and** errorCO/CO_2	**Then** flow
1.	low	verylow	verylow	EH
2.	low	verylow	low	VH
3.	low	verylow	zero	VL
4.	low	verylow	high	VH
5.	low	verylow	veryhigh	VL
6.	low	low	verylow	VH
7.	low	low	low	VH
8.	low	low	zero	B
9.	low	low	high	B
10.	low	low	veryhigh	VL
11.	low	Zero	verylow	VH
12.	low	Zero	low	B
13.	low	Zero	zero	VL
14.	low	Zero	high	VL
15.	low	Zero	veryhigh	EL
16.	low	high	verylow	B
17.	low	high	low	B
18.	low	high	zero	VL
19.	low	high	high	VL
20.	low	high	veryhigh	EL
21.	low	veryhigh	verylow	VH
22.	low	veryhigh	low	B

Sl .No	**If** hp	**and** errorT	**and** errorCO/CO_2	**Then** flow
23.	low	veryhigh	zero	VL
24.	low	veryhigh	high	VL
25.	low	veryhigh	veryhigh	VL
26.	low	verylow	Verylow	EH
27.	low	verylow	low	EH
28.	low	verylow	zero	EH
29.	low	verylow	high	EH
30.	low	verylow	veryhigh	EH
31.	low	low	verylow	EH
32.	low	low	low	EH
33.	low	low	zero	EH
34.	low	low	high	VH
35.	low	low	veryhigh	VL
36.	low	zero	verylow	VH
37.	low	zero	low	B
38.	low	zero	zero	VL
39.	low	zero	high	VL
40.	low	zero	veryhigh	VL
41.	low	veryhigh	verylow	VL

Table 3. Rules for adjusting air flow rate

This developed fuzzy logic controller of MIMO system of gasifier was based on the static model of the gasifier that have been proposed, which can be used in tuning the controller. The controller that has been developed in this manner was implemented in microcontroller [16, 17]. The CO/CO_2 ratio has effectively controlled with fuzzy logic controller by adjusting the frequency of motion to control the biomass residence time inside the reactor. The temperature also effectively controlled with fuzzy logic controller by adjusting the airflow rate. The performance of MIMO system of the gasifier cannot be verified by simulation because doesn't have proper dynamic model of the gasifier. In this paper temperature control system of the gasifier (SISO) has been verified by the simulation in order to prove the efficiency of fuzzy controller in comparison with the conventional controller.

8. Implementation results

Fuzzy logic controller has been implemented for the transfer function model of the gasifier. To prove the efficiency of fuzzy controller in comparison with the conventional controllers, a fuzzy logic controller for the SISO system of gasifier, where flow is the input and temperature is the output has been proposed. The simulation of the fuzzy logic control system is shown in Figure 16. The responses of the process to fuzzy controller for various set points have been shown in Figures 17, 18 and 19. From the responses it is observed that settling time and overshoot have been reduced. The settling time is found to be 51.5 seconds.

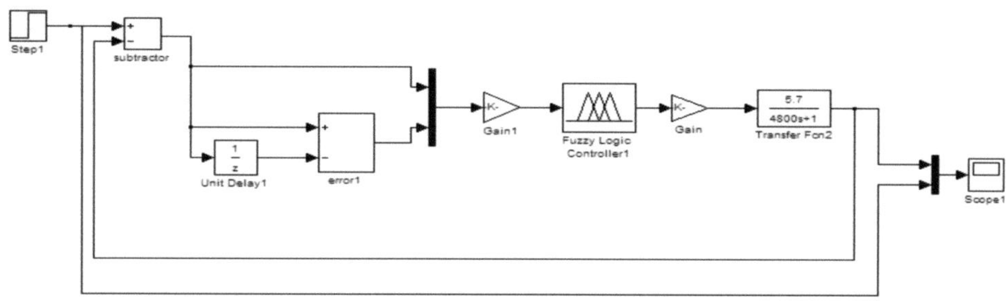

Figure 16. Fuzzy logic controller for transfer function model of the process

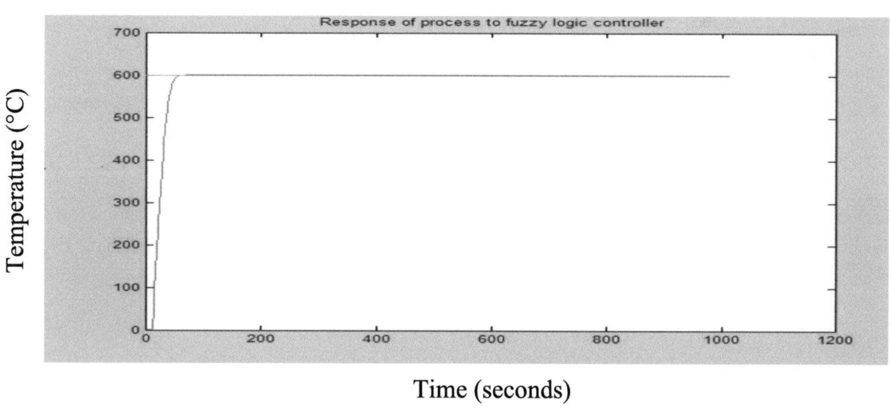

Figure 17. Response of process with fuzzy controller (for set point of 600 ^{0}C)

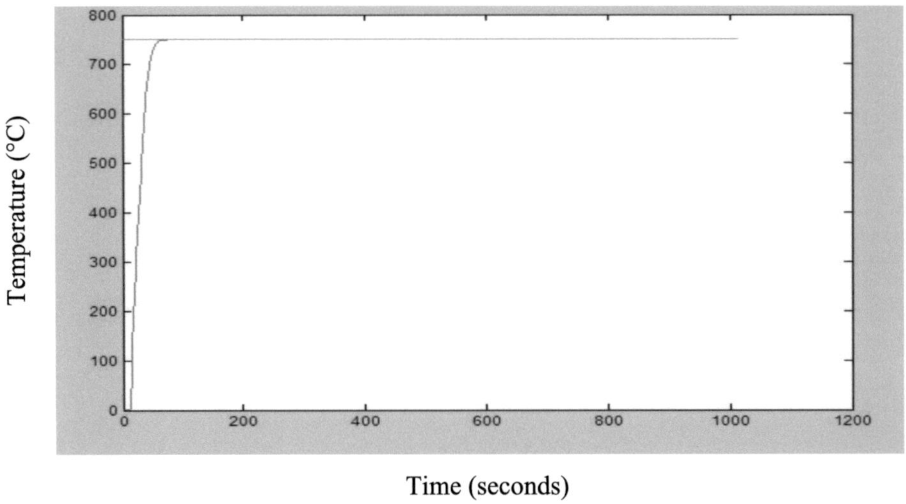

Figure 18. Response of process with fuzzy controller (for set point of 750 ^{0}C)

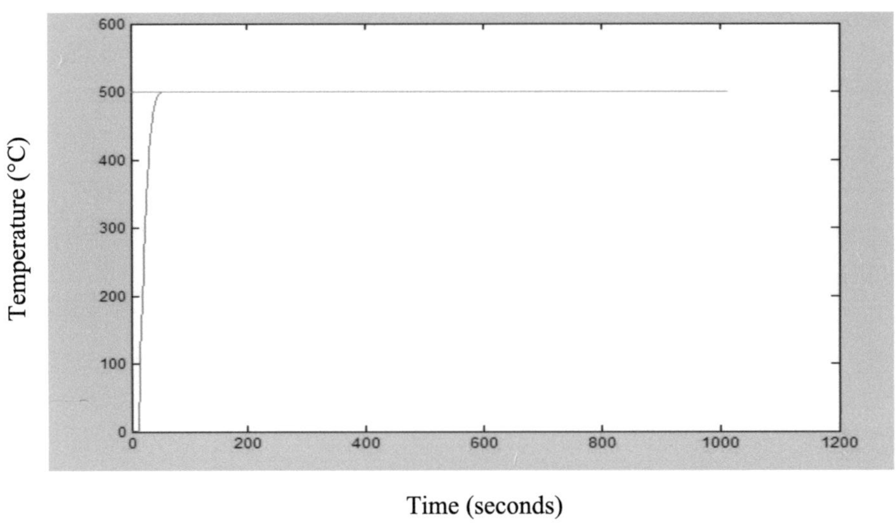

Figure 19. Response of process with fuzzy controller (for set point of 500 ^{0}C)

9. Conclusion

A static model for the gasifier has been developed based on the experimental data of the plant successfully and this could be used to tune the fuzzy logic controller. The implementation results for the gasifier system are found to be better than conventional control and pc based measurement. Fuzzy control is used to control the temperature with less overshoot and settling time . In this paper, it has been observed that fuzzy control gives better performance for the control of biomass gasification.

Author details

A. Sanjeevi Gandhi
Karunya University Coimbatore, Tamilnadu, India

T. Kannadasan
Coimbatore Institute of Tech Coimbatore, Tamilnadu, India

R. Suresh
R.V. College of Engineering,, Bangalore, Karnataka, India

10. References

[1] Milligan J., (1994) Downdraft gasification of biomass, Ph.D. thesis, University of Aston, Birmingham, UK.
[2] Sanjay Gupta.S (2006) Technology of Biomass Gasification, Tata McGraw Hill Publishing Company Private Limited.
[3] Jorapur, R., Rajvanshi, A., (1997) Sugarcane leaf-bagasse gasifiers for industrial heating applications. Biomass and Bioenergy 13 (3), 141– 146.
[4] Garcı´a-Bacaicoa, P., Serrano, S., Berrueco, C., Ceamanos, J., (2004) Study on the gasification of sewage sludge for power production in a dual fueled engine. In: 2nd World Conf. and Technology Exhibition on Biomass for Energy, Industry and Climate Protection, Roma, Italy.
[5] Liu, G.P., Dixon, R., Daley, S., (2000) Multi-objective optimal-tuning Proportional—integral controller design for the Alstom gasifier Problem, Proceedings of the IMechE Part I Journal of Systems and Control Engineering 214 (6), pp.395–404.
[6] Prempain, E., Postlethwaite, I., Sun, X., (2000) Robust control of the gasifier using a mixed-sensitivity h1 approach. Proceedings of the IMechE Part I Journal of Systems and Control Engineering 214 (6), 415–426.
[7] Chern, S. (1989) Equilibrium and kinetic modelling of co-current (Downdraft) moving-bed biomass gasifier, Ph.D. thesis, Kansas State Kansas, EEU.
[8] Claudio (2005) Introduction to Fuzzy Logic,Elec.Energ.vol.18, No.2, pp.319-328.
[9] Driankov D, Hellendoorn H and Reinfrank M,(1996) An Introduction to Fuzzy Control, Narosa Publishing House, New Delhi.
[10] PeterHarriot. (1972), Process Control, Tata McGraw Hill Publishing Company Private Limited
[11] Ronald, R., Yagner, Dimiter. D., Filer (2002), Essentials of Fuzzy Modeling and Control, John Wiley and Sons Inc.
[12] .Sagues, C., Garcia, P-Bacaicoa, and Serrano, S (2007) Automatic control of biomass gasifiers using fuzzy inference system, Bioresource Technology, pp.845-855.
[13] Yamakava. T. (1993) Fuzzy Inference in Nonlinear Analog Mode and its Applications to Fuzzy Logic Conrol, IEEE Trans. Neural Netw., vol.4, pp.496-522.
[14] Zhao, H., Fan, L., Chee, C, and Walawender, W. (1992) Control of a Downdraft gasifier for biomass conversion with fuzzy controller, Powder Technology 69, pp.53–59.

[15] Kevin M Passino and Stephen Yurkovich, (1998) Fuzzy Control, Addison-Wesley Longman, Inc., California.

[16] Svrcek, W.Y., Mahoney, D. and Young, B. (2000) A Real-time Approach to Process Control, Prentice- Hall, Chichester, England, UK.

[17] Sanjeevi Gandhi. A, Kannadasan. T and Suresh.R. (2010) Hardware design of fuzzy controller for biomass downdraft gasifier with labVIEW, IEEE Region 10 Annual International Conference, Proceedings/TENCON, pp.2096-2101.

Section 3

Wastes Gasification

Chapter 6

Supercritical Water Gasification of Municipal Sludge: A Novel Approach to Waste Treatment and Energy Recovery

Jonathan Kamler and J. Andres Soria

Additional information is available at the end of the chapter

http://dx.doi.org/10.5772/51048

1. Introduction

Municipal sewage sludge is often heavily moisture-laden, containing moisture well in excess of 95 w/w% (Fytili & Zabaniotou, 2008; Harrison et al., 2006; Hong et al., 2009). Annual U.S. sludge volume is estimated to be between 500 million and 1.5 billion wet tons, resulting in the need to remove between 130 and 400 billion gallons of water from it for treatment and disposal. This dewatering process expends 80% of the total electricity used by a wastewater treatment facility (Bernardi et al., 2010; Demirbas, 2011b; Fytili & Zabaniotou, 2008; USEPA, 2009), representing an average of 150 billion KWh of electricity each year, at an approximate cost of $6 billion, or ~4% of the total annual U.S. electricity use (CSS, 2011; Kim & Parker, 2008).

Nascent technologies that use the entrained water of the sludge itself have been studied and developed in order to overcome the expense and complexity of dewatering municipal sludges (Savage, 2009). Water, when raised simultaneously to very high temperatures and pressures, becomes one of the most promising solvation media for rapid gasification and complete destruction of aqueous, organic wastes. As temperature and pressure increase, water approaches what is known as the "critical point" (≥374.2°C and 22.1 MPa), above which water becomes "supercritical". This chapter discusses recent developments of using water and elevating its temperature and pressure to near and above supercritical conditions (Figure 1) for the treatment and disposal of municipal sewage sludge.

Supercritical water's unique abilities to quickly dissolve and gasify organic compounds in sludge without dewatering are presented (Kalinci et al., 2009). Furthermore, adding catalysts or oxidants to supercritical water can intensify the reaction, substantially reducing operating costs by creating self-sustaining conditions that can lead to energy recovery and short residence times, as compared to more conventional sludge disposal methods, including incineration.

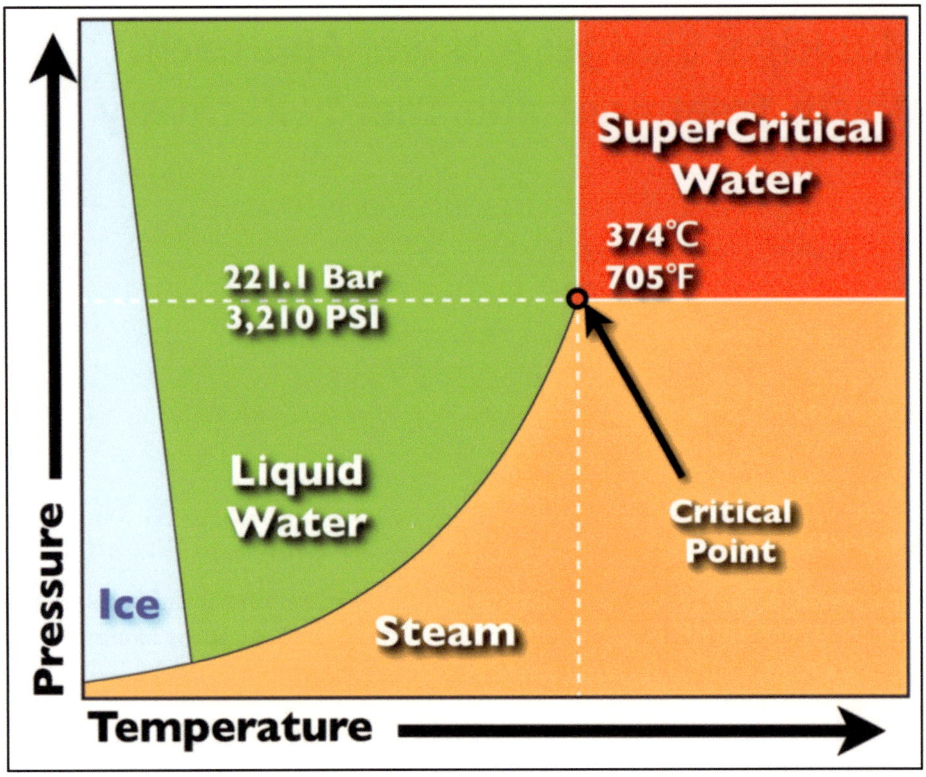

Figure 1. Phases of Water.

The chapter reviews supercritical-water research that addresses various sludge destruction advantages as well as known challenges. The review highlights forays and attempts at commercialization of supercritical water systems for wet-waste destruction and discusses the nascent industrial aspects of the technology and the challenges of creating a commercially viable plant.

1.1. Wastewater sludge

Sewer systems in the U.S. transport over 14.6 trillion gallons of municipal wastewater to ~17 thousand public wastewater facilities each year (CSS, 2011; Fytili & Zabaniotou, 2008). The facilities are designed to collect, remediate, and dispose of human and commercial wastes within an established regulatory framework (Chun et al., 2011; Demirbas, 2011b; Fytili & Zabaniotou, 2008; Svanström et al., 2004; USEPA, 2009). Sewage that enters wastewater treatment facilities gets processed and separated into two products. One is clean water, which is the primary objective of municipal facilities. The other is the leftover waste, generically known as sewage sludge (Abelleira et al., 2011).

Sludge is the most ubiquitous wet waste generated by humans (Abelleira et al., 2011). The U.S. Environmental Protection Agency (EPA) last estimated U.S. sewage-sludge production in 1998 at 6.9 million dry tons (USEPA, 1999b). Unconfirmed estimates dating back as far as 1982, however, put total U.S. sewage sludge volume much higher at nearly 20 million dry tons with an additional comparable amount of other industrial sludges (Gloyna & Li, 1993; Svanström et al., 2004).

All sewage sludge from modern wastewater treatment plants is potentially harmful to human health by design and is designated as a pollutant by the Clean Water Act (Harrison et al., 2006; Mathney, 2011; NASNRC, 1996, 2002; USEPA, 2009). When sewage undergoes treatment, the solids, along with a myriad of entrained hazardous and harmful pollutants and pathogens, are removed from the water and concentrated into sludge (Bernardi et al., 2010; Hong et al., 2009; Snyder, 2005; USEPA, 1999a, 2009). Consequently, the physical properties and chemical constituents of sludges vary widely, depending on the source and treatment of the sewage. Generally, however, sewage sludge is treated as a homogenous, non-standardized slurry of materials, consisting mainly of human metabolic and food wastes as well as varying amounts of industrial, agricultural, and medical wastes (Harrison et al., 2006; Hong et al., 2009).

1.1.1. Sewage sludge composition & regulatory framework

All sewage sludge produced in the U.S. contains varying concentrations of three types of harmful pollutants: 1) heavy metals, 2) hazardous organic compounds, and 3) pathogenic microorganisms. Safely managing these hazardous compounds and pollutants has proven challenging (NASNRC, 2002; USEPA, 2009).

1.1.1.1. Heavy metals

Heavy metals ubiquitously entrained in sludge pose serious and well-documented public health and environmental risks (Babel & del Mundo Dacera, 2006; Bag et al., 1999; Beauchesne et al., 2007; Dimitriou et al., 2006; Fjällborg et al., 2005; Fytianos et al., 1998; Goyal et al., 2003; Hooda, 2003; Kidd et al., 2007; McBride, 2003; Pathak et al., 2009; Reddy et al., 1985; Sánchez-Martín et al., 2007; USEPA, 2009). The EPA, however, limits sludge regulations to only ten (i.e., arsenic, cadmium, chromium, copper, lead, mercury, molybdenum, nickel, selenium, and zinc) of the high-risk, hazardous, bioaccumulating, and leaching metals (Babel & del Mundo Dacera, 2006; Dean & Suess, 1985; Harrison et al., 1999; McBride, 2003; Pathak et al., 2009; Sánchez-Martín et al., 2007; USEPA, 2002b, 2009). Reviews detailing heavy metal prevalence in sludge and related health concerns can be found elsewhere (Babel & del Mundo Dacera, 2006; Bag et al., 1999; Harrison et al., 1999; McBride, 2003; Pathak et al., 2009; Sánchez-Martín et al., 2007; Snyder, 2005).

1.1.1.2. Hazardous organic compounds

Hazardous organic compounds commonly found in sewage sludge matrices are many and varied, including endocrine disrupters, pharmaceuticals, polybrominated fire retardants, polychlorinated biphenyls, carcinogens, pesticides, household chemicals, solvents, and dioxins (Costello & Read, 1994; Gómez et al., 2007; Hale et al., 2001; McBride, 2003; NASNRC, 2002; Qi et al., 2010; Rulkens, 2008; Santos et al., 2010; Sipma et al., 2010; Snyder, 2005; Stasinakis et al., 2008; Zorita et al., 2009). Hazardous pollutants are ubiquitous in sewage sludge. The EPA studied sewage sludges from wastewater facilities across the U.S. and found large amounts of hazardous materials in all of the sludges (USEPA, 2009). Many organic compounds in sludge do not break down quickly in the environment and are often highly mobile, resulting in widespread harmful, organic-compound distribution (Guo et al., 2009; Kulkarni et al., 2008; Leiva et al., 2010; Rulkens, 2008). Consequently, human exposure to some harmful organic compounds from sewage sludge (e.g., dioxins) is considered pervasive and chronic (Kulkarni et al., 2008). Only about 110 organic chemicals (of fewer than 130 total chemicals) are on EPA's antiquated priority pollutant list, and there is no regulatory requirement to monitor any of those in sewage sludge (Clarke & Smith, 2011; Deblonde et al., 2011; Eriksson et al., 2008; Harrison et al., 2006; Hospido et al., 2010; Petrovic´ et al., 2003; Verlicchi et al., 2010).

The proliferation of new pollutants in sewage sludge is also a growing concern. The number of organic chemicals is increasing rapidly, now well in excess of 100 thousand. Very few of the pollutants noted to be commonly present in sludge, including low-grade, radioactive residues in medical wastes, have been studied in detail either in terms of prevalence or harmful effects (Eriksson et al., 2008; Fytili & Zabaniotou, 2008). Even though their effects on environment and human health are largely unknown, these "emerging pollutants" fall outside EPA regulatory status (Deblonde et al., 2011; NASNRC, 2002; Tsai et al., 2009). Furthermore, there have been no major updates to the EPA's priority pollutant list in almost three decades (Harrison et al., 2006; Mathney, 2011; Snyder, 2005).

1.1.1.3. Pathogens

Pathogen loads in sewage sludge are almost universally high and pose a communicable disease hazard (NASNRC, 2002; Reilly, 2001; USEPA, 2009). The pathogens are a result of normal, human metabolic wastes as well as additional loading from medical effluents (Arthurson, 2008; Deblonde et al., 2011; Lewis et al., 2002; Mathney, 2011; Reilly, 2001; Straub et al., 1993; USEPA, 2009; Verlicchi et al., 2010). There are fewer than two dozen pathogens (e.g., fecal coliforms, Salmonella, enteric viruses, and parasites) monitored in sewage sludge (Mathney, 2011; NASNRC, 2002; Reilly, 2001; Snyder, 2005; USEPA, 2000, 2002b, 2003), and many dangerous pathogens (e.g., prions) are neither affected by sewage treatment nor detected by standard analytical methods (Gale & Stanield, 2001; NASNRC, 2002; Peterson et al., 2008b; Saunders et al., 2008; Smith et al., 2011; Snyder, 2005).

Despite considerable controversy surrounding potential sludge hazards, there has been disturbingly little critical inquiry into the environmental effects and human health risks of traditional sludge disposal methods (Deblonde et al., 2011; Mathney, 2011; Nature, 2008; Tollefson, 2008). Nonetheless, some EPA goals (albeit with no specified implementation horizon) indicate that very high destruction requirements (up to 99.9999%) may become standard for some compounds, along with totally enclosed treatment facilities (Lavric et al.,

2005; Veriansyah & Kim, 2007). If such regulatory standards are ever implemented, the feasibility and suitability of conventional sludge disposal techniques will be subject to increased scrutiny (Demirbas, 2011b; Veriansyah & Kim, 2007).

1.2. Sludge processing & disposal

Despite improvements in wastewater cleaning technology and expansion of centralized wastewater services to meet the needs of most of the U.S. population, sludge disposal has historically been, and continues to be, the weak link in the wastewater treatment process (Demirbas et al., 2011; Fytili & Zabaniotou, 2008; Harrison et al., 2006; NASNRC, 1996, 2002). Ocean dumping was a preferred sludge disposal method for the last couple of centuries (Chun et al., 2011; Snyder, 2005), but it was banned in the 1990s by both U.S. and international law due to the high level of harmful pollutants in the sludge and the adverse effect on marine organisms (Abbas et al., 1996; Costello & Read, 1994; Harrison et al., 2006; Snyder, 2005). The loss of ocean-dumping drove most municipalities to embrace either agricultural land application or thermal destruction (viz., incineration) as their primary sludge-disposal routes, with a small percentage using landfilling or composting (Lavric et al., 2006). Current sludge disposal methods, and associated regulations, are outgrowths of the need for municipalities to find a viable solution for treating or disposing large amounts of concentrated harmful pollutants resulting from wastewater treatment. Disposal choice is influenced by economics, public policy, and regional environmental conditions (Cappon, 1991; Rulkens, 2008).

1.2.1. Land application

Agricultural land application is the most commonly used and most controversial of the sludge disposal methods, but has gained favor due to the simple-bottom-line cost. Potential hazards of applying sludge to croplands were noted early on in the adoption of land-application practices. Using material laden with harmful organic compounds in food and forage cultivation makes land application problematic both in terms of operational costs and, more importantly, public health concerns (Borán et al., 2010; CSS, 2011; Demirbas et al., 2011; Eriksson et al., 2008; Fytili & Zabaniotou, 2008; Harrison et al., 2006; NASNRC, 1996, 2002). Specifically, potential food-crop contaminant uptake and subsequent human-food-chain contamination are legitimate concerns (Cappon, 1991). Despite the well-documented, undesirable properties of sewage sludges for agricultural purposes, most communities continue to favor sludge land application over other disposal methods (Beauchesne et al., 2007; Beck et al., 1995; McBride, 2003). The proponents of sludge land application argue that harmful-organic-compound behavior in soils from sludge application is reasonably well understood and that there will be negligible detrimental health and environmental impacts (McBride, 2003).

1.2.2. Thermal destruction

Thermal destruction (i.e., incineration) offers a year-round, all-weather sludge disposal option, albeit an energy-intensive and thus increasingly expensive option. Many large cities in the colder northern climates use incineration, with more than 200 sewage-sludge incinerators (fluidized-bed and multiple hearth configurations) in use nationwide (Sloan et al., 2008). High water content (along with associated high enthalpy demand) poses the main thermodynamic impediment to cost-effective thermal sludge destruction. During the destruction process, all of the energy released from the sludge, and essentially all of the incinerator fuel, is consumed to boil off water (Demirbas, 2011b; Dijkema et al., 2000; Fytili & Zabaniotou, 2008). Furthermore, sludge must initially be dewatered to a "sludge cake" consistency with moisture content below 85% prior to feeding into the incinerator. Once in the incinerator, the sludge cake must be further dewatered thermally to ~35 w/w% moisture before the material itself can actually begin to thermally combust (Abuadala et al., 2010). Dewatering is expensive, and as energy costs continue to rise, drying processes are becoming increasingly prohibitive (Weismantel, 2001).

Dry pyrolysis and gasification face similar thermoeconomic efficiency limitations to incineration, in that high-moisture levels in sludge cause ignition and combustion problems (Demirbas et al., 2011; Dogru et al., 2002). Specifically, traditional gasification technologies encounter operational air:fuel ratio and gas:ventilation mobility problems when the feedstocks exceed 30% moisture content, and sewage-sludge moisture content generally needs to start at less than 15% to serve as a proper feedstock for gasifiers (Dogru et al., 2002). Plus, fuels produced require significant additional cleaning due to the presence of heavy metals and incomplete destruction of harmful organic compounds (Dogru et al., 2002). Indeed, traditional thermal technologies do destroy hazardous organic compounds, but only up to a point. Incineration-derived slag, for example, still contains all of the heavy metals, up to 30% of the original hazardous organic compounds, and additional secondary combustion compounds (Dogru et al., 2002; Fytili & Zabaniotou, 2008). Most contemporary thermal options are prohibitively costly due to high capital investment and increasingly stringent, air-quality permitting and compliance standards (Chun et al., 2011; Fytili & Zabaniotou, 2008). Thermal destruction also meets with considerable, unfavorable public opinion due to the air-borne release of metal emissions and harmful gases (Abbas et al., 1996; Adegoroye et al., 2004; Lavric et al., 2006). Intense public protests of new permits alone have derailed some incinerator permitting efforts (Sloan et al., 2008; Weismantel, 1996).

1.2.3. Landfill disposal

Landfilling (i.e., burial) of sludge is used as a disposal method by many municipalities, often in an effort to avoid expensive regulatory incineration restrictions and to sidestep the greater scrutiny of land application. Nonetheless, landfilling also has a host of problems, including decreased landfill life, increased landfill odor, and increased landfill leachate volume and toxicity. Leachate is a ubiquitous product of landfills, wherein excess water percolates through landfill waste layers, freeing organic compounds from the waste and carrying them away concentrated in leachate. The high water content of sewage sludge is known to escalate leachate volume from landfills (Demirbas et al., 2011). Furthermore, the degradation and conversion of organic compounds in landfilled sludge is usually incomplete (Ejlertsson et al., 2003), and metabolites can be generated that are even more hazardous than their parent compounds, with the secondary organic pollutants also collecting in the leachate (Oleszczuk, 2008). The composition of leachate is complex, environmentally reactive (with very high COD values: above 60K mg/L), and difficult to treat via conventional methods (Wang et al., 2011). Landfill leachate is a noted health and environmental threat, and harmful compounds in sewage sludge exacerbate the problem (Demirbas et al., 2011). A rise in tipping fees, decreased availability of economic landfill sites, and a move toward sustainable solutions has begun to sour municipal fondness for landfilling (Abbas et al., 1996).

1.2.4. Composting

Non-industrial composting of agricultural wastes dates back thousands of years to ancient Rome, Greece, and Israel for agricultural recycling, and has now gained some recent traction as a recycling method for modern organic wastes including sewage sludge (Epstein, 1997; Gajalakshmi & Abbasi, 2008; Hubbe et al., 2010; Kumar, 2011). Industrial composting processes are used to convert sewage sludge into "marketable fertilizer" products and ostensibly reduce sludge volume and organic pollutants (Oleszczuk, 2008). Nonetheless, under U.S. Department of Agriculture (USDA) branding regulations, sludge-derived compost cannot legally be labeled as "Certified Organic", limiting its market potential (USDA, 2011).

There are many composting methods. The simpler composting approaches of mixing sludge with other organic wastes and letting them react with microorganisms are relatively low-tech, inexpensive, slow, odorous, and invariably require large footprints and relatively dry and warm weather conditions for outdoor operations (USEPA, 2002a). More complex approaches often use thermally accelerated, composting processes, commonly known as in-vessel, thermal drying, which produce agricultural "pellets" from sewage sludge at faster processing times in a reduced footprint (Gajalakshmi & Abbasi, 2008; Hubbe et al., 2010; Kumar, 2011; Turovskiy & Mathai, 2005; USEPA, 2002a). A number of municipalities use in-vessel, thermal drying, but the high-temperature, pelletizing process generates secondary, hazardous organic metabolites similar to landfilling, but at a much accelerated rate (Farrell & Jones, 2009; Fytili & Zabaniotou, 2008; Kumar, 2011; Oleszczuk, 2008). High-temperature, in-vessel composting increases mobility and bioavailability of the metabolites, which by extension can significantly contaminate and toxify soil faster (Oleszczuk, 2008). Pellet production costs often exceed $400 per dry ton (and can approach $1,000 per dry ton), but many communities end up landfilling all or part of their pellets due to limited market demand (Sloan et al., 2008). Several reviews have evaluated the advantages and disadvantages of different composting technologies (Farrell & Jones, 2009; Gajalakshmi & Abbasi, 2008; Hubbe et al., 2010; Kumar, 2011; Phillips, 1998; USEPA, 2002a).

1.2.5. Carbonization

Carbonization of the sludge into a solid, fuel-like product is a competing energy recovery option that can be performed for considerably lower cost than compost-pellet production due to elimination of the nuanced need to maintain a marketable fertilizer product. There are a number of competing carbonization conversion processes seeking commercialization that rely on drying and various woody-biomass or coal combinations (Chen et al., 2011; Roy et al., 2011). Some seek stand-alone fuel status, while others function on the expectation of using carbonized sludge as a co-firing fuel supplement with coal at concentrations less than 5 w/w% (Abbas et al., 1996; Roy et al., 2011; Rulkens, 2008). Reviews of sludge-derived, carbonized, solid fuels can be found elsewhere (Maier et al., 2011; Roy et al., 2011).

1.2.6. Regulatory & institutional framework

Municipalities' sludge-disposal difficulties, accompanied by the vexing problems of harmful compound removal, have not been lost on EPA regulators. Historically, regulation has been leniently "tailored" to municipal sludge-disposal needs, only regulating ten metals (i.e., As, Cd, Cr, Cu, Pb, Hg, Mo, Ni, Se, and Zn) and zero organic chemicals (Harrison et al., 1999, 2006; Mathney, 2011; McBride, 2003; Snyder, 2005). Indeed, metal toxin levels legally allowed in sewage sludge applied to croplands or included in sludge compost are several times higher in some cases than levels allowed at superfund sites (Harrison et al., 2006). The EPA has even opted to forgo extending sludge regulations to dioxins (at any level) or any other organic pollutant in sludge (Harrison et al., 2006). Many scientists and other federal agencies point out that EPA assessments for metals, hazardous organic compounds, and pathogens may significantly underestimate risks (Harrison et al., 2006; Mathney, 2011; McBride, 2003; NASNRC, 2002; Nature, 2008; Oleszczuk, 2008; Snyder, 2005; Tollefson, 2008; USDA, 2011). The National Academies of Science, a U.S. District Court, numerous scientists, and the EPA's own Inspector General have openly cast doubt on the quality, objectivity, and integrity of the research upon which the EPA has relied for sludge-disposal policy formulation (Alaimo, 2008; Dominy, 2009; Tollefson, 2008; USEPA, 1999b, 2000, 2002b). No labeling or disclosure is required for compost made from sewage sludge, and very few consumers are aware of the hazards posed by the products (Harrison et al., 2006; USEPA, 2000, 2002b).

Numerous researchers and institutions have noted that in order to fully evaluate sludge safety and risks, it is necessary to go well beyond the EPA's minimalist, chemical analyses and actually combine those with genuine ecotoxicological criteria (Abbas et al., 1996; Chun et al., 2011; Leiva et al., 2010; McBride, 2003; NASNRC, 1996, 2002; Nature, 2008; Oleszczuk, 2008; Snyder, 2005; Tollefson, 2008). There is even growing concern that in the very near future, traditional management options will be unable to handle the increasing sludge quantities (Bernardi et al., 2010). Furthermore, externalized environmental and health costs are beginning to marginalize existing and well-established, sludge disposal methods. In terms of public health risk, the National Academies of Science's National Research Council has expressed concern about the use of sludge-based materials (NASNRC, 1996, 2002). This concern was echoed in a lead editorial in the journal Nature:

> In what can only be called an institutional failure spanning more than three decades … there has been no systematic monitoring program to test what is in the sludge. Nor has

there been much analysis of the potential health effects among local residents – even though anecdotal evidence suggests ample cause for concern (Nature, 2008).

As a result, new processing technologies capable of destroying all organic compounds, including hazardous, pathogenic, and recalcitrant organic compounds, at levels in excess of 99.99%, is of paramount importance for protecting environmental and human health.

2. Hydrothermal decomposition & conversion

Chemical compounds in sludge can be converted by several methods into various forms of energy and energy carriers (Rönnlund et al., 2011). Thermal decomposition methods have been developed over the last two decades into aqueous analogs, namely supercritical water gasification (SCWG), catalytic supercritical water gasification (CSCWG), and supercritical water oxidation (SCWO) (Catallo & Comeaux, 2008; Kruse, 2009; Toor et al., 2011; Veriansyah & Kim, 2007). These hydrothermal decomposition methods use supercritical (≥374.2°C and 22.1 MPa) or near-supercritical states as the destruction medium for sludge. Many of these techniques have been repeatedly demonstrated in laboratory experiments to thoroughly destroy wet wastes such as sewage and oily sludges with efficiencies exceeding 99% (Cao et al., 2011). This high level of destruction is possible due to supercritical water's unique properties that change from standard phases, to allow for solvation of organic substances, diffusivity into solid materials and modified reactivity, leading to the degradation of organic substances into carbon dioxide, carbon monoxide, water, and thermal energy in a single reactor system (Byrd et al., 2008; Savage, 2009; Weiss-Hortala et al., 2010).

Particularly relevant to the destruction of sludge wastes is that water, under supercritical conditions, changes from a polar solvent to a non-polar solvent as the transition between subcritical and supercritical occurs (Byrd et al., 2008; Sato et al., 2003; Savage, 2009). Consequently, hydrocarbons and organics become highly miscible in water above supercritical conditions (Byrd et al., 2008). Due to the increased temperatures and pressures of these systems, the dissolved organics begin thermochemical decomposition above the critical points (Figure 2).

Supercritical water behaves like a "non-ideal" gas, wherein solute molecules in contact with the fluid will interact and react at a faster rate than would be the case with either true liquids or gases (Hyde et al., 2001). Vapor pressure of solutes, based on polarity, will increase for organic molecules and decrease for inorganic compounds, resulting in solvation enhancement for sludge-entrained organics (Sato et al., 2003; Savage, 2009). Local density enhancements resulting from electrostatic and van der Waal effects also have a role in solubility (Brennecke & Chateauneuf, 1999; Marrone et al., 2004). Localized densities surrounding solute increase well above densities throughout most of the fluid due to eddy effects, aggregation, and nucleation around solute molecules (Hyde et al., 2001). Extensive reviews of supercritical water properties can be found elsewhere (Brunner, 2009a; Dinjus & Kruse, 2007; Hauthal, 2001; Hyde et al., 2001; Kruse, 2008; Kruse & Dinjus, 2007a, 2007b;

Loppinet-Serani et al., 2010; Machida et al., 2011; Noyori, 1999). The most important advantage of supercritical-water, hydrothermal destruction systems is that aqueous sludges do not require any pre-treatment or drying steps in order for the thermochemical conversion process to occur, resulting in efficient material transfer and economically beneficial characteristics (Duan & Savage, 2011), which can be further enhanced by modifying the system for production of H_2, the addition of catalysts, or energy recovery.

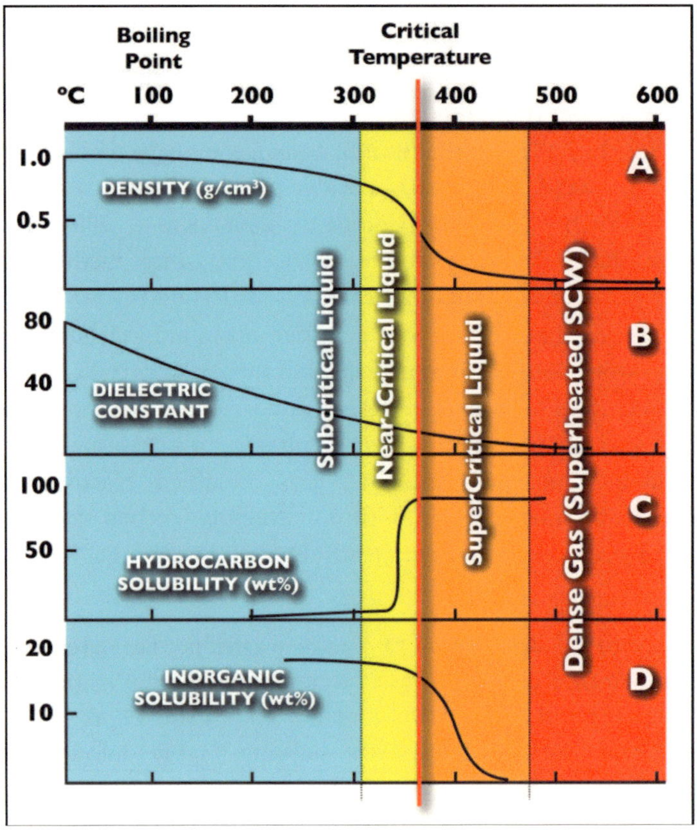

Figure 2. Properties of Water in Subcritical, Near-Critical, and Supercritical Conditions, Adapted from (Zhang et al., 2010).

2.1. Supercritical Water Gasification

Supercritical water gasification (SCWG) technologies have been developed into solutions to wet-waste, wet-biomass, and aqueous-sludge destruction (Savage, 2009). The primary objective of SCWG, however, is similar to conventional thermal gasification, in that SCWG is typically used for the production of fuel or chemicals, with waste stream elimination only a secondary consideration. Nonetheless, the various properties of supercritical water enable supercritical-water gasification to quickly destroy wet biomass and organic aqueous wastes while efficiently producing H_2 and C1 rich gases. Supercritical water gasification product-composition studies using actual sludge are limited and expected supercritical product yields are variable (Afif et al., 2011; Cao et al., 2011). The assumed, basic expected, reaction kinetics (using glucose as a model compound) are represented by the formulas (1-2) listed below (Gasafi et al., 2007; Schmieder et al., 2000):

$$\text{Theoretically: } C_6H_{12}O_6 + 6H_2O \leftrightarrow 6CO_2 + 12H_2 \qquad \Delta H = 158 \text{ kJ mol}^{-1} \qquad (1)$$

$$\text{Experimentally: } 2C_6H_{12}O_6 + 10H_2O \leftrightarrow 11CO_2 + CH_4 + 20H_2 \qquad \Delta H = 152 \text{ kJ mol}^{-1} \qquad (2)$$

However, a review of studies reported that product gas composition under both catalytic and non-catalytic conditions approximate values of H_2: 40%-60%, CO_2: 30%-70%, CH_4: 15%-25%, and CO: 5%-30%, with non-catalytic conditions favoring CO production over CH_4 (Afif et al., 2011). The H_2 and CO gaseous streams can be recombined into liquid hydrocarbon fuels via Fischer-Tropsch or similar catalytic reforming systems (Demirbas, 2007), allowing for sludge to become a second-generation biofuel feedstock (Demirbas, 2011a; Demirbas et al., 2011).

The SCWG process dates back to the late 1970s, with incremental improvements in processing and reactor design, but very little reactor and reaction modeling taking place since (Elliott, 2008; Jessop et al., 1999; Modell, 1977; Modell et al., 1982; Savage, 2009). Like other SCW systems, SCWG can convert wet biomass directly, thereby avoiding high-energy drying processes associated with conventional thermochemical gasification (Hao et al., 2005) leading to similar chemical end-products. Unlike traditional gasification options, most SCWG can demonstrate an energy balance that can yield self-sufficient processing, positively addressing the high-moisture content of very wet wastes such as sewage and oily sludges.

Water simultaneously fulfills multiple roles in the SCWG process. Initially, water serves as the solvent for hydrolysis reactions, which quickly depolymerizes the major biomass sludge components (e.g., polysaccharides and fatty acids) into simpler structures like fructose, glucose, and short-chain organic acids (Di Blasi et al., 2007). The gasification reaction progresses beyond hydrolysis, wherein high-temperature water pyrolyzes those simple sugars and organic products to produce H_2-rich fuel gas and carbon oxides (Elliott, 2008). The H_2 bonds in supercritical water are weak, which means that, during water-gas shift reactions, the water can act as a H_2 donor, thereby increasing H_2 and O_2 availability and the corresponding H_2 yield (Yuan et al., 2006). The increase in O_2 availability can facilitate weak exothermic reactions, which improve process efficiency. Under supercritical conditions, water's hydrolysis solvation characteristics quickly give way to a secondary role as a reactant as well as a H_2 source (Han et al., 2008). Numerous studies have shown the potential role of SCWG for H_2 production from a variety of wet-waste feedstocks including sludge, with H_2 yields increasing by 80% from 330°C subcritical conditions at 380°C supercritical conditions (Demirbas, 2009; Xu et al., 2009; Yan et al., 2009).

Understanding how water molecules interact at supercritical conditions is helpful to predict surface-bound, transition-state species and the reaction energetics (Savage, 2009).

Supercritical water's solvation and dilution characteristics suppress tar and coke formation by preventing polymerization of double-bond intermediates, mainly by spatial distancing and reduced collisions between reactant molecules (Kruse & Dinjus, 2007a).

Surplus H_2 availability also positions supercritical water as the natural upgrading medium for oily sludges, coal pitch, and petroleum coke (Han et al., 2008). Prevailing dogma, however, asserts that H_2-production costs via SCWG of wet biomass (e.g., sludge) are several times higher than the costs of H_2 production via steam CH_4 or natural gas reforming (Balat et al., 2009; Demirbas, 2007). The H_2-production argument is based solely on fuel (viz., H_2) as the cost reference point. When nested within the revenue framework of waste disposal as the primary objective, secondary H_2 fuel conversion costs via SCWG is actually estimated to be a full two magnitudes less than that of natural gas reforming (Gasafi et al., 2008). The conversion costs are constant, regardless of the feedstock origin. However, the bottom-line product production costs are also largely driven by feedstock production and extraction costs, which in the case of natural gas have dropped significantly as a result of shale-based, strata fracturing (a.k.a. "fracking"), albeit fraught with controversy (DiPeso, 2011; Mooney, 2011).

2.2. Catalytically augmented supercritical water

Supercritical-water, fuel-gas production can be catalytically enhanced. The addition of a small quantity of catalyst to the SCWG process enhances gasification efficiency much like in conventional thermochemical gasification, especially at low reaction temperatures (Zhang et al., 2010). Adding catalysts intensifies SCWG reaction kinetics under milder conditions, and in the process, improves the efficiency of the water-gas shift reaction, promoting higher gas yields and a reduced yield of unwanted products (Elliott, 2008; Sınag et al., 2004). Catalysts also intensify hydrolysis liquefaction processes via flash pyrolysis that produces a liquid condensate in the dissolved supercritical water (Penninger & Rep, 2006). The flash-pyrolysis condensate is then readily converted in supercritical water into a H_2-rich gas, which further suppresses char and tar formation and reduces operating costs (Calzavara et al., 2005; Penninger & Rep, 2006; Sınag et al., 2004; Toor et al., 2011).

Catalytic SCWG studies can be divided into two categories based on the types of catalyst used: supported and unsupported catalysis (Lee, 2011). Supported catalysts can include the Noble metals (viz., Ru, Rh, Pd, Ir, and Pt) or lower-cost, common metals (viz., Re, Sn, Pb, W, Mo, Zn, Cr, and Ni) (Chang et al., 1993). Supported catalysts usually consist of various metals (including oxides and ores) dispersed on fixed-bed supports or particles made of ceramic, carbon, or metal oxide (Ding et al., 1996; Lee, 2011). High-performance ceramics (e.g., Al_2O_3, ZrO_2, SiO_2, Si_3N_4, Ce_2O_3, and TiO_2) have been used as supports for catalysts in SCW (Azadi et al., 2011; Ding et al., 1996; Lee, 2011). All of these ceramics are, however, subject to thermal creep at much lower temperatures than when exposed to high-temperature gases, thereby allowing the supported catalyst particles to contact each other, then sinter, weld, or polish, rendering them inactive (Bermejo & Cocero, 2006b; Hyde et al.,

2001). Ceramic supports also serve as nucleation points for salts, which can quickly plug reactors and deactivate the catalysts (Aki & Abraham, 1999; Brunner, 2009a). In some cases, as with silicon-based supports, erosion through solvation in water may occur (Cocero, 2001; Marrone & Hong, 2009). Unsupported catalysts are not fixed in the reactor and can include water-dissolved alkali salts (e.g., KOH, NaOH, Na_2CO_3, and K_2CO_3) in addition to the same metals as those used on fixed supports (Li et al., 2011; Lu et al., 2010; Schmieder et al., 2000). Reactive characteristics of unsupported catalysts are typically higher than supported catalysts (Anglada et al., 2011). An additional advantage of unsupported catalysts is that they can carry salts out of the reactor as the catalyst particles pass through the system (Anglada et al., 2011).

Metal catalysts have been well documented at promoting water-gas shift reactions, methanation, and hydrogenation reactions (Yoshida & Oshima, 2004). Four metals (viz., Ru, Rh, Pt and Ni) have received the greatest amount of attention in the literature. Ruthenium is reported to perform better than either Rh or Pt, in promoting SCWG H_2 production (Azadi & Farnood, 2011; Balat et al., 2009; Byrd et al., 2008; Chakinala et al., 2010; Chang et al., 1993; D'Jesús et al., 2006; Ding et al., 1996; Fang et al., 2008; Hao et al., 2005; Izumizaki et al., 2005; Izumizaki et al., 2008; Krajnc & Levec, 1994; Sato et al., 2011; Sato et al., 2003), especially when supported on TiO_2. Ruthenium commands a lower market price than Rh or Pt, making it an attractive option (Elliott, 2008; Guo et al., 2007; Izumizaki et al., 2005). Ruthenium is also reported to be more easily recovered for reuse than either Rh or Pt (Izumizaki et al., 2005).

Compared to Noble metals, including Ru, Rh, and Pt, Ni is a low cost material capable of catalyzing conversion at high rates with relatively low temperatures without sacrificing H_2 yields (Antal et al., 2000; Calzavara et al., 2005; Matsumura et al., 2002; Xu et al., 1996). Nickel catalysts have been reportedly effective at cracking tar into smaller, volatile fractions and promoting water-gas shift reactions, methanation, and hydrogenation reactions. Nickel can resist deactivation due to polishing and sintering if properly supported on TiO_2 or non-oxide ceramic substrates (viz., silicon carbide or carbon) (Azadi & Farnood, 2011; Marrone & Hong, 2009; Youssef et al., 2010a). Certain forms of Ni, including reduced Ni or skeletal Ni, commonly known as "Raney", have become a primary focus due its high porosity and surface area, which results in a high number of reactive sites and gasification efficiencies above 93% in SCWG laboratory-scale systems (Afif et al., 2011; Elliott, 2008).

Three alternatives to metal and salt catalysts have been identified: carbon catalysts, synergistic catalysts, and *in-situ* catalysts. Carbon can be used as either a catalyst or as a catalyst support (Antal et al., 2000). The conversion efficiencies of carbon due to increased temperature from partial-oxidation reactions can be on par with metals and alkali salts for H_2 and CO_2 production (Kruse et al., 2000; Matsumura et al., 2002; Rönnlund et al., 2011; Xu et al., 2009). Carbon is very stable in supercritical water, especially when H_2 gas is present (Calzavara et al., 2005). Plus, carbonaceous materials are common and relatively inexpensive, meaning that even the need for large catalyst volumes should still be economically feasible (Matsumura et al., 2002). Synergistic catalysts are formed by combining metals with relatively inexpensive alkali salts, creating a highly reactive surface, with less overall catalyst used, resulting in a higher H_2 output with reduced CH_4 production (Bernardi et al., 2010; Elliott, 2008). Carbon and ceramic (e.g., Al_2O_3, ZrO_2, and CeO_2) catalyst supports have also been shown to have significant synergistic effects on catalytic effectiveness, increasing gas yield by as much as five-fold and non-linearly altering the gas fraction (Elliott, 2008; Minowa & Inoue, 1999). Catalytic reactivity is often strongly influenced by characteristics of the dispersion on a support and the support itself (Azadi & Farnood, 2011). The common support for metal catalysts, zirconia, actually doubled H_2 yield from SCWG processing (Guo et al., 2007; Watanabe et al., 2003). Carbon and the rutile form of TiO_2 have shown similar catalytic promoter effects when used as supports for metal catalysts (Chakinala et al., 2010; Elliott, 2008). It is known, however, that soluble salts and insoluble metals catalytically react in different ways (mainly solubilization, mineralization, and oxidation), and a number of researchers have pointed out that there is no straightforward explanation in the literature for the kinetic mechanisms governing these synergistic processes (Azadi & Farnood, 2011; Bernardi et al., 2010; Lu et al., 2010). Catalytic *in-situ* effects resulting from sludge-entrained inorganic species have been studied as alternatives to conventional catalysts (Lee, 2011). This approach makes use of the antisolvent effect of supercritical water, wherein catalytically active salts and metals present in the sludge matrix actually produce catalyst precursors on the fly, such as activated carbon and trace reactive metals (Marques et al., 2011). The *in-situ* propagation of catalysts could rapidly form a supersaturation of nano-scale, semi-homogenous catalyst particles that promote gasification of the matrix (Gadhe & Gupta, 2007; Levy et al., 2006; Sınag et al., 2011).

2.3. Catalyst effect on chars & tars

Employing catalysts pushes total SCWG efficiency up to 98% by converting a high proportion of char and tar to gas products (Calzavara et al., 2005; Xu et al., 2009). Tars and sulfur edicts commonly released from sludge during gasification also present a threat to long-term catalyst stability (Afif et al., 2011; Elliott, 2008; Izumizaki et al., 2005; Yoshida & Oshima, 2004; Zhang et al., 2011). Tarry deposits and sulfur poisons dramatically impact metal catalyst effectiveness and resulting gas yield volume, even though they have little discernable effect on gas fraction and composition (Afif et al., 2011; Lee, 2011). Some catalyst regeneration was evident with the flushing of sulfur-poisoned catalysts with subcritical water (at 250-300°C), which removed up to 75% of the sulfur (Elliott, 2008). Water can activate or deactivate metal-catalyzed reactions via autoxidation (García-Verdugo et al., 2004). Regeneration effects can be augmented via the addition of an oxidant, such as H_2O_2 (Elliott, 2008). Deactivation effects, however, can be extensive and irreversible due to unique interactions between the catalyst and water (Ding et al., 1996).

2.4. Catalyst stability & sintering

Catalysts under harsh SCW conditions are subject to numerous morphological challenges affecting reactivity, lifecycle, stability, and economical operation (Ding et al., 1996; Elliott,

2008; van Rossum et al., 2009). Harsh SCW conditions demand more durable catalyst materials than gaseous operations (Ding et al., 1996). Significant loss of catalyst surface area, interstitial space, and chemically active sites result from numerous phenomena, including hydrothermal sintering, friction welding, friction polishing, thermal glazing, support creep, and aqueous dissolution (Aki & Abraham, 1999; Hao et al., 2005).

High temperatures alone are insufficient to cause significant catalyst sintering problems (Hao et al., 2005). Raney Ni shows a high resistance to heat in a gaseous atmosphere (Afif et al., 2011; Hao et al., 2005). Nickel has been noted to resist deactivation due to polishing and sintering when supported on TiO_2 or non-oxide ceramics such as silicon carbide or carbon (Azadi & Farnood, 2011; Marrone & Hong, 2009; Youssef et al., 2010a). If there is no proper support, as with Raney Ni, sintering can occur even after short-term operation (Hao et al., 2005; Lee, 2011). Raney Ni deactivates due to accretional crystal growth resulting from hydrothermal sintering (Afif et al., 2011; Elliott, 2008). Furthermore, the effect of the hydrothermal sintering was measured to be six-fold higher under exposure to hydrothermal treatment (380°C) than was observed at the same temperature in a gas atmosphere (Afif et al., 2011). Even when stabilized by Ru doping, Raney Ni sintered rapidly at 400°C (Elliott, 2008). Development of hydrothermally stable supports continues to be an area of active research and development efforts (Xu et al., 2009).

2.5. Supercritical Water Oxidation (SCWO)

Supercritical water oxidation (SCWO) is closely related to SCWG, both in terms of kinetics and technology. The objective of SCWO, however, is to oxidatively destroy organic compounds in water (Bermejo & Cocero, 2006b; Jing et al., 2008). The technology was originally developed nearly 30 years ago at the Massachusetts Institute of Technology for NASA, back when it was thought there would be a human colony on the moon and a need for a single system to treat and purify water was a priority (Bubenheim & Wydeven, 1994; Modell, 1977; Modell et al., 1982; Slavin & Oleson, 1991; Sloan et al., 2008; Svanström et al., 2004; Svanström et al., 2005).

Applications of SCWO technology ensued in the defense industry, where it was perfected as a destruction method for the most dangerous organic compounds in the world (Crooker et al., 2000; Onwudili & Williams, 2006; Savage, 2009; Veriansyah & Kim, 2007). The U.S. Department of Defense (DOD) developed SCWO into a viable technology specifically in support of the 1993 International Chemical Weapon Convention (Marrone et al., 2005; Savage, 2009; Veriansyah & Kim, 2007; Veriansyah et al., 2007). Currently, SCWO is used on a regular basis by the U.S. Army and U.S. Air Force to destroy nerve gas, biological weapons, and other dangerous munitions (Crooker et al., 2000; Onwudili & Williams, 2006; Savage, 2009; Veriansyah et al., 2005). The U.S. Navy has developed compact SCWO units for ship-board, hazardous waste treatment in order to comply with national and international waste discharge standards (Crooker et al., 2000; Veriansyah & Kim, 2007; Veriansyah et al., 2005).

The SCWO technology functions as a hydrothermal analog to incineration, thermochemically destroying wet wastes, such as sewage sludge, that are rich in organic compounds and residues (Onwudili & Williams, 2006). The SCWO process functions in much the same way as SCWG, but it is taken a step further by adding a strong oxidant (e.g., O_2, H_2O_2, or $KMnO_4$) in order to completely oxidize organic compounds dissolved in the supercritical water (Anglada et al., 2011; Castello & Fiori, 2011; Guo et al., 2007; Youssef et al., 2010a). Under highly oxidizing conditions, carbon compounds are quickly converted into carbon oxides (CO_2 and CO), H_2 is converted to H_2O, and the active oxidation process results in the exothermic release of energy (Abelleira et al., 2011; Castello & Fiori, 2011; Guo et al., 2007; Mahmood & Elliott, 2006; Sınag et al., 2004; Svanström et al., 2004). Notably, the biomass destruction rate efficiency using H_2O_2 in supercritical water has been shown to be 16-fold higher (based on free molar O_2 mass) than injected O_2, which by extension makes the effective oxidative cost of H_2O_2 less than $1/5^{th}$ that of injected O_2 (D'Jesús et al., 2005).

The unique reaction media provided by SCW is important, because it induces almost zero, inter-phase mass transfer limitations (Byrd et al., 2008; Letellier et al., 2010). Consequently, SCW operates in a homogeneous phase where O_2 (or other oxidants) availability becomes high. Oxygen, therefore, dissolves faster in supercritical water than in subcritical water. As water transitions into the supercritical state, it becomes a strong oxidant further enhancing the process. Depending on the quantity of oxidants introduced, partial-oxidation reactions occur in the working fluid, actually heating itself *in-situ* rather than relying on an external reactor heater. The resultant internal heating by the working fluid itself (i.e., water) dramatically lowers the transport phenomena resistance, and thus produces high efficiencies for heat-transfer and gasification processes inside the reactor (Calzavara et al., 2005). High transfer efficiencies are the primary drivers behind the very short residence times (i.e., <1sec) and smaller reactor volumes characteristic of SCWO systems (Letellier et al., 2010). The overall chemical transformations achieve complete organic destruction (>99.99%) while producing essentially no char, tar, or NO_x (Du et al., 2010; Mahmood & Elliott, 2006).

2.6. Effects of temperature, pressure, & residence time

Temperature, pressure, and residence time have been noted to be the most important variables for modifying supercritical reaction conditions (Brunner, 2009a; Elliott, 2008). Optimal supercritical conditions can be experimentally derived and aided by models to induce the ideal combination of temperature, pressure, and residence time (Soria et al., 2008). System optimization, however, involves maximizing the desired output (energy or organic destruction), while reducing reaction times to minutes or seconds versus the hours required for similar results in subcritical water (Gloyna & Li, 1993).

Temperature is considered the most sensitive variable in SCWG processes, with 600°C serving as an often-cited, optimal target temperature due to associated high conversion rates (D'Jesús et al., 2006; Elliott, 2008; Susanti et al., 2010). When temperature was increased in SCWG, for example, from 601°C to 676°C, CH_4 yield was reduced and H_2 yield doubled

(Susanti et al., 2010). A similar, inverse effect was observed as temperature declined. A drop in temperature from 600°C to 500°C during the SCWG process resulted in an overall decline in gasification efficiency from 98% to 51% (Elliott, 2008). Substantial changes either side of 600°C were evident in CSCWG as well, suggesting that it too has a narrow effective temperature range (Antal et al., 1995; Brunner, 2009a; Izumizaki et al., 2008; Jessop et al., 1999). The CSCWG process achieved unacceptably low efficiencies when temperatures declined far below 600°C, and carbon catalyst decomposition occurred when temperatures increased far above 600°C (Antal et al., 1995; Xu et al., 2011; Xu et al., 1996).

Short residence times (<1 min) and high organics destruction efficiencies (>99.99%) occur during gasification and oxidative reactions at supercritical operating conditions above 600°C (Cao et al., 2011; Du et al., 2010). Furthermore, when temperatures are above 600°C, reactions can take as little as a few milliseconds (Augustine & Tester, 2009; Bermejo et al., 2011; Cabeza et al., 2011; Narayanan et al., 2008; Wellig et al., 2009). Longer residence time can improve gasification thoroughness, but there is also an inverse relationship between temperature and reaction completeness, dropping from a few minutes below 600°C to a few seconds above 600°C (Cao et al., 2011). The optimal temperature threshold for SCWG (i.e., 600°C) has been shown to be on the low side of the rapid-conversion range for higher concentration biomass in the absence of a catalyst or strong oxidant (Afif et al., 2011; Antal et al., 1995; Xu et al., 2011; Xu et al., 1996). Without a catalyst or oxidant, temperatures more in the range of 800°C are required for rapid conversion (Afif et al., 2011; Guo et al., 2007; Izumizaki et al., 2008). Conversely, water just below the critical temperature (375°C) has been shown to be highly effective for gasification when performed with active catalysts when primarily targeting CH_4 (vs H_2) production (Elliott, 2008). However, at temperatures more than about 20°C below the critical temperature, all gasification ceases, with or without catalyst, resulting in only hydrolysis and solvation reactions (Elliott, 2008).

While both temperature and reaction times seem to consistently be straightforward influences on reducing organic content, several studies indicate that pressure variations have more subtle and complex effects on conversion efficiency and gas product fraction (Brunner, 2009a; Cui et al., 2009; D'Jesús et al., 2006; Guo et al., 2007). Supercritical water reactions have been demonstrated to be very stable. Temperature profiles represent a quasi-constant plateau near the critical conditions, and there is little or no reaction effect from pressure variations above a threshold point (Dutournie & Mercadier, 2005). Nonetheless, the limited reaction effect from pressure could be attributed to the fact that high pressure stabilizes reaction energetics (Dutournie & Mercadier, 2005). Similarly, the potential for complex pressure effects should not be ignored because water properties including density, dielectric constant, and ion product increase with pressure (Guo et al., 2007). Higher ionic reaction rates can restrain free-radical reactions (Guo et al., 2007). These complex pressure effects can be used to fine tune the chemical composition of the solvent and control gas composition and yield (Savage, 2009). Specifically, pressure has little or no influence on reaction rate, but it does affect solvent density (Brunner, 2009a). Density also has little effect on gasification efficiency above the critical point, but can have significant affects on gas fraction characteristics (Brunner, 2009a). High pressures, and correspondingly higher densities, favor CH_4 production and inhibit H_2 production (Brunner, 2009a).

2.7. Char & tar formation

Substantial amounts of char and condensable volatile tars form during hydrothermal decomposition of sludge, especially in the absence of catalysts or oxidants (Afif et al., 2011; Azadi et al., 2011). Sewage sludge is highly prone to char and tar formation due to the presence of high levels of condensable volatile materials, which favor the production of cyclic compounds (Adegoroye et al., 2004; Dogru et al., 2002; Onwudili & Williams, 2006). Char and tar formation can severely impair carbon gasification efficiency, which is a common and persistent problem with both traditional, dry gasification and hydrothermal gasification (Chuntanapum & Matsumura, 2010). Even under supercritical conditions, if the thermal kinetics are not high enough, wet biomass (including sludges) can dehydrate and then polymerize into tarry condensates prior to hydropyrolytic liquefaction (Azadi & Farnood, 2011; Matsumura et al., 2005; Onwudili & Williams, 2006). The exact influence of SCW reaction kinetics on tar formed during biomass gasification is largely unknown beyond the general benefits of higher temperatures and higher heating rates (Adegoroye et al., 2004; Kruse, 2009; Matsumura et al., 2005).

Chars and tars are difficult to gasify and once formed, act as persistent barriers to complete gasification (Afif et al., 2011; Calzavara et al., 2005; Chuntanapum & Matsumura, 2010). If not properly handled, chars and tars can quickly plug SCW reactors in as little as one hour of operation (Calzavara et al., 2005; Chuntanapum & Matsumura, 2010; Jin et al., 2010). Slow, reaction-heating rates and low reaction temperatures accelerate char and tar formation (Azadi & Farnood, 2011; Jin et al., 2010). Thus, in the absence of catalytic promoters, char and tar formation is especially problematic during process startup, wherein the reaction relies on external heating (Azadi & Farnood, 2011; Jin et al., 2010). The preheating of reactor and heat-up zones where feedstocks first enter has been noted as a possible solution to char and tar formation, buildup, and plugging (Antal et al., 2000; Elliott, 2008). Despite the fact that SCWG processes produce less char and tar, the lower reactor volume and small diameter typical of SCW systems are still vulnerable to plugging (Calzavara et al., 2005). Even if plugging is avoided, char formation can still cause a cascading loss of carbon gasification efficiency (Chuntanapum & Matsumura, 2010).

Nevertheless, char and tar formation in SCW is usually considerably less than that in low-pressure processes, largely due to higher water solubility, intensified kinetics, high heat, and mass transport properties (Byrd et al., 2008; Calzavara et al., 2005; Chuntanapum & Matsumura, 2010). Plus, hydrolysis and hydropyrolysis reactions in SCW quickly dissolve sludge educts before they can dehydrate, thus suppressing polymerization of cleavage products and tar formation (Gasafi et al., 2007). Although small quantities of an oxidant can produce partial oxidation, catalysts appear to be the key for reliably achieving both char and tar avoidance and selectivity for efficient H_2 production (Balat et al., 2009; Calzavara et al., 2005; Ding et al., 1996).

3. Reactor, kinetics, & design considerations

Supercritical-water reaction kinetics and effective reactor engineering are inextricably linked. New reactors, able to withstand harsh SCW operating conditions, are needed for SCWG technology to advance from laboratory and emerging status to genuine commercial operations (Yoshida et al., 2003). Most commercial and industrial applications require that engineering designs and materials overcome corrosion and plugging problems and that systems operate on a continuous duty cycle (Azadi et al., 2011; Elliott, 2008; Guo et al., 2007). Some common, SCW-reactor considerations are presented here, while extensive and detailed reviews can be found elsewhere (Bermejo et al., 2005a, 2005b, 2009; Calzavara et al., 2004; Fauvel et al., 2003, 2005; Lieball et al., 2001; Machida et al., 2011; Marrone & Hong, 2009; Mitton et al., 2000; Peter, 2004; Tan et al., 2011; Veriansyah et al., 2009; Wellig et al., 2005; Yoshida & Matsumura, 2009).

Continuous-flow reactor systems provide the most suitable options for real-world applications, because they offer plant expansion flexibility and versatile, industrial scale-up (Guo et al., 2007; Veriansyah & Kim, 2007). There are several categories of reactor designs, including in-line tubular systems, transpiring-wall (both tubular and vessel), and pressure vessel setups (Azadi et al., 2011; Elliott, 2008). Basic choices of system configuration require a complete understanding of how water molecules interact with each other at supercritical conditions and how reactants influence catalyst-surface adsorption and desorption events (Feng et al., 2004; Matsumura et al., 2005; Savage, 2009).

Even when water's transport properties can be predicted, thermodynamic phase equilibria are still handicapped by varying real-world, compositions of reactant educts and the presence of inorganic salts (Bermejo & Cocero, 2006b). Furthermore, the sequential and simultaneous progression of hydrolysis, pyrolysis, steam-reforming, and water-gas shift reactions in supercritical gasification chemistry are complex and have yet to be comprehensively described beyond speculative assumptions based largely on limited observations and first-order kinetics (Calzavara et al., 2005; Kruse, 2009; Matsumura et al., 2005; Sato et al., 2004; Vogel et al., 2005).

Designing reactor energy flows requires clearly definable equilibrium relationships (Feng et al., 2004; Gassner & Maréchal, 2009). The characterization of inherently arbitrary reactor feed equilibria, however, is complicated by thermodynamic mechanics of fluid mechanics, heat transfer, mass transfer, kinetics, and phase behavior (Hodes et al., 2004). Modeling, predicting, and defining these thermodynamic mechanisms is difficult, and there is no straightforward explanation in the literature for SCW reaction kinetic mechanisms (Azadi & Farnood, 2011; Bermejo & Cocero, 2006b; Bernardi et al., 2010; Hodes et al., 2004; Lu et al., 2010). The assumed, basic reaction kinetics are represented by the formulas (3-8) listed below (Chun et al., 2011; White et al., 2011):

$$\text{Hydrolysis: } Organics \overset{K_1}{\rightarrow} Sugars \overset{K_2}{\rightarrow} Degradation\ Products \tag{3}$$

$$\text{Pyrolysis: } Organics \overset{K_1}{\rightarrow} \text{"Active" Organics} \overset{K_2}{\rightarrow} Volatiles\ or\ Chars \tag{4}$$

$$\text{Steam reforming: } Organics + H_2O \leftrightarrow CO + H_2 \tag{5}$$

$$\text{Steam gasification: } C + H_2O \leftrightarrow CO + H_2 \qquad \Delta H\ 298\text{K} = 132\ \text{kJ mol}^{-1} \tag{6}$$

$$CH_4 \text{ gasification: } CH_4 + H_2O \leftrightarrow CO + 3H_2 \qquad \Delta H\ 298\text{K} = 206{:}1\ \text{kJ mol}^{-1} \tag{7}$$

$$H_2O\text{-}CO \text{ shift: } CO + H_2O \leftrightarrow CO_2 + H_2 \qquad \Delta H\ 298\text{K} = \text{-}41{:}5\ \text{kJ mol}^{-1} \tag{8}$$

All of the reactions are assumed to use a first-order rate constant that obeys the Arrhenius equation (9) in which k_{io} serves as a pre-exponential factor, with A as acid concentration (wt%) raised to the power m_i, E_i as the activation energy, and R and T as gas constant and temperature, respectively (Jacobsen & Wyman, 2000):

$$k_i = k_{io} \times A^{m_i} \times e^{-\frac{E_i}{RT}} \tag{9}$$

Attempts to develop detailed understanding of the reaction kinetics have so far been limited and isolated, relying primarily on *in-situ* diagnostics gleaned through direct visual/optical observations and indirect nuclear radiographic observations. Visually observing reactions is advantageous when compared to drawing surrogate reactant samples, in that direct observations support real-time, kinetic diagnostics and operational integrity (Hunter et al., 1996). Direct observations of small-scale, transparent reactors (e.g., diamond anvil cells or quartz capillary tubes) allow reactions to be seen, photographed, and quickly halted if necessary (Azadi & Farnood, 2011; Fang et al., 2008; Hashaikeh et al., 2007; Maharrey & Miller, 2001; Peterson et al., 2008a; Sasaki et al., 2000; Vogel et al., 2005). Larger scale systems have been directly observed via optical, laser Raman spectroscopy through sapphire reactor viewing ports in order to capture finite details of the reaction progress, fluid mechanics, reactant destruction completeness, and oxidation efficiencies (Chuntanapum & Matsumura, 2010; García-Verdugo et al., 2004; Hunter et al., 1996; Koda et al., 2001; Rice et al., 1996). Indirect, nuclear radiography accomplishes the same result as optical Raman spectroscopy, but does not require viewing-port reactor modifications (Peterson et al., 2008a, 2008b, 2010). Consequently, radiography is more flexible than direct observation, because reaction observations can be made from different angles independent of reactor design (Peterson et al., 2008b).

Supercritical-water reaction educts will ultimately be determined exclusively by thermodynamic kinetics (Savage, 2009). The exact influence of SCW reaction kinetics is largely unknown beyond the general benefits of higher temperatures and higher heating rates (Adegoroye et al., 2004; Kruse, 2009; Matsumura et al., 2005; Vogel et al., 2005). Newer reaction-observation techniques show promise for developing an understanding of the missing, critical kinetics needed for comprehensive modeling of SCWG reactions (Vogel et al., 2005). Reaction observation techniques (particularly Raman spectroscopy), nonetheless, are not widely used, are limited to methodological studies, and have no comprehensive kinetics models based on them (Hunter et al., 1996; Rice et al., 1996). Existing observation studies have, however, partially confirmed the assumptions that endothermic, acid-catalyzed hydrolysis reactions quickly dissolve sludge educts before they can dehydrate, resulting in complete solubilization and liquefaction early in the process (Brunner, 2009b; Koda et al., 2001; Peterson et al., 2008a). The rapid endothermic, hydrothermal-pyrolytic decomposition of liquefied organic materials appears to progress concurrently with hydrolysis, reaching completion within seconds to minutes (Brunner, 2009b; Peterson et al., 2008a). Partial oxidation of the pyrolyzed compounds drives the pyrolysis and gasification reaction exothermically (Koda et al., 2001; Kruse & Vogel, 2010; Peterson et al., 2008a; Vogel et al., 2005). The disintegration of sludge under SCW conditions results in the formation of hydrolysis products, including volatile fatty acids, phosphorous compounds, dissolved biodegradable organics, gases (i.e., CO and CO_2), and H_2O (Rulkens, 2008). Due to the lack of a well-established Equation of State (EOS) for SCW and any form of biomass, very few studies have systematically investigated the complexities of reaction progress or even heat transfer to reactants in supercritical reactors, and a comprehensive description of reaction kinetics is unlikely to evolve in the absence of an EOS (Azadi et al., 2011; Bermejo et al., 2007; Yoshida et al., 2004).

Supercritical-water reaction kinetics and effective reactor engineering may very well be inextricably linked, but the connections are largely unknown. Consequently, rather than designing systems to accommodate any particular reaction progression or kinetics, progress in SCW reaction-kinetics engineering has largely relied on trial-and-error to solve corrosion and scaling problems (Hodes et al., 2004; Marrone et al., 2004). Efforts to design SCW systems continue in the absence of clearly defined models of reaction kinetics, and progress is reflected by the many successful industrial applications of SCW (notably, General Atomics, Foster Wheeler, and Chematur Engineering) (Bermejo & Cocero, 2006b).

Mixtures of supercritical water and sludge can be thought of as dynamic systems, wherein regions may predictably or transiently exist (Savage, 2009). An increase in organic content, for example, shifts the critical point of the mixture further from that of pure water (Savage, 2009). Despite well-documented SCW effectiveness for gasification, data is very limited for phase behavior of sludge-decomposition (Fang et al., 2008). Consequently, a number of broad assumptions and logical leaps must be made to model supercritical reactor conditions (based on either *Peng-Robinson* or *Anderko-Pitzer* EOSs), including volume translation corrections to reproduce densities (Bermejo & Cocero, 2006b; Bermejo et al., 2007; Vogel et al., 2005).

3.1. Corrosion influence on reactor design

Corrosion has historically impeded SCW commercialization due to limited reactor life (Barner et al., 1992; Hodes et al., 2004). Metal corrosion in SCW systems is driven, in part, by water's own natural solvation characteristics and is largely localized to areas where water drops below the critical point (Marrone & Hong, 2009). Water in the near-critical region actually exhibits maximum corrosion effects (Marrone & Hong, 2009). Just below the critical point, water's fast kinetics from high temperatures, high pressures, and natural acidity are particularly taxing on metals (Marrone & Hong, 2009). Escalated corrosive severity of near-critical water means that components used in preheating and cool-down are typically more susceptible to corrosion than the reactor itself (Marrone & Hong, 2009).

Reactor-specific corrosion problems often result from the fact that supercritical water cannot solvate charged (polar) species. Precipitation of polar species (i.e., inorganic salt) can form subcritical-water "microenvironments" between salt deposits on reactor walls and the reactor's internal metal surface (Hodes et al., 2004; Marrone & Hong, 2009). The highly saline and acidic, subcritical water in the microenvironments ultimately leads to severe and localized, reactor-wall corrosion (Hodes et al., 2004; Marrone & Hong, 2009). Consequently, reactor corrosion is a particular concern when alkali-salt catalysts are used or when high-salt-content sludges are processed (Lee, 2011).

Oxidative, metal corrosion results from supercritical water's high O_2 availability and correspondingly high electrochemical potential (Marrone & Hong, 2009). Corrosion prevention often requires the use of expensive alloys capable of withstanding high temperatures and pressures (Toor et al., 2011). Materials such as advanced Ni-based alloys (Inconel 625 or Hastelloy C276) and Ti alloys can suppress corrosion losses. Nickel-based alloys resist aqueous corrosion by forming a passivated and impermeable oxide surface coating that prevents corrosive solvent contact with the underlying metal, which protects the metal from further corrosion (Lee, 2011).

A separate, but related issue, is that corrosion-resistant, high-Ni-content reactor alloys (viz., Inconel and Hastelloy alloys) exert a catalytic influence on gasification chemistry (Afif et al., 2011; Antal et al., 2000; Chakinala et al., 2010; D'Jesús et al., 2006). This phenomenon has become commonly known as the "wall effect" (Sınag et al., 2004), wherein the reactor-wall alloys promote water-gas shift activity in SCW conditions (Chakinala et al., 2010). There has been considerable work with reactor alloys in an attempt to control and promote these catalytic effects (Afif et al., 2011). The fabrication of a fixed-bed catalyst from the same Inconel material used in the reactor, for example, increased gasification efficiency four-fold (Ding et al., 1998).

Corrosion in SCW is species-specific, targeting and selectively dissolving chromium's passivating oxide layer, thereby exposing the underlying alloy to further attack (D'Jesús et al., 2006; Marrone & Hong, 2009). Specific corrosive activity is discernable with process effluent analysis. Effluent laden with Ni, Cr, and Mo would indicate that corrosion is stripping those metals from the reactor wall (Sınag et al., 2004). One advantage of corrosion dynamics in SCW processes is that, even if corrosion occurs, the gas products are almost completely free of any corrosive substances, because the corrosive educts remain in the liquid phase (Kruse, 2009). Consequently, unlike dry processes, extensive cleaning of the SCW-produced fuel gases is typically not necessary (Kruse, 2009). Also, the metal embrittlement resulting from H_2 exposure in dry gasification is not a major problem in SCW processes (Kruse, 2009).

Transpiring-wall reactors are recent developments designed to avoid high temperatures near the walls and flush away corrosive salts with a thin film of subcritical water (Lavric et al., 2006). The reaction chamber consists of a porous inner wall through which clean water continuously flows, creating a thin film of subcritical water (Figure 3). A second outer wall contains high pressure water that is never exposed to the extreme temperatures or corrosive effects of the reaction working fluid (Bermejo & Cocero, 2006a). The transpiring-wall approach allows one reactor wall to contain the pressure while the other wall endures exposure to the corrosive effects. This arrangement potentially allows for lower temperature operating conditions, less-extensive and costly containment alloys, and lower capital costs (Elliott, 2008).

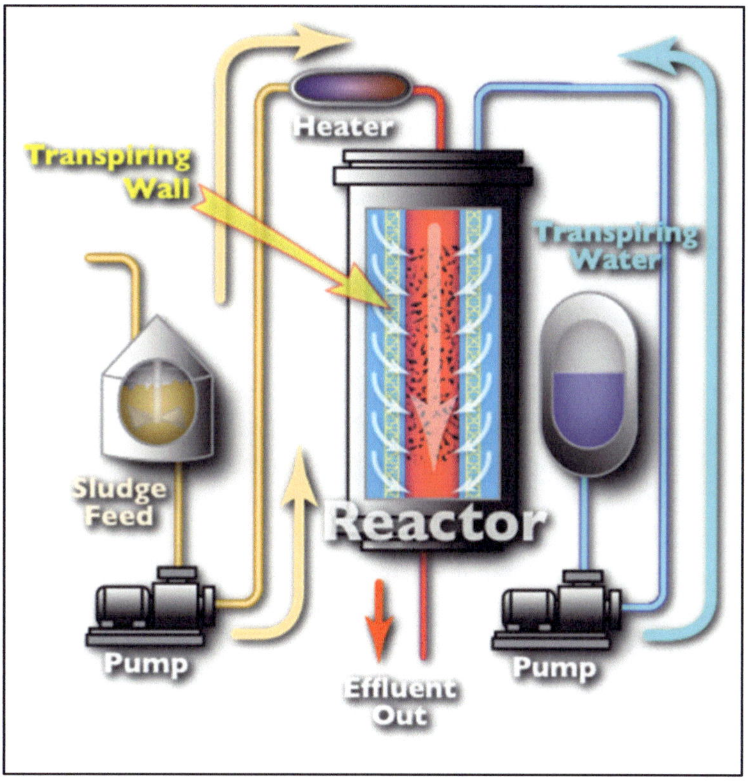

Figure 3. Transpiring Wall Reactor.

3.2. Salt precipitation & scaling influence on reactor design

Salts are pervasive and abundant in sludge (Brunner, 2009b; Elliott, 2008; Kruse et al., 2010). Waste-dissolved salts can precipitate and, if not controlled, eventually block SCW reactors (Brunner, 2009a, 2009b; Demirbas, 2011a; Marrone et al., 2004). Salt precipitation persistently complicates SCW systems (Du et al., 2010), thereby impeding widespread commercialization due to inherent practical difficulties of scale buildup, fouling, and corrosion (Cocero et al., 2003; Hodes et al., 2004). Salt handling represents one of the greatest remaining technical challenges for development of SCW biomass gasification processes at commercial scales (Hodes et al., 2004; Kruse et al., 2010). Salt precipitation and control are briefly discussed here, but reviews of the subject can be found elsewhere (Hodes et al., 2004; Kruse et al., 2010; Leusbrock et al., 2010; Marrone et al., 2004; Príkopský et al., 2007; Schubert et al., 2010a, b, 2012; Xu et al., 2010).

The low, dielectric constant of SCW reduces the salt-dissolving power of SCW to nearly zero (Figure 2), which results in the formation of solid precipitates (Brunner, 2009a, 2009b; Demirbas, 2011a; Marrone et al., 2004). Salt precipitation is particularly common in the preheating sections of SCW systems, due to steep concentration gradients as liquid water transitions to SCW, and mineral ions release from sludge matrices (Bermejo & Cocero, 2006b; Hodes et al., 2004; Penninger & Rep, 2006). Sludge-entrained, acidic solutions also precipitate educt salts during neutralization at the end of the SCW process (Hodes et al., 2004). Therefore, low solubility results in rapid salt precipitation immediately after sludge enters, and salts also precipitate as the effluent exits (Elliott, 2008; Marrone et al., 2004). Salt precipitation and plugging is particularly challenging with fixed-catalyst beds, which serve as both bottlenecks to velocity and ready nucleation sites for salt (Elliott, 2008; Kruse, 2009).

Numerous reactor designs, some paired with dedicated salt-separation equipment, have been proposed and studied (Brunner, 2009a; Du et al., 2010). Transpiring-wall reactors provide a nuanced use of liquid water to flush salt from the system (Bermejo & Cocero, 2006a; Lavric et al., 2006). A more common method, however, is a brute-force approach using high fluid velocities in tubular reactors such as AquaCat® and AquaCritox® processes developed by Chematur (Bermejo & Cocero, 2006b; Marrone et al., 2004). Tubular reactors are designed with small tube diameters in order to maintain high fluid velocity (Bermejo & Cocero, 2006b). The high fluid velocities overcome salt nucleation and agglomerate adhesion via high shear forces, in combination with scouring effects of entrained inorganic solids such as sand (Marrone et al., 2004). Tubular reactors have become the overwhelming technology of choice for commercial applications in part because of their salt control advantages (Barner et al., 1992; Bermejo & Cocero, 2006b; Brunner, 2009a; Cabeza et al., 2011; Matsumura et al., 2005). Well over 80% of the industrial applications of supercritical treatment of industrial wastewaters use tubular reactors coupled with a countercurrent heat exchanger for increased efficiency (Vadillo et al., 2011; Vogel et al., 2005).

Salt-control, maintenance duty has become a significant SCW implementation consideration, requiring frequent and costly shutdowns for cleaning (Hodes et al., 2004; Marrone et al., 2004). Some commercial entities (e.g., Chematur and SCFI) have developed a design workaround to the shutdown problem with parallel redundant systems, so that one unit can be in operation while a second unit is in cleaning mode (Bermejo & Cocero, 2006b). In addition to this reactor design solution, a variety of additive "magic sauces" have been developed to mitigate salt precipitation. These mixtures are comprised of "Type-1" salts (e.g., K_3PO_4 or KNO_3) that are sometimes supplemented with very finely ground abrasives (Marrone et al., 2004; Yoshida et al., 2003). Under supercritical water conditions, the Type-1 salts remain liquid and are not very "sticky" to metal, thereby acting as both solvating and nucleating agents for "Type-2" sticky salts such as NaCl and Na_2SO_4 (Marrone & Hong, 2009; Yoshida et al., 2003). Entrained inert solids also act as additional nucleating media for

Type-2 salts and as a scouring and polishing media for reactor walls (Kruse, 2009; Marrone et al., 2004). The resulting mixture forms a eutectic with a melting point less than the operating temperature at the system pressure (Bermejo & Cocero, 2006b; Hodes et al., 2004). The resulting molten-salt blend will flow more easily through the reactor. Abrasive solids have very high surface areas (1,000 times higher than polished reactor walls) and exert much higher van der Waal and electrostatic attraction on the salt than the metal walls themselves (Marrone et al., 2004).

The liquid-salt phases are consistent with density eddies or "local density augmentations" often observed in supercritical fluids and are related to the isothermal compressibility of the supercritical state (Brennecke & Chateauneuf, 1999; Hyde et al., 2001). The localized densities cluster together and are perpetuated by van der Waals forces (Brennecke & Chateauneuf, 1999; Hyde et al., 2001; Marrone et al., 2004). Consequently, a dense, liquid slurry of "good" salts forms and abrasives constantly "clean" and clear the reactor of the "bad, sticky" salts.

3.3. Sludge dewatering & water retention

A critical consideration for the apparatus setup is the degree to which the sludge can be pumped into the reactor, which is primarily limited by viscosity from biomass solids content. Preheating sludge slurries allows higher solid concentrations, because increased temperature and pressure, even when well below the critical point, decrease sludge viscosity (Abelleira et al., 2011; Xu et al., 1996). Reduced viscosity is likely due to biomass liquefaction as a result of accelerated hydrolysis (Abelleira et al., 2011).

Beyond reaction kinetics, sewage sludge is mostly water, and dewatering has been a key focus of transportation logistics, with less water reflecting lower transport costs (Weismantel, 2001). Most estimates put the moisture content of sewage sludge at well above 95% for liquid sludge and nearly 90% for dewatered, semi-solid sludge cake (Fytili & Zabaniotou, 2008; Harrison et al., 2006; Hong et al., 2009). One reason for the difficulty in dewatering sludge is the presence of macromolecules and extracellular polymeric substances, as well as large quantities of cellular bacteria (Abelleira et al., 2011; Xu et al., 1996). Polymeric, cellular substances impede ion movement and thus promote water retention (Abelleira et al., 2011). When pre-heated to about 150°C at relatively low pressure (about 10 bar), walls of sludge-entrained cellular bodies are destroyed, thus decreasing sludge viscosity and making the cell contents more available to catalytic, oxidative, and thermal degradation (Abelleira et al., 2011). Preheating the wet organic feedstock with heat recycled from the hot reaction gases is also important in terms of reaching a self-sustaining, process threshold (Abelleira et al., 2011; Cocero, 2001; Cocero et al., 2002; Guan et al., 2011; Matsumura et al., 2005).

3.4. Energy recovery

The advantage of SCW is that much of the process energy investment can be recovered from the hot effluent at supercritical temperatures and reused to preheat the wet, organic feedstock (Guan et al., 2011). Recycling heat from the hot effluent through heat exchangers back to the incoming, pressurized feedstock achieves a positive energy balance and acceptable system efficiency, which is critical to overcoming the high enthalpy of water (Abelleira et al., 2011; Cocero, 2001; Cocero et al., 2002; Matsumura et al., 2005). Preheating temperature (to well above the critical point 375°C) can also be a very important optimization (or simulation) consideration, especially when dealing with dilute or low-heating-value feedstocks (Barner et al., 1992; Cocero, 2001; Cocero et al., 2002). High preheating demand can be offset by supplementing the feed stream with a liquid, high-heating-value "fuel" such as waste oils or discarded organic solvents from industrial processes (Barner et al., 1992; Cocero, 2001; Cocero et al., 2002).

Even though supercritical water exhibits excellent heat and mass transfer properties, making use of those properties is a much higher technical challenge than other oxidation processes such as incineration (Bermejo & Cocero, 2006a; Cocero, 2001). External preheating through heat exchangers is usually necessary to initiate the reaction process, but can be discontinued once oxidative, exothermic reaction kinetics occur (Abelleira et al., 2011; Bermejo & Cocero, 2006a). Heat exchanger inefficiencies often negate high heat production, thereby undermining overall process efficiency (Yoshida et al., 2003). Heating rate is strongly related to reactor flow rate, which by extension defines heating duty length (Azadi et al., 2011). As flow rates decline, an increase in external heating has little impact on the temperature profile along the reactor (Azadi et al., 2011).

3.5. Potential

Until a sustainable sludge destruction solution is found, the sludge disposal problem will continue to grow with increasing population, rather than dissipate (Lavric et al., 2006). Application of sewage sludge on agricultural land has become socially unacceptable due to the fact that it is increasingly regarded as an unsafe and insecure handling route (Eriksson et al., 2008; Fytili & Zabaniotou, 2008; Harrison et al., 2006; Mathney, 2011; McBride, 2003; Snyder, 2005; USEPA, 2000, 2002b). Relying on agricultural and horticultural options as a disposal route for sewage sludge is simply not a valid, long-term solution by a whole host of sustainability and safety measures (Alaimo, 2008; Angenent et al., 2004; Arthurson, 2008; Booth et al., 2010; Costello & Read, 1994; CSS, 2011; Deblonde et al., 2011; Dijkema et al., 2000; Duić et al., 2011; Farré & Barceló, 2003; Fodor & Klemes, 2011; García-Serna et al., 2007; Harrison et al., 1999, 2006; McBride, 2003; NASNRC, 2002; Nature, 2008; Phillips, 1998; Smith et al., 2011; Snyder, 2005; Tollefson, 2008; USEPA, 2002a, 2009). Furthermore, public policy and regulations governing sludge disposal methods are beginning to reflect the growing public recognition that sewage sludge is more appropriately treated as hazardous waste than as fertilizer (Veriansyah & Kim, 2007; Youssef et al., 2011). Beyond the fact that sludge disposal regulations are becoming increasingly stringent worldwide, land available for waste disposal has also become more limited (Youssef et al., 2011). Driven by all these issues, sewage sludge is prime for capitalizing on a paradigm shift in the municipal wastewater industry.

Adoption of a new paradigm requires critically questioning the benefits of continuing with existing sludge disposal methods, and doing so will ultimately lead to dramatic technological developments, with SCW processes providing a viable alternative (Dijkema et al., 2000; Domínguez et al., 2006; Youssef et al., 2011). The hydrothermal decomposition of municipal wastewater solids would shift the view of sewage sludge as a costly disposal problem to that of a valuable, sustainable energy source. The U.S. annually consumes 4% of total national electricity just separating water from sludge (CSS, 2011; NASNRC, 2002; Phillips, 1998). Destroying sewage sludge *in-situ* would mean that municipalities would not have to dewater the sludge, thereby gaining massive, immediate financial savings, while simultaneously addressing a critical and vexing wastewater management problem (Fytili & Zabaniotou, 2008).

3.6. Commercial forays & missteps

Supercritical water processes have been used in the defense industry for over 30 years, but their use in the biofuels and wastewater industries dates back only a decade. Current SCW commercialization efforts focus in two main areas: 1) biomass (including coal) gasification and 2) municipal waste destruction. Applications, aside from the military, have been limited so far to demonstration units. Detailed performance reports from those demonstration units are scarce, with correspondingly even fewer review articles (Brunner, 2009a; Crooker et al., 2000; Onwudili & Williams, 2006; Savage, 2009; Veriansyah & Kim, 2007; Veriansyah et al., 2005; Xu et al., 2012). The engineering challenges discussed previously (viz., corrosion and salt precipitation) have been limiting factors to SCW commercialization (Brunner, 2009a). Efforts to commercialize SCW have been made by only a handful of companies, including Foster Wheeler, General Atomics, EcoWaste Technologies, Chematur Engineering, HydroProcessing, SuperWater Solutions, and Supercritical Fluids International (Bermejo & Cocero, 2006b; Bermejo et al., 2009; Marrone et al., 2004; Xu et al., 2012). Notably, much of the work by those companies has been limited to SCWO and funded by defense-industry contracts, with only more recent developments focused on municipal waste and energy (Bermejo et al., 2009).

Foster Wheeler developed several full-scale SCWO projects for multiple branches of the U.S. Armed Forces. Their systems were based on transpiring-wall reactor designs for the destruction of U.S. Army munitions. Sandia National Laboratories continues to operate one of Foster Wheeler's systems (Veriansyah & Kim, 2007). Foster Wheeler successfully tested the same system on halogenated solvents for the U.S. Navy as well as nerve agent hydrolysates (e.g., HD, GB, and VX) and propellants (Veriansyah & Kim, 2007).

General Atomics tested very similar systems to those of Foster Wheeler for comparable defense-industry purposes (Veriansyah & Kim, 2007). General Atomics took their designs a step further with a full-scale design for chemical weapons demilitarization as well as operational, compact SCWO systems for U.S. Navy shipboard-waste destruction.

EcoWaste Technologies designed and built the world's first commercial SCWO plant for Huntsman Chemical in 1994. The plant was a tubular reactor system constructed for the destruction of organic wastes produced on-site at Huntsman's Austin Research Laboratories (Bermejo & Cocero, 2006b; Veriansyah & Kim, 2007; Xu et al., 2012). The system was able to operate at about half the cost of incineration, even without integrated energy recovery (Veriansyah & Kim, 2007).

Chematur Engineering acquired a world-wide licensing agreement in 1995 for the EcoWaste SCWO process developed for Huntsman Chemical, in order to further develop the process in Europe and elsewhere (Bermejo & Cocero, 2006b; Bermejo et al., 2009; Mahmood & Elliott, 2006; Veriansyah & Kim, 2007). Chematur developed the SCWO process into an integrated energy production and waste destruction system, which they marketed under the trade name AquaCritox®. Chematur built two pilot plants, one in Europe and one in Japan for Shinko Pantec (Bermejo & Cocero, 2006b; Bermejo et al., 2009; Mahmood & Elliott, 2006; Veriansyah & Kim, 2007). Chematur constructed a full-scale SCWO facility for Johnson Matthey in the UK (Bermejo et al., 2011; Bermejo & Cocero, 2006b).

HydroProcessing built a full-scale, tubular SCWO unit in 2001 specifically for sewage sludge destruction in Harlingen, Texas (Bermejo & Cocero, 2006b; Bermejo et al., 2009; Griffith & Raymond, 2002; Marrone et al., 2004; Sloan et al., 2008; Svanström et al., 2004; Veriansyah & Kim, 2007; Weismantel, 2001; Xu et al., 2012). The Harlingen system was touted as the first U.S. SCWO system dedicated to sewage sludge and operated for about four years achieving very high destruction efficiencies (Bermejo & Cocero, 2006b; Bermejo et al., 2009; Griffith & Raymond, 2002; Marrone et al., 2004; Sloan et al., 2008; Svanström et al., 2004; Veriansyah & Kim, 2007; Weismantel, 2001; Xu et al., 2012). However, the system ultimately failed due to inadequate pump durability and insufficient flow velocities, which resulted in salt precipitation, corrosion, and plugging (Bermejo & Cocero, 2006b; Bermejo et al., 2009; Marrone et al., 2005; Sloan et al., 2008; Xu et al., 2012).

SuperWater Solutions began jointly developing a pilot-scale, SCWO system with the City of Orlando, Florida in 2007. The high-velocity, tubular reactor system is designed to destroy sewage sludge (Sloan et al., 2008). Reported pilot runs of the system indicate that high velocities have been able to keep solids in suspension, thus eliminating clogging caused by inorganic solids (Sloan et al., 2008). Development apparently is continuing, and additional details are not available.

Supercritical Fluids International purchased all intellectual property rights to the AquaCritox® technology from Chematur in 2007 (Bermejo et al., 2009; Marrone & Hong, 2009). Supercritical Fluids International has further refined the process into a sewage sludge destruction system with integrated, electrical generation and value-added production (e.g., CO_2, phosphorus, and silica) (Bermejo et al., 2009). The AquaCritox® system appears to be the only fully integrated and turn-key system currently available for commercial sale. The basic flow schematic for the AquaCritox® system is shown (Figure 4).

Supercritical reactor sizes (sufficient for 99.99% decomposition of biomass) can be commercially scaled down by orders of magnitude, making such systems suitable for much smaller physical footprints as long as viable salt precipitation control mechanisms are deployed. Scalability could facilitate the location of conversion systems at the source of wet waste production, enabling on-site use of the energy contained in the waste streams while simultaneously avoiding waste storage and transportation. The successful process intensification of wet-biomass conversion into an energy-efficient and sustainable pathway has many potential markets and public benefits.

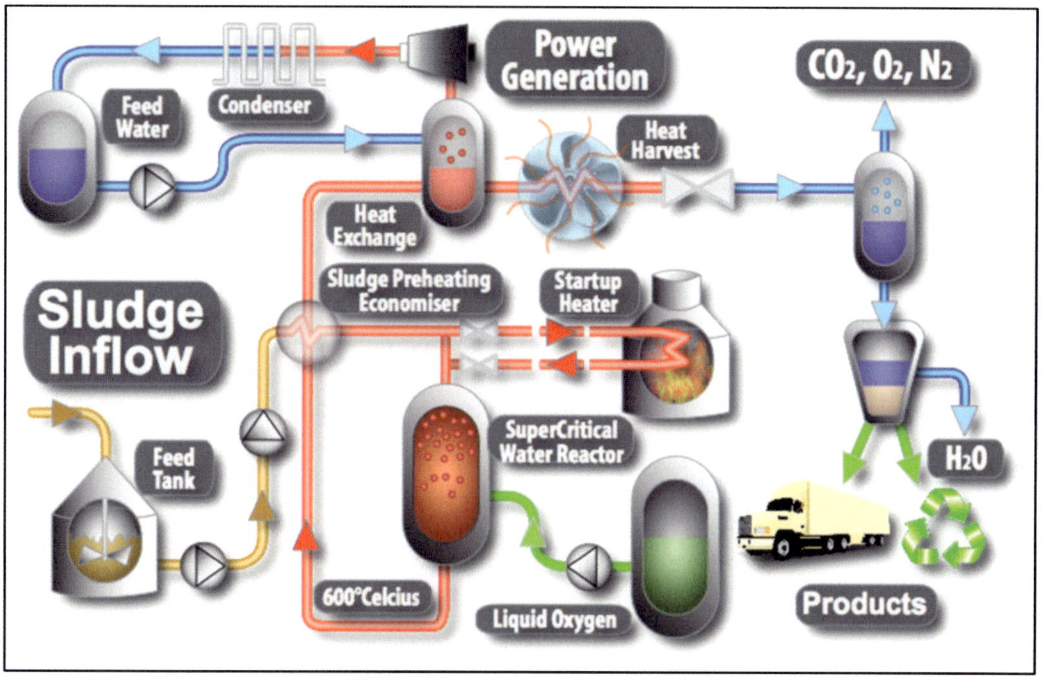

Figure 4. AquaCritox® System Flow Schematic, Adapted from (SCFI, 2012).

4. Conclusion

Solutions to contemporary, waste-management and energy problems are becoming universally multi-objective (Klemes & Stehlík, 2006). Important considerations include reductions in energy consumption and associated costs, while simultaneously stemming harmful pollutant releases. Ideally, wastewater treatment facilities would produce clean water effluent, operate at net-plus energy, and have near-zero pollutant releases to the environment (Münster & Lund, 2010; Villar et al., 2012; Weismantel, 1996). Currently, wastewater treatment facilities do produce clean water, but only at immense energetic and environmental expense. Municipalities collectively spend about $6 billion annually in the U.S. just reducing the moisture content of sewage sludge (CSS, 2011; Kim & Parker, 2008; USEPA, 2009). After moisture reduction, the millions of tons of still-moisture-laden sludge are then typically dried, at great expense, prior to disposal by conventional methods (Abuadala et al., 2010; Borán et al., 2010; CSS, 2011; Demirbas et al., 2011; Eriksson et al., 2008; Fytili & Zabaniotou, 2008; Harrison et al., 2006; Kruse et al., 2000; NASNRC, 1996, 2002). The disposal of the increasingly pollutant-laden and voluminous leftover sludge is also expensive and is becoming more problematic. Sewage sludge contains heavy metals, harmful organic compounds, and pathogens that are potentially hazardous to human and environmental health. Consequently, current disposal methods are increasingly unpopular with the public. It is likely that climbing wastewater-related energy costs, coupled with more stringent environmental regulations, will ultimately lead to the adoption of new wastewater processing and sludge disposal techniques (Dutournie & Mercadier, 2005).

Given that sewage sludge is mostly water, hydrothermal decomposition via supercritical water gasification (SCWG) is viewed as a promising technology for sustainable sludge disposal. The unique properties of supercritical water enable SCWG to quickly destroy wet biomass such as sludge while efficiently producing marketable byproducts (e.g., heat, H_2, and CO-rich fuel gases) (Kalinci et al., 2009). Adding catalysts or oxidants to SCWG (i.e., CSCWG or SCWO) can further reduce operating costs by creating self-sustaining reactions under milder conditions with even shorter residence times.

As a research area, the number of SCW-related studies has grown exponentially during the last 10 years. Much of this research is focused on the technological SCW limitations, notably reactor corrosion and plugging, which seems to be nearly resolved (Bermejo et al., 2009; Calzavara et al., 2005; Marrone & Hong, 2009). Many supercritical-water techniques have been repeatedly shown to thoroughly destroy wet wastes such as sewage and oily sludges. Most SCW technologies, nonetheless, remain relatively new and untested beyond laboratory applications, relegating SCW systems to emerging-technology status (Youssef et al., 2010b). Likewise, SCW designs are somewhat niche-based, with few flexible enough to deal with the wide range of mixed wastes generated by municipalities and industry. Commercial viability sets a very high bar for an emerging technology to clear, and, so far, no SCW systems have yet reached the bar of true commercialization in the wastewater industry (Savage, 2009).

The destruction of sewage *in-situ*, without separating the water from the sludge offers the best potential for sustainable wastewater processing. Furthermore, sewage has a relatively high energy content in the form of organic matter and can be used to produce renewable energy (Chen et al., 2011; Fodor & Klemes, 2011). The annual energy entrained in U.S. sewage sludge is estimated at 1.5 Quads. The *in-situ* destruction of sewage sludge via SCW processes would help municipalities gain massive and immediate financial savings, while simultaneously addressing a critical and vexing wastewater management problem (Fytili & Zabaniotou, 2008). The widespread use of hydrothermal decomposition of municipal wastewater solids would shift the view of sewage sludge as a costly disposal problem to that of a valuable, sustainable energy source (Tsai et al., 2009). Supercritical-water technologies (particularly SCWO) can be integrated into existing wastewater treatment facilities as bolt-on, end-of-pipe systems (Bermejo et al., 2009; Brunner, 2009a; Cocero Alonso et al., 2002; Gloyna & Li, 1993), and might very well be the panacea for municipal sewage sludge.

Author details

Jonathan Kamler
University of Alaska Fairbanks, School of Natural Resources and Agricultural Sciences, USA

J. Andres Soria
University of Alaska Fairbanks, School of Natural Resources and Agricultural Sciences,
University of Alaska Anchorage, School of Engineering, USA

Acknowledgement

The authors would like to acknowledge the University of Alaska Fairbanks, Agricultural and Forestry Experiment Station, part of the School of Natural Resources and Agricultural Sciences for their financial support.

Addendum

Research publication trends

The annual trend in the number of published studies provides a suitable proxy of the importance of promising research areas (Savage, 2009). The amount of supercritical water (SCW) research has grown at a near-exponential rate since 1988 (Figure 5A). Data were collected using a Web of Science search for "supercritical water". The data were plotted biennially, and a trend line was fitted to the growth curve. Despite the fact that SCW research dates back to the 1970s, there were only a total of 22 hits from the Web of Science database prior to 1988.

The amount of supercritical water gasification (SCWG) research has grown at a near-exponential rate since 1996, and shows no signs of slowing (Figure 5B). The proportion for the larger category of supercritical water research is still relatively small (23.1%), but is poised to lead supercritical water research in the next two or three years. Data were collected using a Web of Science search for "supercritical water" and "gasif*". The data were plotted biennially, and a trend line was fitted to the growth curve. There were a total of ten search hits from the Web of Science database prior to 1998. Catalytic supercritical water gasification (CSCWG) research is one of the newest areas of study within the scope of supercritical water science. Interest in CSCWG has grown considerably over the past two decades, with most of the research taking place in the last ten years (Azadi & Farnood, 2011). Indeed, the amount of CSCWG research has grown exponentially since 2000, and shows no signs of slowing down (Figure 5C). The data were collected using a Web of Science search of articles with the terms "supercritical water", "gasif*", and "cataly*". The data were plotted biennially, and a trend line was fitted to the growth curve. There were a total of only five search hits from the Web of Science database prior to 2000. Catalytic gasification now makes up a full two thirds of the SCWG research, and if the growth rate continues, it is likely that CSCWG will increase its current 15.4% proportion of all supercritical water research. The most prevalent message from recent works is that CSCWG shows great scientific promise as well as notable technologic and commercialization feasibility for hydrogen production from biomass (Calzavara et al., 2005). Supercritical water oxidation (SCWO) research dates back a bit further than either SCWG or CSCWG, but most of the research occurred as greater interest developed at universities, national laboratories, and government agencies in the late 1980s (Barner et al., 1992). Despite its early beginnings, SCWO now has only a slightly larger percentage of overall supercritical water research (27.8%) than SCWG (23.1%). The amount of SCWO research grew rapidly from about 1988, and continues to grow (Figure 5D) but at a much more moderate rate than either SCWG or CSCWG. The data were collected using a Web of Science search for "supercritical water" and "oxida*". The data were plotted biennially, and a trend line was fitted to the growth curve. There were a total of only three search hits from the Web of Science database prior to 1988.

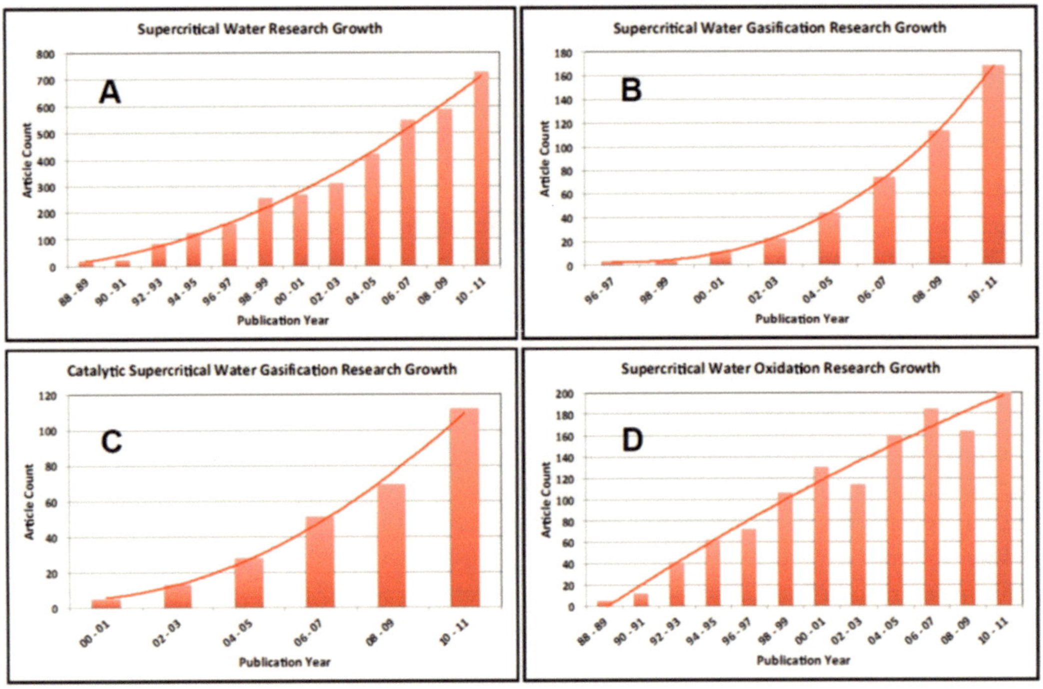

Figure 5. Supercritical Water Research Publication Trends.

5. References

Abbas, T., Costen, P., De Soete, G., Glaser, K., Hassan, S., & Lockwood, F.C. (1996). The Energy and Environmental Implications of Using Sewage Sludge as a Co-Fired Fuel Applied to Boilers. *Symposium (International) on Combustion*, Vol.26, No.2, pp. 2487-2493, ISSN 0082-0784.

Abelleira, J., Pérez-Elvirab, S.I., Sánchez-Oneto, J., Portela, J.R., & Nebot, E. (2011). Advanced Thermal Hydrolysis of Secondary Sewage Sludge: A Novel Process

Combining Thermal Hydrolysis and Hydrogen Peroxide Addition. *Resources, Conservation and Recycling*, ISSN 0921-3449.

Abuadala, A., Dincer, I., & Naterer, G.F. (2010). Exergy Analysis of Hydrogen Production from Biomass Gasification. *International Journal of Hydrogen Energy*, Vol.35, No.10, pp. 4981-4990, ISSN 0360-3199.

Adegoroye, A., Paterson, N., Li, X., Morgan, T., Herod, A.A., Dugwell, D.R., & Kandiyoti, R. (2004). The Characterisation of Tars Produced During the Gasification of Sewage Sludge in a Spouted Bed Reactor. *Fuel*, Vol.83, No.14–15, pp. 1949-1960, ISSN 0016-2361.

Afif, E., Azadi, P., & Farnood, R. (2011). Catalytic Hydrothermal Gasification of Activated Sludge. *Applied Catalysis B: Environmental*, Vol.105, No.1-2, pp. 136-143, ISSN 0926-3373.

Aki, S., & Abraham, M.A. (1999). Catalytic Supercritical Water Oxidation of Pyridine: Comparison of Catalysts. *Industrial & Engineering Chemistry Research*, Vol.38, No.2, pp. 358-367, ISSN 0888-5885.

Alaimo, A.A. (2008). *R.A. McElmurray, III, R.A. McElmurray, Jr., Richard P. McElmurray, and Earl D. McElmurray, Plaintiffs, V. United States Department of Agriculture, Defendant.*, in: United States District Court, S.D.G., Augusta Division. (Ed.). United States District Court, Augusta, Georgia.

Angenent, L.T., Karim, K., Al-Dahhan, M.H., Wrenn, B.A., & Domíguez-Espinosa, R. (2004). Production of Bioenergy and Biochemicals from Industrial and Agricultural Wastewater. *Trends in Biotechnology*, Vol.22, No.9, pp. 477-485, ISSN 0167-7799.

Anglada, Á., Urtiaga, A., Ortiz, I., Mantzavinos, D., & Diamadopoulos, E. (2011). Treatment of Municipal Landfill Leachate by Catalytic Wet Air Oxidation: Assessment of the Role of Operating Parameters by Factorial Design. *Waste Management*, Vol.31, No.8, pp. 1833-1840, ISSN 0956-053X.

Antal, M.J., Allen, S.G., Schulman, D., Xu, X., & Divilio, R.J. (2000). Biomass Gasification in Supercritical Water. *Industrial & Engineering Chemistry Research*, Vol.39, No.11, pp. 4040-4053, ISSN 0888-5885.

Antal, M.J.J., Matsumurn, Y., Xu, X., Stenberg, J., & Lipnik, P. (1995). Catalytic Gasification of Wet Biomass in Supercritical Water. *Preprint Papers - American Chemical Society, Division of Fuel Chemistry*, pp. 304–307.

Arthurson, V. (2008). Proper Sanitization of Sewage Sludge: A Critical Issue for a Sustainable Society. *Applied and Environmental Microbiology*, Vol.74, No.17, pp. 5267-5276.

Augustine, C., & Tester, J.W. (2009). Hydrothermal Flames: From Phenomenological Experimental Demonstrations to Quantitative Understanding. *The Journal of Supercritical Fluids*, Vol.47, No.3, pp. 415-430, ISSN 0896-8446.

Azadi, P., & Farnood, R. (2011). Review of Heterogeneous Catalysts for Sub- and Supercritical Water Gasification of Biomass and Wastes. *International Journal of Hydrogen Energy*, Vol.36, No.16, pp. 9529-9541, ISSN 0360-3199.

Azadi, P., Farnood, R., & Vuillardot, C. (2011). Estimation of Heating Time in Tubular Supercritical Water Reactors. *The Journal of Supercritical Fluids*, Vol.55, No.3, pp. 1038-1045, ISSN 0896-8446.

Babel, S., & del Mundo Dacera, D. (2006). Heavy Metal Removal from Contaminated Sludge for Land Application: A Review. *Waste Management*, Vol.26, No.9, pp. 988-1004, ISSN 0956-053X.

Bag, S., Vora, T., Ghatak, R., Nilufer, I., D'Mello, D., Pereira, L., Pereira, J., Cutinho, C., & Rao, V. (1999). A Study of Toxic Effects of Heavy Metal Contaminants from Sludge-Supplemented Diets on Male Wistar Rats. *Ecotoxicology and Environmental Safety*, Vol.42, No.2, pp. 163-170, ISSN 0147-6513.

Balat, M., Balat, M., Kirtay, E., & Balat, H. (2009). Main Routes for the Thermo-Conversion of Biomass into Fuels and Chemicals. Part 2: Gasification Systems. *Energy Conversion and Management*, Vol.50, No.12, pp. 3158-3168, ISSN 0196-8904.

Barner, H.E., Huang, C.Y., Johnson, T., Jacobs, G., Martch, M.A., & Killilea, W.R. (1992). Supercritical Water Oxidation: An Emerging Technology. *Journal of Hazardous Materials*, Vol.31, No.1, pp. 1-17, ISSN 0304-3894.

Beauchesne, I., Cheikh, R.B., Mercier, G., Blais, J.-F., & Ouarda, T. (2007). Chemical Treatment of Sludge: In-Depth Study on Toxic Metal Removal Efficiency, Dewatering Ability and Fertilizing Property Preservation. *Water Research*, Vol.41, No.9, pp. 2028-2038, ISSN 0043-1354.

Beck, A.J., Alcock, R.E., Wilson, S.C., Wang, M.-J., Wild, S.R., Sewart, A.P., & Jones, K.C. (1995). Long-Term Persistence of Organic Chemicals in Sewage Sludge-Amended Agricultural Land: A Soil Quality Perspective, in: Donald, L.S. (Ed.), *Advances in Agronomy*. Academic Press, pp. 345-391.

Bermejo, M.D., Cabeza, P., Queiroz, J.P.S., Jiménez, C., & Cocero, M.J. (2011). Analysis of the Scale up of a Transpiring Wall Reactor with a Hydrothermal Flame as a Heat Source for the Supercritical Water Oxidation. *The Journal of Supercritical Fluids*, Vol.56, No.1, pp. 21-32, ISSN 0896-8446.

Bermejo, M.D., & Cocero, M.J. (2006a). Destruction of an Industrial Wastewater by Supercritical Water Oxidation in a Transpiring Wall Reactor. *Journal of Hazardous Materials*, Vol.137, No.2, pp. 965-971, ISSN 0304-3894.

Bermejo, M.D., & Cocero, M.J. (2006b). Supercritical Water Oxidation: A Technical Review. *AIChE Journal*, Vol.52, No.11, pp. 3933-3951, ISSN 1547-5905.

Bermejo, M.D., Fernández-Polanco, F., & Cocero, M.J. (2005a). Effect of the Transpiring Wall on the Behavior of a Supercritical Water Oxidation Reactor: Modeling and Experimental Results. *Industrial & Engineering Chemistry Research*, Vol.45, No.10, pp. 3438-3446, ISSN 0888-5885.

Bermejo, M.D., Fernández-Polanco, F., & Cocero, M.J. (2005b). Modeling of a Transpiring Wall Reactor for the Supercritical Water Oxidation Using Simple Flow Patterns: Comparison to Experimental Results. *Industrial & Engineering Chemistry Research*, Vol.44, No.11, pp. 3835-3845, ISSN 0888-5885.

Bermejo, M.D., Martín, A., & Cocero, M.J. (2007). Application of the Anderko–Pitzer EOS to the Calculation of Thermodynamical Properties of Systems Involved in the Supercritical Water Oxidation Process. *The Journal of Supercritical Fluids*, Vol.42, No.1, pp. 27-35, ISSN 0896-8446.

Bermejo, M.D., Rincon, D., Martin, A., & Cocero, M.J. (2009). Experimental Performance and Modeling of a New Cooled-Wall Reactor for the Supercritical Water Oxidation. *Industrial & Engineering Chemistry Research*, Vol.48, No.13, pp. 6262-6272, ISSN 0888-5885.

Bernardi, M., Cretenot, D., Deleris, S., Descorme, C., Chauzy, J., & Besson, M. (2010). Performances of Soluble Metallic Salts in the Catalytic Wet Air Oxidation of Sewage Sludge. *Catalysis Today*, Vol.157, No.1-4, pp. 420-424, ISSN 0920-5861.

Booth, S., Barnett, J., Burman, K., Hambrick, J., & Westby, R. (2010). *Net Zero Energy Military Installations: A Guide to Assessment and Planning*, in: Energy, U.S.D.O. (Ed.). U.S. Department of Commerce, National Technical Information Service, 5285 Port Royal Road, Springfield, VA 22161, Golden, Colorado, p. 49.

Borán, J., Houdková, L., & Elsäßer, T. (2010). Processing of Sewage Sludge: Dependence of Sludge Dewatering Efficiency on Amount of Flocculant. *Resources, Conservation and Recycling*, Vol.54, No.5, pp. 278-282, ISSN 0921-3449.

Brennecke, J.F., & Chateauneuf, J.E. (1999). Homogeneous Organic Reactions as Mechanistic Probes in Supercritical Fluids. *Chemical Reviews*, Vol.99, No.2, pp. 433-452, ISSN 0009-2665.

Brunner, G. (2009a). Near and Supercritical Water. Part II: Oxidative Processes. *The Journal of Supercritical Fluids*, Vol.47, No.3, pp. 382-390, ISSN 0896-8446.

Brunner, G. (2009b). Near Critical and Supercritical Water. Part I. Hydrolytic and Hydrothermal Processes. *The Journal of Supercritical Fluids*, Vol.47, No.3, pp. 373-381, ISSN 0896-8446.

Bubenheim, D.L., & Wydeven, T. (1994). Approaches to Resource Recovery in Controlled Ecological Life Support Systems. *Advances in Space Research*, Vol.14, No.11, pp. 113-123, ISSN 0273-1177.

Byrd, A.J., Pant, K.K., & Gupta, R.B. (2008). Hydrogen Production from Glycerol by Reforming in Supercritical Water over Ru/Al_2O_3 Catalyst. *Fuel*, Vol.87, No.13-14, pp. 2956-2960, ISSN 0016-2361.

Cabeza, P., Bermejo, M.D., Jiménez, C., & Cocero, M.J. (2011). Experimental Study of the Supercritical Water Oxidation of Recalcitrant Compounds Under Hydrothermal Flames Using Tubular Reactors. *Water Research*, Vol.45, No.8, pp. 2485-2495, ISSN 0043-1354.

Calzavara, Y., Joussot-Dubien, C., Boissonnet, G., & Sarrade, S. (2005). Evaluation of Biomass Gasification in Supercritical Water Process for Hydrogen Production. *Energy Conversion and Management*, Vol.46, No.4, pp. 615-631, ISSN 0196-8904.

Calzavara, Y., Joussot-Dubien, C., Turc, H.A., Fauvel, E., & Sarrade, S. (2004). A New Reactor Concept for Hydrothermal Oxidation. *The Journal of Supercritical Fluids*, Vol.31, No.2, pp. 195-206, ISSN 0896-8446.

Cao, C., Guo, L., Chen, Y., Guo, S., & Lu, Y. (2011). Hydrogen Production from Supercritical Water Gasification of Alkaline Wheat Straw Pulping Black Liquor in Continuous Flow System. *International Journal of Hydrogen Energy*, Vol.36, No.21, pp. 13528-13535, ISSN 0360-3199.

Cappon, C.J. (1991). Sewage Sludge as a Source of Environmental Selenium. *Science of The Total Environment*, Vol.100, pp. 177-205, ISSN 0048-9697.

Castello, D., & Fiori, L. (2011). Supercritical Water Gasification of Biomass: Thermodynamic Constraints. *Bioresource Technology*, Vol.102, No.16, pp. 7574-7582, ISSN 0960-8524.

Catallo, W.J., & Comeaux, J.L. (2008). Reductive Hydrothermal Treatment of Sewage Sludge. *Waste Management*, Vol.28, No.11, pp. 2213-2219, ISSN 0956-053X.

Chakinala, A.G., Brilman, D.W.F., van Swaaij, W.P.M., & Kersten, S.R.A. (2010). Catalytic and Non-Catalytic Supercritical Water Gasification of Microalgae and Glycerol. *Industrial & Engineering Chemistry Research*, Vol.49, No.3, pp. 1113-1122, ISSN 0888-5885.

Chang, K.-C., Li, L., & Gloyna, E.F. (1993). Supercritical Water Oxidation of Acetic Acid by Potassium Permanganate. *Journal of Hazardous Materials*, Vol.33, No.1, pp. 51-62, ISSN 0304-3894.

Chen, W.-S., Chang, F.-C., Shen, Y.-H., & Tsai, M.-S. (2011). The Characteristics of Organic Sludge/Sawdust Derived Fuel. *Bioresource Technology*, Vol.102, No.9, pp. 5406-5410, ISSN 0960-8524.

Chun, Y.N., Kim, S.C., & Yoshikawa, K. (2011). Pyrolysis Gasification of Dried Sewage Sludge in a Combined Screw and Rotary Kiln Gasifier. *Applied Energy*, Vol.88, No.4, pp. 1105-1112, ISSN 0306-2619.

Chuntanapum, A., & Matsumura, Y. (2010). Char Formation Mechanism in Supercritical Water Gasification Process: A Study of Model Compounds. *Industrial & Engineering Chemistry Research*, Vol.49, No.9, pp. 4055-4062, ISSN 0888-5885.

Clarke, B.O., & Smith, S.R. (2011). Review of Emerging Organic Contaminants in Biosolids and Assessment of International Research Priorities for the Agricultural Use of Biosolids. *Environment International*, Vol.37, No.1, pp. 226-247, ISSN 0160-4120.

Cocero Alonso, M.J., Alonso Sánchez, E., & Fernandez-Polanco, F. (2002). Supercritical Water Oxidation of Wastewater and Sludges – Design Considerations. *Engineering in Life Sciences*, Vol.2, No.7, pp. 195-200, ISSN 1618-2863.

Cocero, M.J. (2001). Supercritical Water Oxidation (SCWO). Application to Industrial Wastewater Treatment, in: Bertucco, A., & Vetter, G. (Eds.), *Industrial Chemistry Library*. Elsevier, pp. 509-526.

Cocero, M.J., Alonso, E., Sanz, M.T., & Fdz-Polanco, F. (2002). Supercritical Water Oxidation Process Under Energetically Self-Sufficient Operation. *The Journal of Supercritical Fluids*, Vol.24, No.1, pp. 37-46, ISSN 0896-8446.

Cocero, M.J., Martín, A., Bermejo, M.D., Santos, M., Rincón, D., Alonso, E., & Fdez-Polanco, F. (2003). Supercritical Water Oxidation of Industrial Waste Water from Pilot to Demonstration Plant, in: Brunner, G.H., & Kikic, I., & Perrut, M. (Eds.), *6th International Symposium on Supercritical Fluids*. International Society for the Advancement of Supercritical Fluids, Versailles, France.

Costello, M.J., & Read, P. (1994). Toxicity of Sewage Sludge to Marine Organisms: A Review. *Marine Environmental Research*, Vol.37, No.1, pp. 23-46, ISSN 0141-1136.

Crooker, P.J., Ahluwalia, K.S., Fan, Z., & Prince, J. (2000). Operating Results from Supercritical Water Oxidation Plants. *Industrial & Engineering Chemistry Research*, Vol.39, No.12, pp. 4865-4870, ISSN 0888-5885.

CSS. (2011). *U.S. Wastewater Treatment Factsheet*. Center for Sustainable Systems, University of Michigan.

Cui, B., Cui, F., Jing, G., Xu, S., Huo, W., & Liu, S. (2009). Oxidation of Oily Sludge in Supercritical Water. *Journal of Hazardous Materials*, Vol.165, No.1-3, pp. 511-517, ISSN 0304-3894.

D'Jesús, P., Artiel, C., Boukis, N., Kraushaar-Czarnetzki, B., & Dinjus, E. (2005). Influence of Educt Preparation on Gasification of Corn Silage in Supercritical Water. *Industrial & Engineering Chemistry Research*, Vol.44, No.24, pp. 9071-9077, ISSN 0888-5885.

D'Jesús, P., Boukis, N., Kraushaar-Czarnetzki, B., & Dinjus, E. (2006). Gasification of Corn and Clover Grass in Supercritical Water. *Fuel*, Vol.85, No.7-8, pp. 1032-1038, ISSN 0016-2361.

Dean, R.B., & Suess, M.J. (1985). The Risk to Health of Chemicals in Sewage Sludge Applied to Land. *Waste Management & Research*, Vol.3, No.3, pp. 251-278, ISSN 0734-242X.

Deblonde, T., Cossu-Leguille, C., & Hartemann, P. (2011). Emerging Pollutants in Wastewater: A Review of the Literature. *International Journal of Hygiene and Environmental Health*, ISSN 1438-4639.

Demirbas, A. (2007). Progress and Recent Trends in Biofuels. *Progress in Energy and Combustion Science*, Vol.33, No.1, pp. 1-18, ISSN 0360-1285.

Demirbas, A. (2009). Biofuels Securing the Planet's Future Energy Needs. *Energy Conversion and Management*, Vol.50, No.9, pp. 2239-2249, ISSN 0196-8904.

Demirbas, A. (2011a). Competitive Liquid Biofuels from Biomass. *Applied Energy*, Vol.88, No.1, pp. 17-28, ISSN 0306-2619.

Demirbas, A. (2011b). Waste Management, Waste Resource Facilities and Waste Conversion Processes. *Energy Conversion and Management*, Vol.52, No.2, pp. 1280-1287, ISSN 0196-8904.

Demirbas, M.F., Balat, M., & Balat, H. (2011). Biowastes-to-Biofuels. *Energy Conversion and Management*, Vol.52, No.4, pp. 1815-1828, ISSN 0196-8904.

Di Blasi, C., Branca, C., Galgano, A., Meier, D., Brodzinski, I., & Malmros, O. (2007). Supercritical Gasification of Wastewater from Updraft Wood Gasifiers. *Biomass and Bioenergy*, Vol.31, No.11-12, pp. 802-811, ISSN 0961-9534.

Dijkema, G.P.J., Reuter, M.A., & Verhoef, E.V. (2000). A New Paradigm for Waste Management. *Waste Management*, Vol.20, No.8, pp. 633-638, ISSN 0956-053X.

Dimitriou, I., Eriksson, J., Adler, A., Aronsson, P., & Verwijst, T. (2006). Fate of Heavy Metals after Application of Sewage Sludge and Wood, Ash Mixtures to Short-Rotation Willow Coppice. *Environmental Pollution*, Vol.142, No.1, pp. 160-169, ISSN 0269-7491.

Ding, Z.Y., Frisch, M.A., Li, L., & Gloyna, E.F. (1996). Catalytic Oxidation in Supercritical Water. *Industrial & Engineering Chemistry Research*, Vol.35, No.10, pp. 3257-3279, ISSN 0888-5885.

Ding, Z.Y., Li, L., Wade, D., & Gloyna, E.F. (1998). Supercritical Water Oxidation of NH_3 Over a MnO_2/CeO_2 Catalyst. *Industrial & Engineering Chemistry Research*, Vol.37, No.5, pp. 1707-1716, ISSN 0888-5885.

Dinjus, E., & Kruse, A. (2007). Applications of Supercritical Water, *High Pressure Chemistry*. Wiley-VCH Verlag GmbH, pp. 422-446.

DiPeso, J. (2011). Natural Gas: Fueling Our Future? *Environmental Quality Management*, Vol.21, No.2, pp. 97-104, ISSN 1520-6483.

Dogru, M., Midilli, A., & Howarth, C.R. (2002). Gasification of Sewage Sludge Using a Throated Downdraft Gasifier and Uncertainty Analysis. *Fuel Processing Technology*, Vol.75, No.1, pp. 55-82, ISSN 0378-3820.

Domínguez, A., Menéndez, J.A., Inguanzo, M., & Pís, J.J. (2006). Production of Bio-Fuels by High Temperature Pyrolysis of Sewage Sludge Using Conventional and Microwave Heating. *Bioresource Technology*, Vol.97, No.10, pp. 1185-1193, ISSN 0960-8524.

Dominy, M. (2009). *Lewis Et Al. V. Walker Et Al, Deposition of Madolyn Dominy*. U. S. District Court, Middle District of Georgia, Athens, Georgia.

Du, X., Zhang, R., Gan, Z., & Bi, J. (2010). Treatment of High Strength Coking Wastewater by Supercritical Water Oxidation. *Fuel*, ISSN 0016-2361.

Duan, P., & Savage, P.E. (2011). Upgrading of Crude Algal Bio-Oil in Supercritical Water. *Bioresource Technology*, Vol.102, No.2, pp. 1899-1906, ISSN 0960-8524.

Duić, N., Guzović, Z., & Lund, H. (2011). Sustainable Development of Energy, Water and Environment Systems. *Energy*, Vol.36, No.4, pp. 1839-1841, ISSN 0360-5442.

Dutournie, P., & Mercadier, J. (2005). Unsteady Behaviour of Hydrothermal Oxidation Reactors: Theoretical and Numerical Studies near the Critical Point. *The Journal of Supercritical Fluids*, Vol.35, No.3, pp. 247-253, ISSN 0896-8446.

Ejlertsson, J., Karlsson, A., Lagerkvist, A., Hjertberg, T., & Svensson, B.H. (2003). Effects of Co-Disposal of Wastes Containing Organic Pollutants with Municipal Solid Waste—A Landfill Simulation Reactor Study. *Advances in Environmental Research*, Vol.7, No.4, pp. 949-960, ISSN 1093-0191.

Elliott, D.C. (2008). Catalytic Hydrothermal Gasification of Biomass. *Biofuels, Bioproducts and Biorefining*, Vol.2, No.3, pp. 254-265, ISSN 1932-1031.

Epstein, E. (1997). *The Science of Composting*. CRC Press, Boca Raton, Florida.

Eriksson, E., Christensen, N., Ejbye Schmidt, J., & Ledin, A. (2008). Potential Priority Pollutants in Sewage Sludge. *Desalination*, Vol.226, No.1-3, pp. 371-388, ISSN 0011-9164.

Fang, Z., Minowa, T., Fang, C., Smith, J.R.L., Inomata, H., & Kozinski, J.A. (2008). Catalytic Hydrothermal Gasification of Cellulose and Glucose. *International Journal of Hydrogen Energy*, Vol.33, No.3, pp. 981-990, ISSN 0360-3199.

Farré, M., & Barceló, D. (2003). Toxicity Testing of Wastewater and Sewage Sludge by Biosensors, Bioassays and Chemical Analysis. *TrAC Trends in Analytical Chemistry*, Vol.22, No.5, pp. 299-310, ISSN 0165-9936.

Farrell, M., & Jones, D.L. (2009). Critical Evaluation of Municipal Solid Waste Composting and Potential Compost Markets. *Bioresource Technology*, Vol.100, No.19, pp. 4301-4310, ISSN 0960-8524.

Fauvel, E., Joussot-Dubien, C., Pomier, E., Guichardon, P., Charbit, G., Charbit, F., & Sarrade, S. (2003). Modeling of a Porous Reactor for Supercritical Water Oxidation by a Residence Time Distribution Study. *Industrial & Engineering Chemistry Research*, Vol.42, No.10, pp. 2122-2130, ISSN 0888-5885.

Fauvel, E., Joussot-Dubien, C., Tanneur, V., Moussière, S., Guichardon, P., Charbit, G., & Charbit, F. (2005). A Porous Reactor for Supercritical Water Oxidation: Experimental Results on Salty Compounds and Corrosive Solvents Oxidation. *Industrial & Engineering Chemistry Research*, Vol.44, No.24, pp. 8968-8971, ISSN 0888-5885.

Feng, W., van der Kooi, H.J., & de Swaan Arons, J. (2004). Biomass Conversions in Subcritical and Supercritical Water: Driving Force, Phase Equilibria, and Thermodynamic Analysis. *Chemical Engineering and Processing: Process Intensification*, Vol.43, No.12, pp. 1459-1467, ISSN 0255-2701.

Fjällborg, B., Ahlberg, G., Nilsson, E., & Dave, G. (2005). Identification of Metal Toxicity in Sewage Sludge Leachate. *Environment International*, Vol.31, No.1, pp. 25-31, ISSN 0160-4120.

Fodor, Z., & Klemes, J.J. (2011). Waste as Alternative Fuel - Minimising Emissions and Effluents by Advanced Design. *Process Safety and Environmental Protection*, ISSN 0957-5820.

Fytianos, K., Charantoni, E., & Voudrias, E. (1998). Leaching of Heavy Metals from Municipal Sewage Sludge. *Environment International*, Vol.24, No.4, pp. 467-475, ISSN 0160-4120.

Fytili, D., & Zabaniotou, A. (2008). Utilization of Sewage Sludge in EU Application of Old and New Methods—A Review. *Renewable and Sustainable Energy Reviews*, Vol.12, No.1, pp. 116-140, ISSN 1364-0321.

Gadhe, J.B., & Gupta, R.B. (2007). Hydrogen Production by Methanol Reforming in Supercritical Water: Catalysis by In-Situ-Generated Copper Nanoparticles. *International Journal of Hydrogen Energy*, Vol.32, No.13, pp. 2374-2381, ISSN 0360-3199.

Gajalakshmi, S., & Abbasi, S.A. (2008). Solid Waste Management by Composting: State of the Art. *Critical Reviews In Environmental Science And Technology*, Vol.38, No.5, pp. 311-400.

Gale, P., & Stanield, G. (2001). Towards a Quantitative Risk Assessment for BSE in Sewage Sludge. *Journal of Applied Microbiology*, Vol.91, No.3, pp. 563-569.

García-Serna, J., Pérez-Barrigón, L., & Cocero, M.J. (2007). New Trends for Design Towards Sustainability in Chemical Engineering: Green Engineering. *Chemical Engineering Journal*, Vol.133, No.1-3, pp. 7-30, ISSN 1385-8947.

García-Verdugo, E., Venardou, E., Thomas, W.B., Whiston, K., Partenheimer, W., Hamley, P.A., & Poliakoff, M. (2004). Is It Possible to Achieve Highly Selective Oxidations in Supercritical Water? Aerobic Oxidation of Methylaromatic Compounds. *Advanced Synthesis & Catalysis*, Vol.346, No.2-3, pp. 307-316, ISSN 1615-4169.

Gasafi, E., Meyer, L., & Schebek, L. (2007). Exergetic Efficiency and Options for Improving Sewage Sludge Gasification in Supercritical Water. *International Journal of Energy Research*, Vol.31, No.4, pp. 346-363, ISSN 1099-114X.

Gasafi, E., Reinecke, M.-Y., Kruse, A., & Schebek, L. (2008). Economic Analysis of Sewage Sludge Gasification in Supercritical Water for Hydrogen Production. *Biomass and Bioenergy*, Vol.32, No.12, pp. 1085-1096, ISSN 0961-9534.

Gassner, M., & Maréchal, F. (2009). Methodology for the Optimal Thermo-Economic, Multi-Objective Design of Thermochemical Fuel Production from Biomass. *Computers & Chemical Engineering*, Vol.33, No.3, pp. 769-781, ISSN 0098-1354.

Gloyna, E.F., & Li, L. (1993). Supercritical Water Oxidation: An Engineering Update. *Waste Management*, Vol.13, No.5-7, pp. 379-394, ISSN 0956-053X.

Gómez, M.J., Martínez Bueno, M.J., Lacorte, S., Fernández-Alba, A.R., & Agüera, A. (2007). Pilot Survey Monitoring Pharmaceuticals and Related Compounds in a Sewage Treatment Plant Located on the Mediterranean Coast. *Chemosphere*, Vol.66, No.6, pp. 993-1002, ISSN 0045-6535.

Goyal, N., Jain, S.C., & Banerjee, U.C. (2003). Comparative Studies on the Microbial Adsorption of Heavy Metals. *Advances in Environmental Research*, Vol.7, No.2, pp. 311-319, ISSN 1093-0191.

Griffith, J.W., & Raymond, D.H. (2002). The First Commercial Supercritical Water Oxidation Sludge Processing Plant. *Waste Management*, Vol.22, No.4, pp. 453-459.

Guan, Q., Savage, P.E., & Wei, C. (2011). Gasification of Alga *Nannochloropsis* Sp. In Supercritical Water. *The Journal of Supercritical Fluids*, ISSN 0896-8446.

Guo, L., Zhang, B., Xiao, K., Zhang, Q., & Zheng, M. (2009). Levels and Distributions of Polychlorinated Biphenyls in Sewage Sludge of Urban Wastewater Treatment Plants. *Journal of Environmental Sciences*, Vol.21, No.4, pp. 468-473, ISSN 1001-0742.

Guo, L.J., Lu, Y.J., Zhang, X.M., Ji, C.M., Guan, Y., & Pei, A.X. (2007). Hydrogen Production by Biomass Gasification in Supercritical Water: A Systematic Experimental and Analytical Study. *Catalysis Today*, Vol.129, No.3-4, pp. 275-286, ISSN 0920-5861.

Hale, R.C., La Guardia, M.J., Harvey, E.P., Gaylor, M.O., Mainor, T.M., & Duff, W.H. (2001). Flame Retardants: Persistent Pollutants in Land-Applied Sludges. *Nature*, Vol.412, No.6843, pp. 140-141, ISSN 0028-0836.

Han, L.-n., Zhang, R., & Bi, J.-c. (2008). Upgrading of Coal-Tar Pitch in Supercritical Water. *Journal of Fuel Chemistry and Technology*, Vol.36, No.1, pp. 1-5, ISSN 1872-5813.

Hao, X., Guo, L., Zhang, X., & Guan, Y. (2005). Hydrogen Production from Catalytic Gasification of Cellulose in Supercritical Water. *Chemical Engineering Journal*, Vol.110, No.1-3, pp. 57-65, ISSN 1385-8947.

Harrison, E.Z., McBride, M.B., & Bouldin, D.R. (1999). Land Application of Sewage Sludges: An Appraisal of the US Regulations. *International Journal of Environment and Pollution*, Vol.11, No.1, pp. 1-36.

Harrison, E.Z., Oakes, S.R., Hysell, M., & Hay, A. (2006). Organic Chemicals in Sewage Sludges. *Science of The Total Environment*, Vol.367, No.2-3, pp. 481-497, ISSN 0048-9697.

Hashaikeh, R., Fang, Z., Butler, I.S., Hawari, J., & Kozinski, J.A. (2007). Hydrothermal Dissolution of Willow in Hot Compressed Water as a Model for Biomass Conversion. *Fuel*, Vol.86, No.10-11, pp. 1614-1622, ISSN 0016-2361.

Hauthal, W.H. (2001). Advances with Supercritical Fluids [Review]. *Chemosphere*, Vol.43, No.1, pp. 123-135, ISSN 0045-6535.

Hodes, M., Marrone, P.A., Hong, G.T., Smith, K.A., & Tester, J.W. (2004). Salt Precipitation and Scale Control in Supercritical Water Oxidation, Part A: Fundamentals and Research. *The Journal of Supercritical Fluids*, Vol.29, No.3, pp. 265-288, ISSN 0896-8446.

Hong, J., Hong, J., Otaki, M., & Jolliet, O. (2009). Environmental and Economic Life Cycle Assessment for Sewage Sludge Treatment Processes in Japan. *Waste Management*, Vol.29, No.2, pp. 696-703, ISSN 0956-053X.

Hooda, P.S. (2003). A Special Issue on Heavy Metals in Soils: Editorial Foreword. *Advances in Environmental Research*, Vol.8, No.1, pp. 1-3, ISSN 1093-0191.

Hospido, A., Carballa, M., Moreira, M., Omil, F., Lema, J.M., & Feijoo, G. (2010). Environmental Assessment of Anaerobically Digested Sludge Reuse in Agriculture: Potential Impacts of Emerging Micropollutants. *Water Research*, Vol.44, No.10, pp. 3225-3233, ISSN 0043-1354.

Hubbe, M.A., Nazhad, M., & Sanchez, C. (2010). Composting as a Way to Convert Cellulosic Biomass and Organic Waste into High-Value Soil Amendments: A Review. *Bioresources*, Vol.5, No.4, pp. 2808-2854.

Hunter, T.B., Rice, S.F., & Hanush, R.G. (1996). Raman Spectroscopic Measurement of Oxidation in Supercritical Water. 2. Conversion of Isopropyl Alcohol to Acetone. *Industrial & Engineering Chemistry Research*, Vol.35, No.11, pp. 3984-3990, ISSN 0888-5885.

Hyde, J.R., Licence, P., Carter, D., & Poliakoff, M. (2001). Continuous Catalytic Reactions in Supercritical Fluids. *Applied Catalysis A: General*, Vol.222, No.1–2, pp. 119-131, ISSN 0926-860X.

Izumizaki, Y., Park, K.C., Tachibana, Y., Tomiyasu, H., & Fujii, Y. (2005). Organic Decomposition in Supercritical Water by an Aid of Ruthenium (IV) Oxide as a Catalyst-Exploitation of Biomass Resources for Hydrogen Production. *Progress in Nuclear Energy*, Vol.47, No.1–4, pp. 544-552, ISSN 0149-1970.

Izumizaki, Y., Park, K.C., Yamamura, T., Tomiyasu, H., Goda, B., & Fujii, Y. (2008). Exothermic Hydrogen Production System in Supercritical Water from Biomass and Usual Domestic Wastes with an Exploitation of RuO_2 Catalyst. *Progress in Nuclear Energy*, Vol.50, No.2-6, pp. 438-442, ISSN 0149-1970.

Jacobsen, S., & Wyman, C. (2000). Cellulose and Hemicellulose Hydrolysis Models for Application to Current and Novel Pretreatment Processes. *Applied Biochemistry and Biotechnology*, Vol.84-86, No.1, pp. 81-96, ISSN 0273-2289.

Jessop, P.G., Ikariya, T., & Noyori, R. (1999). Homogeneous Catalysis in Supercritical Fluids. *Chemical Reviews*, Vol.99, No.2, pp. 475-494, ISSN 0009-2665.

Jin, H., Lu, Y., Guo, L., Cao, C., & Zhang, X. (2010). Hydrogen Production by Partial Oxidative Gasification of Biomass and Its Model Compounds in Supercritical Water. *International Journal of Hydrogen Energy*, Vol.35, No.7, pp. 3001-3010, ISSN 0360-3199.

Jing, G., Huo, W., Cui, B., & Zhao, T. (2008). *Supercritical Water Oxidation of Oilfield Sludge*, Bioinformatics and Biomedical Engineering, 2008. ICBBE 2008. The 2nd International Conference on pp. 4096 - 4099.

Kalinci, Y., Hepbasli, A., & Dincer, I. (2009). Biomass-Based Hydrogen Production: A Review and Analysis. *International Journal of Hydrogen Energy*, Vol.34, No.21, pp. 8799-8817, ISSN 0360-3199.

Kidd, P.S., Domínguez-Rodríguez, M.J., Díez, J., & Monterroso, C. (2007). Bioavailability and Plant Accumulation of Heavy Metals and Phosphorus in Agricultural Soils Amended by Long-Term Application of Sewage Sludge. *Chemosphere*, Vol.66, No.8, pp. 1458-1467, ISSN 0045-6535.

Kim, Y., & Parker, W. (2008). A Technical and Economic Evaluation of the Pyrolysis of Sewage Sludge for the Production of Bio-Oil. *Bioresource Technology*, Vol.99, No.5, pp. 1409-1416, ISSN 0960-8524.

Klemes, J.J., & Stehlík, P. (2006). Recent Advances on Heat, Chemical and Process Integration, Multiobjective and Structural Optimisation. *Applied Thermal Engineering*, Vol.26, No.13, pp. 1339-1344, ISSN 1359-4311.

Koda, S., Kanno, N., & Fujiwara, H. (2001). Kinetics of Supercritical Water Oxidation of Methanol Studied in a CSTR by Means of Raman Spectroscopy. *Industrial & Engineering Chemistry Research*, Vol.40, No.18, pp. 3861-3868, ISSN 0888-5885.

Krajnc, M., & Levec, J. (1994). Catalytic Oxidation of Toxic Organics in Supercritical Water. *Applied Catalysis B: Environmental*, Vol.3, pp. L101-L107.

Kruse, A. (2008). Supercritical Water Gasification. *Biofuels, Bioproducts and Biorefining*, Vol.2, No.5, pp. 415-437, ISSN 1932-1031.

Kruse, A. (2009). Hydrothermal Biomass Gasification. *The Journal of Supercritical Fluids*, Vol.47, No.3, pp. 391-399, ISSN 0896-8446.

Kruse, A., & Dinjus, E. (2007a). Hot Compressed Water as Reaction Medium and Reactant: 2. Degradation Reactions. *The Journal of Supercritical Fluids*, Vol.41, No.3, pp. 361-379, ISSN 0896-8446.

Kruse, A., & Dinjus, E. (2007b). Hot Compressed Water as Reaction Medium and Reactant: Properties and Synthesis Reactions. *The Journal of Supercritical Fluids*, Vol.39, No.3, pp. 362-380, ISSN 0896-8446.

Kruse, A., Forchheim, D., Gloede, M., Ottinger, F., & Zimmermann, J. (2010). Brines in Supercritical Biomass Gasification: 1. Salt Extraction by Salts and the Influence on Glucose Conversion. *The Journal of Supercritical Fluids*, Vol.53, No.1-3, pp. 64-71, ISSN 0896-8446.

Kruse, A., Meier, D., Rimbrecht, P., & Schacht, M. (2000). Gasification of Pyrocatechol in Supercritical Water in the Presence of Potassium Hydroxide. *Industrial & Engineering Chemistry Research*, Vol.39, No.12, pp. 4842-4848, ISSN 0888-5885.

Kruse, A., & Vogel, G.H. (2010). Chemistry in Near- and Supercritical Water, *Handbook of Green Chemistry*. Wiley-Verlag. pp. 457-475.

Kulkarni, P.S., Crespo, J.G., & Afonso, C.A.M. (2008). Dioxins Sources and Current Remediation Technologies—A Review. *Environment International*, Vol.34, No.1, pp. 139-153, ISSN 0160-4120.

Kumar, S. (2011). Composting of Municipal Solid Waste. *Critical Reviews In Biotechnology*, Vol.31, No.2, pp. 112-136, ISSN 0738-8551.

Lavric, E.D., Weyten, H., De Ruyck, J., Plesu, V., & Lavric, V. (2005). Delocalized Organic Pollutant Destruction Through a Self-Sustaining Supercritical Water Oxidation Process. *Energy Conversion and Management*, Vol.46, No.9-10, pp. 1345-1364, ISSN 0196-8904.

Lavric, E.D., Weyten, H., De Ruyck, J., Plesu, V., & Lavric, V. (2006). Supercritical Water Oxidation Improvements Through Chemical Reactors Energy Integration. *Applied Thermal Engineering*, Vol.26, No.13, pp. 1385-1392, ISSN 1359-4311.

Lee, I.-G. (2011). Effect of Metal Addition to Ni/Activated Charcoal Catalyst on Gasification of Glucose in Supercritical Water. *International Journal of Hydrogen Energy*, Vol.36, No.15, pp. 8869-8877, ISSN 0360-3199.

Leiva, C., Ahumada, I., Sepúlveda, B., & Richter, P. (2010). Polychlorinated Biphenyl Behavior in Soils Amended with Biosolids. *Chemosphere*, Vol.79, No.3, pp. 273-277, ISSN 0045-6535.

Letellier, S., Marias, F., Cezac, P., & Serin, J.P. (2010). Gasification of Aqueous Biomass in Supercritical Water: A Thermodynamic Equilibrium Analysis. *The Journal of Supercritical Fluids*, Vol.51, No.3, pp. 353-361, ISSN 0896-8446.

Leusbrock, I., Metz, S.J., Rexwinkel, G., & Versteeg, G.F. (2010). The Solubilities of Phosphate and Sulfate Salts in Supercritical Water. *The Journal of Supercritical Fluids*, Vol.54, No.1, pp. 1-8, ISSN 0896-8446.

Levy, C., Watanabe, M., Aizawa, Y., Inomata, H., & Sue, K. (2006). Synthesis of Nanophased Metal Oxides in Supercritical Water: Catalysts for Biomass Conversion. *International Journal of Applied Ceramic Technology*, Vol.3, No.5, pp. 337-344, ISSN 1744-7402.

Lewis, D., Gattie, D., Novak, M., Sanchez, S., & Pumphrey, C. (2002). Interactions of Pathogens and Irritant Chemicals in Land-Applied Sewage Sludges (Biosolids). *BMC Public Health*, Vol.2, No.1, pp. 11, ISSN 1471-2458.

Li, H., Hurley, S., & Xu, C. (2011). Liquefactions of Peat in Supercritical Water with a Novel Iron Catalyst. *Fuel*, Vol.90, No.1, pp. 412-420, ISSN 0016-2361.

Lieball, K., Wellig, B., & von Rohr, P.R. (2001). Operating Conditions for a Transpiring Wall Reactor for Supercritical Water Oxidation. *Chemie Ingenieur Technik*, Vol.73, No.6, pp. 658-658, ISSN 1522-2640.

Loppinet-Serani, A., Aymonier, C., & Cansell, F. (2010). Supercritical Water for Environmental Technologies. *Journal of Chemical Technology & Biotechnology*, Vol.85, No.5, pp. 583-589, ISSN 1097-4660.

Lu, Y., Li, S., Guo, L., & Zhang, X. (2010). Hydrogen Production by Biomass Gasification in Supercritical Water over Ni/γ Al_2O_3 and Ni/CeO_2-γ Al_2O_3 Catalysts. *International Journal of Hydrogen Energy*, Vol.35, No.13, pp. 7161-7168, ISSN 0360-3199.

Machida, H., Takesue, M., & Smith Jr., R.L. (2011). Green Chemical Processes with Supercritical Fluids: Properties, Materials, Separations and Energy. *The Journal of Supercritical Fluids*, Vol.60, ISSN 0896-8446.

Maharrey, S.P., & Miller, D.R. (2001). Quartz Capillary Microreactor for Studies of Oxidation in Supercritical Water. *AIChE Journal*, Vol.47, No.5, pp. 1203-1211, ISSN 1547-5905.

Mahmood, T., & Elliott, A. (2006). A Review of Secondary Sludge Reduction Technologies for the Pulp and Paper Industry. *Water Research*, Vol.40, No.11, pp. 2093-2112, ISSN 0043-1354.

Maier, J., Gerhardt, A., & Dunnu, G. (2011). Experiences on Co-Firing Solid Recovered Fuels in the Coal Power Sector Solid Biofuels for Energy, in: Grammelis, P. (Ed.). Springer London, pp. 75-94.

Marques, R.R.N., Stüber, F., Smith, K.M., Fabregat, A., Bengoa, C., Font, J., Fortuny, A., Pullket, S., Fowler, G.D., & Graham, N.J.D. (2011). Sewage Sludge Based Catalysts for Catalytic Wet Air Oxidation of Phenol: Preparation, Characterisation and Catalytic Performance. *Applied Catalysis B: Environmental*, Vol.101, No.3-4, pp. 306-316, ISSN 0926-3373.

Marrone, P.A., Cantwell, S.D., & Dalton, D.W. (2005). SCWO System Designs for Waste Treatment: Application to Chemical Weapons Destruction. *Industrial & Engineering Chemistry Research*, Vol.44, No.24, pp. 9030-9039, ISSN 0888-5885.

Marrone, P.A., Hodes, M., Smith, K.A., & Tester, J.W. (2004). Salt Precipitation and Scale Control in Supercritical Water Oxidation, Part B: Commercial/Full-Scale Applications. *The Journal of Supercritical Fluids*, Vol.29, No.3, pp. 289-312, ISSN 0896-8446.

Marrone, P.A., & Hong, G.T. (2009). Corrosion Control Methods in Supercritical Water Oxidation and Gasification Processes. *The Journal of Supercritical Fluids*, Vol.51, No.2, pp. 83-103, ISSN 0896-8446.

Mathney, J.M.J. (2011). A Critical Review of the U.S. EPA's Risk Assessment for the Land Application of Sewage Sludge. *New Solutions: A Journal of Environmental and Occupational Health Policy*, Vol.21, No.1, pp. 43-45.

Matsumura, Y., Minowa, T., Potic, B., Kersten, S.R.A., Prins, W., van Swaaij, W.P.M., van de Beld, B., Elliott, D.C., Neuenschwander, G.G., Kruse, A., & Jerry Antal Jr., M. (2005). Biomass Gasification in Near- and Super-Critical Water: Status and Prospects. *Biomass and Bioenergy*, Vol.29, No.4, pp. 269-292, ISSN 0961-9534.

Matsumura, Y., Urase, T., Yamamoto, K., & Nunoura, T. (2002). Carbon Catalyzed Supercritical Water Oxidation of Phenol. *The Journal of Supercritical Fluids*, Vol.22, No.2, pp. 149-156, ISSN 0896-8446.

McBride, M.B. (2003). Toxic Metals in Sewage Sludge-Amended Soils: Has Promotion of Beneficial Use Discounted the Risks? *Advances in Environmental Research*, Vol.8, No.1, pp. 5-19, ISSN 1093-0191.

Minowa, T., & Inoue, S. (1999). Hydrogen Production from Biomass by Catalytic Gasification in Hot Compressed Water. *Renewable Energy*, Vol.16, No.1-4, pp. 1114-1117, ISSN 0960-1481.

Mitton, D.B., Yoon, J.H., Cline, J.A., Kim, H.S., Eliaz, N., & Latanision, R.M. (2000). Corrosion Behavior of Nickel-Based Alloys in Supercritical Water Oxidation Systems. *Industrial & Engineering Chemistry Research*, Vol.39, No.12, pp. 4689-4696, ISSN 0888-5885.

Modell, M. (1977). Reforming of Glucose and Wood at Critical Conditions of Water. *Mechanical Engineering*, Vol.99, No.10, pp. 108.

Modell, M., Gaudet, G.G., Simon, M., Hong, G.T., & Biemann, K. (1982). Supercritical Water Testing Reveals New Process Holds Promise. *Solid Wastes Management*, Vol.25, No.8, pp. 26-28.

Mooney, C. (2011). The Truth About Fracking. *Scientific American*, Vol.305, No.5, pp. 80-85, ISSN 00368733.

Münster, M., & Lund, H. (2010). Comparing Waste-to-Energy Technologies by Applying Energy System Analysis. *Waste Management*, Vol.30, No.7, pp. 1251-1263, ISSN 0956-053X.

Narayanan, C., Frouzakis, C., Boulouchos, K., Príkopský, K., Wellig, B., & Rudolf von Rohr, P. (2008). Numerical Modelling of a Supercritical Water Oxidation Reactor Containing a Hydrothermal Flame. *The Journal of Supercritical Fluids*, Vol.46, No.2, pp. 149-155, ISSN 0896-8446.

NASNRC. (1996). *Use of Reclaimed Water and Sludge in Food Crop Production*, National Academies of Science, National Research Council, pp. 1-178.

NASNRC. (2002). *Biosolids Applied to Land*, National Academies of Science, National Research Council.

Nature. (2008). Stuck in the Mud. *Nature*, Vol.453, No.7193, pp. 258.

Noyori, R. (1999). Supercritical Fluids: Introduction. *Chemical Reviews*, Vol.99, No.2, pp. 353-354, ISSN 0009-2665.

Oleszczuk, P. (2008). The Toxicity of Composts from Sewage Sludges Evaluated by the Direct Contact Tests Phytotoxkit and Ostracodtoxkit. *Waste Management*, Vol.28, No.9, pp. 1645-1653, ISSN 0956-053X.

Onwudili, J.A., & Williams, P.T. (2006). Flameless Incineration of Pyrene Under Sub-Critical and Supercritical Water Conditions. *Fuel*, Vol.85, No.1, pp. 75-83, ISSN 0016-2361.

Pathak, A., Dastidar, M.G., & Sreekrishnan, T.R. (2009). Bioleaching of Heavy Metals from Sewage Sludge: A Review. *Journal of Environmental Management*, Vol.90, No.8, pp. 2343-2353, ISSN 0301-4797.

Penninger, J.M.L., & Rep, M. (2006). Reforming of Aqueous Wood Pyrolysis Condensate in Supercritical Water. *International Journal of Hydrogen Energy*, Vol.31, No.11, pp. 1597-1606, ISSN 0360-3199.

Peter, K. (2004). Corrosion in High-Temperature and Supercritical Water and Aqueous Solutions: A Review. *The Journal of Supercritical Fluids*, Vol.29, No.1–2, pp. 1-29, ISSN 0896-8446.

Peterson, A.A., Tester, J.W., & Vogel, F. (2010). Water-in-Water Tracer Studies of Supercritical-Water Reversing Jets Using Neutron Radiography. *The Journal of Supercritical Fluids*, Vol.54, No.2, pp. 250-257, ISSN 0896-8446.

Peterson, A.A., Vogel, F., Lachance, R.P., Froling, M., Antal, J.M.J., & Tester, J.W. (2008a). Thermochemical Biofuel Production in Hydrothermal Media: A Review of Sub- and Supercritical Water Technologies. *Energy & Environmental Science*, Vol.1, No.1, pp. 32-65, ISSN 1754-5692.

Peterson, A.A., Vontobel, P., Vogel, F., & Tester, J.W. (2008b). In Situ Visualization of the Performance of a Supercritical-Water Salt Separator Using Neutron Radiography. *The Journal of Supercritical Fluids*, Vol.43, No.3, pp. 490-499, ISSN 0896-8446.

Petrovic', M., Gonzalez, S., & Barcelo, D. (2003). Analysis and Removal of Emerging Contaminants in Wastewater and Drinking Water. *TrAC Trends in Analytical Chemistry*, Vol.22, No.10, pp. 685-696, ISSN 0165-9936.

Phillips, J.A. (1998). *Managing America's Solid Waste*. U.S. Department of Energy, National Technical Information Service (NTIS), Golden, Colorado, p. 162.

Príkopský, K., Wellig, B., & von Rohr, P.R. (2007). SCWO of Salt Containing Artificial Wastewater Using a Transpiring-Wall Reactor: Experimental Results. *The Journal of Supercritical Fluids*, Vol.40, No.2, pp. 246-257, ISSN 0896-8446.

Qi, Y., Yue, Q., Han, S., Yue, M., Gao, B., Yu, H., & Shao, T. (2010). Preparation and Mechanism of Ultra-Lightweight Ceramics Produced from Sewage Sludge. *Journal of Hazardous Materials*, Vol.176, No.1-3, pp. 76-84, ISSN 0304-3894.

Reddy, C.S., Dorn, C.R., Lamphere, D.N., & Powers, J.D. (1985). Municipal Sewage Sludge Application on Ohio Farms: Tissue Metal Residues and Infections. *Environmental Research*, Vol.38, No.2, pp. 360-376, ISSN 0013-9351.

Reilly, M. (2001). The Case Against Land Application of Sewage Sludge Pathogens. *Canadian Journal of Infectious Diseases*, Vol.12, No.4, pp. 205-207.

Rice, S.F., Hunter, T.B., Rydén, Å.C., & Hanush, R.G. (1996). Raman Spectroscopic Measurement of Oxidation in Supercritical Water. 1. Conversion of Methanol to Formaldehyde. *Industrial & Engineering Chemistry Research*, Vol.35, No.7, pp. 2161-2171, ISSN 0888-5885.

Rönnlund, I., Myréen, L., Lundqvist, K., Ahlbeck, J., & Westerlund, T. (2011). Waste to Energy by Industrially Integrated Supercritical Water Gasification—Effects of Alkali Salts in Residual By-Products from the Pulp and Paper Industry. *Energy*, Vol.36, No.4, pp. 2151-2163, ISSN 0360-5442.

Roy, M.M., Dutta, A., Corscadden, K., Havard, P., & Dickie, L. (2011). Review of Biosolids Management Options and Co-Incineration of a Biosolid-Derived Fuel. *Waste Management*, Vol.31, No.11, pp. 2228-2235, ISSN 0956-053X.

Rulkens, W. (2008). Sewage Sludge as a Biomass Resource for the Production of Energy: Overview and Assessment of the Various Options. *Energy & Fuels*, Vol.22, No.1, pp. 9-15, ISSN 0887-0624.

Sánchez-Martín, M.J., García-Delgado, M., Lorenzo, L.F., Rodríguez-Cruz, M.S., & Arienzo, M. (2007). Heavy Metals in Sewage Sludge Amended Soils Determined by Sequential Extractions as a Function of Incubation Time of Soils. *Geoderma*, Vol.142, No.3-4, pp. 262-273, ISSN 0016-7061.

Santos, L.H.M.L.M., Araújo, A.N., Fachini, A., Pena, A., Delerue-Matos, C., & Montenegro, M.C.B.S.M. (2010). Ecotoxicological Aspects Related to the Presence of Pharmaceuticals in the Aquatic Environment. *Journal of Hazardous Materials*, Vol.175, No.1-3, pp. 45-95, ISSN 0304-3894.

Sasaki, M., Fang, Z., Fukushima, Y., Adschiri, T., & Arai, K. (2000). Dissolution and Hydrolysis of Cellulose in Subcritical and Supercritical Water. *Industrial & Engineering Chemistry Research*, Vol.39, No.8, pp. 2883-2890, ISSN 0888-5885.

Sato, T., Inda, K., & Itoh, N. (2011). Gasification of Bean Curd Refuse with Carbon Supported Noble Metal Catalysts in Supercritical Water. *Biomass and Bioenergy*, Vol.35, No.3, pp. 1245-1251, ISSN 0961-9534.

Sato, T., Kurosawa, S., Smith Jr, R.L., Adschiri, T., & Arai, K. (2004). Water Gas Shift Reaction Kinetics Under Noncatalytic Conditions in Supercritical Water. *The Journal of Supercritical Fluids*, Vol.29, No.1-2, pp. 113-119, ISSN 0896-8446.

Sato, T., Osada, M., Watanabe, M., Shirai, M., & Arai, K. (2003). Gasification of Alkylphenols with Supported Noble Metal Catalysts in Supercritical Water. *Industrial & Engineering Chemistry Research*, Vol.42, No.19, pp. 4277-4282, ISSN 0888-5885.

Saunders, S.E., Bartelt-Hunt, S.L., & Bartz, J.C. (2008). Prions in the Environment—Occurrence, Fate and Mitigation. *Prion*, Vol.2, No.4, pp. 162-169.

Savage, P.E. (2009). A Perspective on Catalysis in Sub- and Supercritical Water. *The Journal of Supercritical Fluids*, Vol.47, No.3, pp. 407-414, ISSN 0896-8446.

SCFI. (2012). *What Is Super Critical Water Oxidation?* http://www.scfi.eu/products/.

Schmieder, H., Abeln, J., Boukis, N., Dinjus, E., Kruse, A., Kluth, M., Petrich, G., Sadri, E., & Schacht, M. (2000). Hydrothermal Gasification of Biomass and Organic Wastes. *The Journal of Supercritical Fluids*, Vol.17, No.2, pp. 145-153, ISSN 0896-8446.

Schubert, M., Aubert, J., Müller, J.B., & Vogel, F. (2012). Continuous Salt Precipitation and Separation from Supercritical Water. Part 3: Interesting Effects in Processing Type 2 Salt Mixtures. *The Journal of Supercritical Fluids*, Vol.61, pp. 45-54, ISSN 0896-8446.

Schubert, M., Regler, J.W., & Vogel, F. (2010a). Continuous Salt Precipitation and Separation from Supercritical Water. Part 1: Type 1 Salts. *The Journal of Supercritical Fluids*, Vol.52, No.1, pp. 99-112, ISSN 0896-8446.

Schubert, M., Regler, J.W., & Vogel, F. (2010b). Continuous Salt Precipitation and Separation from Supercritical Water. Part 2. Type 2 Salts and Mixtures of Two Salts. *The Journal of Supercritical Fluids*, Vol.52, No.1, pp. 113-124, ISSN 0896-8446.

Sınag, A., Kruse, A., & Rathert, J. (2004). Influence of the Heating Rate and the Type of Catalyst on the Formation of Key Intermediates and on the Generation of Gases During Hydropyrolysis of Glucose in Supercritical Water in a Batch Reactor. *Industrial & Engineering Chemistry Research*, Vol.43, No.2, pp. 502-508, ISSN 0888-5885.

Sınag, A., Yumak, T., Balci, V., & Kruse, A. (2011). Catalytic Hydrothermal Conversion of Cellulose Over SnO_2 and ZnO Nanoparticle Catalysts. *The Journal of Supercritical Fluids*, Vol.56, No.2, pp. 179-185, ISSN 0896-8446.

Sipma, J., Osuna, B., Collado, N., Monclús, H., Ferrero, G., Comas, J., & Rodriguez-Roda, I. (2010). Comparison of Removal of Pharmaceuticals in MBR and Activated Sludge Systems. *Desalination*, Vol.250, No.2, pp. 653-659, ISSN 0011-9164.

Slavin, T.J., & Oleson, M.W. (1991). Technology Tradeoffs Related to Advanced Mission Waste Processing. *Waste Management & Research*, Vol.9, No.5, pp. 401-414, ISSN 0734-242X.

Sloan, D.S., Pelletier, R.A., & Modell, M. (2008). Sludge Management in the City of Orlando—It's Supercritical! *Florida Water Resources Journal*, No.June, pp. 46-54.

Smith, C.B., Booth, C.J., & Pederson, J.A. (2011). Fate of Prions in Soil: A Review. *Journal of Environmental Quality*, Vol.40, No.2, pp. 449-461.

Snyder, C. (2005). The Dirty Work of Promoting "Recycling" of America's Sewage Sludge. *International Journal Of Occupational And Environmental Health*, Vol.11, No.4, pp. 415-427.

Soria, J.A., McDonald, A.G., & Shook, S.R. (2008). Wood Solubilization and Depolymerization Using Supercritical Methanol. Part 1: Process Optimization and Analysis of Methanol Insoluble Components (Bio-Char). *Holzforschung*, Vol.62, No.4, pp. 402–408.

Stasinakis, A.S., Gatidou, G., Mamais, D., Thomaidis, N.S., & Lekkas, T.D. (2008). Occurrence and Fate of Endocrine Disrupters in Greek Sewage Treatment Plants. *Water Research*, Vol.42, No.6-7, pp. 1796-1804, ISSN 0043-1354.

Straub, T.M., Pepper, I.L., & Gerba, C.P. (1993). Hazards from Pathogenic Microorganisms in Land-Disposed Sewage Sludge. *Reviews of Environmental Contamination & Toxicology*, Vol.132, pp. 55-91.

Susanti, R.F., Veriansyah, B., Kim, J.-D., Kim, J., & Lee, Y.-W. (2010). Continuous Supercritical Water Gasification of Isooctane: A Promising Reactor Design. *International Journal of Hydrogen Energy*, Vol.35, No.5, pp. 1957-1970, ISSN 0360-3199.

Svanström, M., Fröling, M., Modell, M., Peters, W.A., & Tester, J. (2004). Environmental Assessment of Supercritical Water Oxidation of Sewage Sludge. *Resources, Conservation and Recycling*, Vol.41, No.4, pp. 321-338, ISSN 0921-3449.

Svanström, M., Modell, M., & Tester, J. (2005). Direct Energy Recovery from Primary and Secondary Sludges by Supercritical Water Oxidation. *ChemInform*, Vol.36, No.19, pp. 201-208, ISSN 1522-2667.

Tan, L., Allen, T.R., & Yang, Y. (2011). Corrosion of Austenitic Stainless Steels and Nickel-Base Alloys in Supercritical Water and Novel Control Methods, *Green Corrosion Chemistry and Engineering*. Wiley-VCH Verlag GmbH & Co. KGaA, pp. 211-242.

Tollefson, J. (2008). Raking through Sludge Exposes a Stink. *Nature*, Vol.453, No.15 May, pp. 262.

Toor, S.S., Rosendahl, L., & Rudolf, A. (2011). Hydrothermal Liquefaction of Biomass: A Review of Subcritical Water Technologies. *Energy*, Vol.36, No.5, pp. 2328-2342, ISSN 0360-5442.

Tsai, W.-T., Chang, J.-H., Hsien, K.-J., & Chang, Y.-M. (2009). Production of Pyrolytic Liquids from Industrial Sewage Sludges in an Induction-Heating Reactor. *Bioresource Technology*, Vol.100, No.1, pp. 406-412, ISSN 0960-8524.

Turovskiy, I.S., & Mathai, P.K. (2005). Frontmatter, *Wastewater Sludge Processing*. John Wiley & Sons, Inc., pp. i-xii.

USDA. (2011). *Guidance: Allowance of Green Waste in Organic Production Systems*. United States Department of Agriculture Agricultural Marketing Service National Organic Program, July 22, 2011.

USEPA. (1999a). *Biosolids Generation, Use, and Disposal in the United States*. United States Environmental Protection Agency, Municipal and Industrial Solid Waste Division, Office of Solid Waste.

USEPA. (1999b). *Diagnostic Evaluation of Sludge Facilities for Messerly Wastewater Treatment Plant Augusta, Georgia*. Enforcement and Investigations Branch, July 01, 1999.

USEPA. (2000). *Biosolids Management and Enforcement*. Office of Inspector General, March 20, 2000.

USEPA. (2002a). *Biosolids Technology Fact Sheet: Use of Composting for Biosolids Management*, Office of Water, September 2002.

USEPA. (2002b). *Land Application of Biosolids*. Office of Inspector General, March 28, 2000.

USEPA. (2003). *Environmental Regulations and Technology: Control of Pathogens and Vector Attraction in Sewage Sludge*. United States Environmental Protection Agency, Office of Research and Development National Risk Management Research Laboratory Center for Environmental Research Information Cincinnati, OH 45268.

USEPA. (2009). *Biosolids: Targeted National Sewage Sludge Survey Report*. United States Environmental Protection Agency, Municipal and Industrial Solid Waste Division, Office of Solid Waste.

Vadillo, V., García-Jarana, M.B., Sánchez-Oneto, J., Portela, J.R., & de la Ossa, E.J.M. (2011). Supercritical Water Oxidation of Flammable Industrial Wastewaters: Economic Perspectives of an Industrial Plant. *Journal of Chemical Technology & Biotechnology*, Vol.86, No.8, pp. 1049-1057, ISSN 1097-4660.

van Rossum, G., Potic, B., Kersten, S.R.A., & van Swaaij, W.P.M. (2009). Catalytic Gasification of Dry and Wet Biomass. *Catalysis Today*, Vol.145, No.1-2, pp. 10-18, ISSN 0920-5861.

Veriansyah, B., & Kim, J.-D. (2007). Supercritical Water Oxidation for the Destruction of Toxic Organic Wastewaters: A Review. *Journal of Environmental Sciences*, Vol.19, No.5, pp. 513-522, ISSN 1001-0742.

Veriansyah, B., Kim, J.-D., & Lee, J.-C. (2007). Destruction of Chemical Agent Simulants in a Supercritical Water Oxidation Bench-Scale Reactor. *Journal of Hazardous Materials*, Vol.147, No.1-2, pp. 8-14, ISSN 0304-3894.

Veriansyah, B., Kim, J.-D., & Lee, J.-C. (2009). A Double Wall Reactor for Supercritical Water Oxidation: Experimental Results on Corrosive Sulfur Mustard Simulant Oxidation. *Journal of Industrial and Engineering Chemistry*, Vol.15, No.2, pp. 153-156, ISSN 1226-086X.

Veriansyah, B., Kim, J.-D., Lee, J.-C., & Lee, Y.-W. (2005). OPA Oxidation Rates in Supercritical Water. *Journal of Hazardous Materials*, Vol.124, No.1-3, pp. 119-124, ISSN 0304-3894.

Verlicchi, P., Galletti, A., Petrovic, M., & Barceló, D. (2010). Hospital Effluents as a Source of Emerging Pollutants: An Overview of Micropollutants and Sustainable Treatment Options. *Journal of Hydrology*, Vol.389, No.3-4, pp. 416-428, ISSN 0022-1694.

Villar, A., Arribas, J., & Parrondo, J. (2012). Waste-to-Energy Technologies in Continuous Process Industries. *Clean Technologies and Environmental Policy*, Vol.14, No.1, pp. 29-39, ISSN 1618-954X.

Vogel, F., Blanchard, J.L.D., Marrone, P.A., Rice, S.F., Webley, P.A., Peters, W.A., Smith, K.A., & Tester, J.W. (2005). Critical Review of Kinetic Data for the Oxidation of Methanol in Supercritical Water. *The Journal of Supercritical Fluids*, Vol.34, No.3, pp. 249-286, ISSN 0896-8446.

Wang, S., Guo, Y., Chen, C., Zhang, J., Gong, Y., & Wang, Y. (2011). Supercritical Water Oxidation of Landfill Leachate. *Waste Management*, Vol.31, No.9–10, pp. 2027-2035, ISSN 0956-053X.

Watanabe, M., Inomata, H., Osada, M., Sato, T., Adschiri, T., & Arai, K. (2003). Catalytic Effects of NaOH and ZrO_2 for Partial Oxidative Gasification of N-Hexadecane and Lignin in Supercritical Water. *Fuel*, Vol.82, pp. 545-552.

Weismantel, G. (1996). Supercritical Water Oxidation Treats Toxic Organics in Sludge. *Environmental Technology*, No.September/October, pp. 30-34.

Weismantel, G. (2001). What's New in Sewage Sludge Separation and Processing? *Filtration and Separation*, Vol.38, No.5, pp. 22-25, ISSN 0015-1882.

Weiss-Hortala, E., Kruse, A., Ceccarelli, C., & Barna, R. (2010). Influence of Phenol on Glucose Degradation During Supercritical Water Gasification. *The Journal of Supercritical Fluids*, Vol.53, No.1–3, pp. 42-47, ISSN 0896-8446.

Wellig, B., Lieball, K., & Rudolf von Rohr, P. (2005). Operating Characteristics of a Transpiring-Wall SCWO Reactor with a Hydrothermal Flame as Internal Heat Source. *The Journal of Supercritical Fluids*, Vol.34, No.1, pp. 35-50, ISSN 0896-8446.

Wellig, B., Weber, M., Lieball, K., Príkopský, K., & von Rohr, P.R. (2009). Hydrothermal Methanol Diffusion Flame as Internal Heat Source in a SCWO Reactor. *The Journal of Supercritical Fluids*, Vol.49, No.1, pp. 59-70, ISSN 0896-8446.

White, J.E., Catallo, W.J., & Legendre, B.L. (2011). Biomass Pyrolysis Kinetics: A Comparative Critical Review with Relevant Agricultural Residue Case Studies. *Journal of Analytical and Applied Pyrolysis*, Vol.91, No.1, pp. 1-33, ISSN 0165-2370.

Xu, D., Wang, S., Hu, X., Chen, C., Zhang, Q., & Gong, Y. (2009). Catalytic Gasification of Glycine and Glycerol in Supercritical Water. *International Journal of Hydrogen Energy*, Vol.34, No.13, pp. 5357-5364, ISSN 0360-3199.

Xu, D., Wang, S., Tang, X., Gong, Y., Guo, Y., Wang, Y., & Zhang, J. (2012). Design of the First Pilot Scale Plant of China for Supercritical Water Oxidation of Sewage Sludge. *Chemical Engineering Research and Design*, Vol.90, No.2, pp. 288–297, ISSN 0263-8762.

Xu, D.H., Wang, S.Z., Gong, Y.M., Guo, Y., Tang, X.Y., & Ma, H.H. (2010). A Novel Concept Reactor Design for Preventing Salt Deposition in Supercritical Water. *Chemical Engineering Research and Design*, Vol.88, No.11, pp. 1515-1522, ISSN 0263-8762.

Xu, L., Brilman, D.W.F., Withag, J.A.M., Brem, G., & Kersten, S. (2011). Assessment of a Dry and a Wet Route for the Production of Biofuels from Microalgae: Energy Balance Analysis. *Bioresource Technology*, Vol.102, No.8, pp. 5113-5122, ISSN 0960-8524.

Xu, X., Matsumura, Y., Stenberg, J., & Antal, M.J. (1996). Carbon-Catalyzed Gasification of Organic Feedstocks in Supercritical Water. *Industrial & Engineering Chemistry Research*, Vol.35, No.8, pp. 2522-2530, ISSN 0888-5885.

Yan, B., Wu, J., Xie, C., He, F., & Wei, C. (2009). Supercritical Water Gasification with Ni/ZrO_2 Catalyst for Hydrogen Production from Model Wastewater of Polyethylene Glycol. *The Journal of Supercritical Fluids*, Vol.50, No.2, pp. 155-161, ISSN 0896-8446.

Yoshida, T., & Matsumura, Y. (2009). Reactor Development for Supercritical Water Gasification of 4.9 wt% Glucose Solution at 673 K by Using Computational Fluid Dynamics. *Industrial & Engineering Chemistry Research*, Vol.48, No.18, pp. 8381-8386, ISSN 0888-5885.

Yoshida, T., & Oshima, Y. (2004). Partial Oxidative and Catalytic Biomass Gasification in Supercritical Water: A Promising Flow Reactor System. *Industrial & Engineering Chemistry Research*, Vol.43, No.15, pp. 4097-4104, ISSN 0888-5885.

Yoshida, T., Oshima, Y., & Matsumura, Y. (2004). Gasification of Biomass Model Compounds and Real Biomass in Supercritical Water. *Biomass and Bioenergy*, Vol.26, No.1, pp. 71-78, ISSN 0961-9534.

Yoshida, Y., Dowaki, K., Matsumura, Y., Matsuhashi, R., Li, D., Ishitani, H., & Komiyama, H. (2003). Comprehensive Comparison of Efficiency and CO_2 Emissions between Biomass Energy Conversion Technologies—Position of Supercritical Water Gasification in Biomass Technologies. *Biomass and Bioenergy*, Vol.25, No.3, pp. 257-272, ISSN 0961-9534.

Youssef, E.A., Chowdhury, M.B.I., Nakhla, G., & Charpentier, P. (2010a). Effect of Nickel Loading on Hydrogen Production and Chemical Oxygen Demand (COD) Destruction from Glucose Oxidation and Gasification in Supercritical Water. *International Journal of Hydrogen Energy*, Vol.35, No.10, pp. 5034-5042, ISSN 0360-3199.

Youssef, E.A., Elbeshbishy, E., Hafez, H., Nakhla, G., & Charpentier, P. (2010b). Sequential Supercritical Water Gasification and Partial Oxidation of Hog Manure. *International Journal of Hydrogen Energy*, Vol.35, No.21, pp. 11756-11767, ISSN 0360-3199.

Youssef, E.A., Nakhla, G., & Charpentier, P.A. (2011). Oleic Acid Gasification Over Supported Metal Catalysts in Supercritical Water: Hydrogen Production and Product Distribution. *International Journal of Hydrogen Energy*, Vol.36, No.8, pp. 4830-4842, ISSN 0360-3199.

Yuan, P.-Q., Cheng, Z.-M., Zhang, X.-Y., & Yuan, W.-K. (2006). Catalytic Denitrogenation of Hydrocarbons Through Partial Oxidation in Supercritical Water. *Fuel*, Vol.85, No.3, pp. 367-373, ISSN 0016-2361.

Zhang, L., Champagne, P., & Xu, C. (2011). Supercritical Water Gasification of an Aqueous By-Product from Biomass Hydrothermal Liquefaction with Novel Ru Modified Ni Catalysts. *Bioresource Technology*, Vol.102, No.17, pp. 8279-8287, ISSN 0960-8524.

Zhang, L., Xu, C., & Champagne, P. (2010). Overview of Recent Advances in Thermo-Chemical Conversion of Biomass. *Energy Conversion and Management*, Vol.51, No.5, pp. 969-982, ISSN 0196-8904.

Zorita, S., Mårtensson, L., & Mathiasson, L. (2009). Occurrence and Removal of Pharmaceuticals in a Municipal Sewage Treatment System in the South of Sweden. *Science of The Total Environment*, Vol.407, No.8, pp. 2760-2770, ISSN 0048-9697.

Chapter 7

Thermal Plasma Gasification of Municipal Solid Waste (MSW)

Youngchul Byun, Moohyun Cho, Soon-Mo Hwang and Jaewoo Chung

Additional information is available at the end of the chapter

http://dx.doi.org/10.5772/48537

1. Introduction

Rapid economic development has led to an annual increase in municipal solid waste (MSW) production. According to the US Environmental Protection Agency (US EPA), MSW generation has increased by a factor of 2.6 since 1960 [1]. The US EPA endorsed the concept of integrated waste management that could be tailored to fit particular community's needs. Sustainable and successful treatment of MSW should be safe, effective, and environmentally friendly. The primary components of the philosophy are (a) source reduction including reuse of products and on-site composting of yard trimmings, (b) recycling, including off-site (or community) composting, (c) combustion with energy recovery, and (d) disposal through landfill. Among them, landfill has been the practice most widely adopted. There are two main drawbacks of landfill. One is that surrounding areas of landfills are often heavily polluted since it is difficult to keep dangerous chemicals from leaching out into the surrounding land [2]. The other is that landfill can increase chances of global warming by releasing CH_4, which is 20 times more dangerous as a greenhouse gas than CO_2. Therefore, we must find a more environmentally friendly alternative to treat MSW.

A plasma is defined as a quasineutral gas of charged and neutral particles which exhibits collective behavior [3]. Plasma can be classified into non-thermal and thermal plasmas according to the degree of ionization and the difference of temperature between heavy particles and electrons [4, 5]. Thermal plasma can be characterized by approximate equality between heavy particle and electron temperatures and have numerous advantages including high temperature and high energy density [6]. Electrically generated thermal plasma can reach temperature of ~10,000 ^{o}C or more, whereas only an upper temperature limit of 2,000 ^{o}C can be achieved by burning fossil fuels [7]. For this reason, thermal plasma has been traditionally used in high temperature and large enthalpy processes [8-11].

Thermal plasma technology has been applied in various industrial applications such as cutting, welding, spraying, metallurgy, mass spectroscopy, nano-sized particle synthesis, powder spheroidization, and waste treatment [12-15]. Over the past decade, thermal plasma process has also been regarded as a viable alternative to treat highly toxic wastes, such as air pollutant control (APC) residues, radioactive, and medical wastes [16-25]. It has also been demonstrated that the thermal plasma process is environmentally friendly, producing only inert slag and minimal air pollutants that are well within regional regulations. Recently, a thermal plasma process for a gasification of MSW has been planned and constructed as a pilot program in commercial plants. The thermal plasma process employs extremely high temperatures in the absence or near-absence of O_2 to treat MSW containing organics and other materials. The MSW is dissociated into its constituent chemical elements, transformed into other materials some of which are valuable products. The organic components are transformed into syngas, which is mainly composed of H_2 and CO and inorganic components are vitrified into inert glass-like slag.

We constructed thermal plasma plants for the recovery of high purity H_2 (> 99.99%) from paper mill waste at 3 TPD (ton/day) and the gasification of MSW at 10 TPD [26, 27]. For the recovery of high purity H_2, gases emitted from a gasification furnace equipped with a non-transferred thermal plasma torch were purified using a bag-filter and wet scrubber. Thereafter, the gases, which contained syngas, were introduced into a H_2 recovery system, consisting largely of a water gas shift (WGS) unit for the conversion of CO to H_2 and a pressure swing adsorption (PSA) unit for the separation and purification of H_2. It was successfully demonstrated that the thermal plasma process for solid wastes gasification, combined with the WGS and PSA, produced high purity H_2 (20 Nm^3/hr (400 H_2-Nm^3/ton), up to 99.99%) using a plasma torch that used 1.6 MWh/ton of electricity. For the treatment of MSW, we developed a gasification commercial plant for the direct treatment of municipal solid waste (MSW) with a capacity of 10 TPD, using an integrated furnace equipped with two non-transferred thermal plasma torches. It was successfully demonstrated that the thermal plasma process converted MSW into innocuous slag, with much lower levels of environmental air pollutant emissions and the syngas (287 Nm^3/ton for H_2 and 395 Nm^3/ton for CO), using 1.14 MWh/ton of electricity (thermal plasma torch (0.817 MWh/ton) + utilities (0.322 MWh/ton)) and 7.37 Nm^3/ton of liquefied petroleum gas (LPG).

Such a plant is currently operating in Cheongsong, Korea. The 3.5 years' worth of data obtained from this plant has given us the insight into the economics and design parameters for extending capacity to 100 TPD. In this chapter, we describe the past operational performances of 10 TPD thermal plasma gasification plant for MSW treatment, evaluate the economics, and suggest the design parameters for extending capacity to 100 TPD with brief discussion on recent achievements in thermal plasma technology for the treatment of solid wastes on the basis of selected scientific and technical literatures.

2. Characteristics of thermal plasma process for the treatment of solid wastes

Thermal plasma for wastes treatment has received great attention recently to meet the contemporary needs to solve problems with increasing environmental pollutions.

Compared with commonly used combustion methods for waste treatment, thermal plasma provides the following advantages; (1) high energy density and temperatures, and the correspondingly fast reaction times, offer the potential for a large throughput with a small furnace. (2) High heat flux densities at the furnace boundaries lead to fast attainment of steady state conditions. This allows rapid start-up and shutdown times compared with other thermal treatments such as incineration. (3) Only a small amount of oxidant is necessary to generate syngas, therefore, the gas volume produced is much smaller than with conventional combustion processes and so is easier and less expensive to manage. These characteristics make thermal plasma process an ideal alternative to conventional methods of solid waste treatment.

There are three kinds of processes inside the thermal plasma furnace for solid waste treatment. First is pyrolysis (without O_2) of gaseous, liquid, and solid waste in a thermal plasma furnace with plasma torches. Second is gasification (O_2-starved) of solid waste containing organic compounds to produce syngas (H_2 + CO). Last is vitrification of solid wastes by transferred, non-transferred, or hybrid arc plasma torch according to electric conductivity of substrate. Processes being considered importantly for the treatment of solid wastes are gasification and vitrification; this is due to the energy recovery and volume reduction. The gasification process is an old industrial process that uses heat in an O_2-starved environment to break down carbon based materials into fuel gases. It is closely related to combustion and pyrolysis, but there are important distinctions between them. Gasification is similar to starved-air burning because O_2 is strictly controlled and limited so that the feedstock is not allowed to be completely burned as heat is applied. Instead of combusting, the raw materials go through the progress of pyrolysis, producing char and tar. The char and tar are broken down into syngas, mainly composed of H_2 and CO, as the gasification process continues. The global gasification reaction is written as follows; waste material is described by its ultimate analysis (CHxOy) [28]:

$$CH_xO_y + wH_2O + mO_2 + 3.76mN_2 \rightarrow aH_2 + bCO + cCO_2 + dH_2O + eCH_4 + fN_2 + gC \quad (1)$$

where w is the amount of water per mole of waste material, m is the amount of O_2 per mole of waste, a, b, c, d, e, f and g are the coefficients of the gaseous products and soot (all stoichiometric coefficients in moles). This overall equation has also been used for the calculation of chemical equilibrium occurring in the thermal plasma gasification with input electrical energy [28]. The concentrations of each gas have been decided depending on the amount of injected O_2, H_2O, and input thermal plasma enthalpy. The detailed main reactions are as follows [28, 29]:

$$CH_4 + H_2O \rightarrow CO + 3H_2 \text{ (}CH_4\text{ decomposition-endothermic)} \quad (2)$$

$$CO + H_2O \rightarrow CO_2 + H_2 \text{ (water gas shift reaction-exthermic)} \quad (3)$$

$$C + H_2O \rightarrow CO + H_2 \text{ (Heterogeneous water gas shift reaction-endothermic)} \quad (4)$$

$$C + CO_2 \rightarrow 2CO \text{ (Boudouard equilibrium-endothermic)} \quad (5)$$

$$2C + O_2 \rightarrow CO \quad (6)$$

The H_2 and CO generated during the gasification process can be a fuel source. Therefore, plasma gasification process has been combined with many other technologies to recover energy from the syngas. Representatives include a combination with the integrated gasification combined cycle (IGCC), fuel cells, and the production of high purity H_2 [26, 30, 31].

Heberlein and Murphy [32] described that a typical plasma system for the treatment of solid wastes consists of (a) the plasma furnace, with a metal and the slag collection at the bottom that periodically tapped and cast into some usable form and power supply, cooling water supplies, gas supplies, and control and data acquisition equipment; (b) a secondary combustion chamber for allowing sufficient residence time at elevated temperatures to assure complete reactions and gasification of soot; this secondary combustion chamber can be fired either by a burner or by a low power non-transferred plasma torch; (c) depending on the waste, a quenching chamber (usually water quencher) to avoid formation of dioxins and furans; (d) a cyclone or bag-house for particulate removal; (e) a scrubber for eliminating acidic gases; (f) if necessary a hydrogen sulfide absorber; (g) high efficiency filters or precipitators for small particulate removal; (h) an activated carbon filter for removal of heavy metals; (i) finally a fan for generating sub-atmospheric pressure in the entire installation. Additionally, various forms of waste preparation and feeding systems have to be integrated with the furnace. Therefore, to operate such a plant, many careful considerations are necessary. Figure 1 shows the necessary technologies boundary. Initially, total process control and safety management systems are necessary. A thermal plasma plant consists of a number of unit processes. To make each process connect with others efficiently, a total control system is essential. In addition, a safety management system is also necessary to protect workers. Based on these two fundamental systems, solid waste pretreatment, plasma torch and furnace, waste heat recovery, power generation, air pollutant control, and syngas utilization systems are necessary.

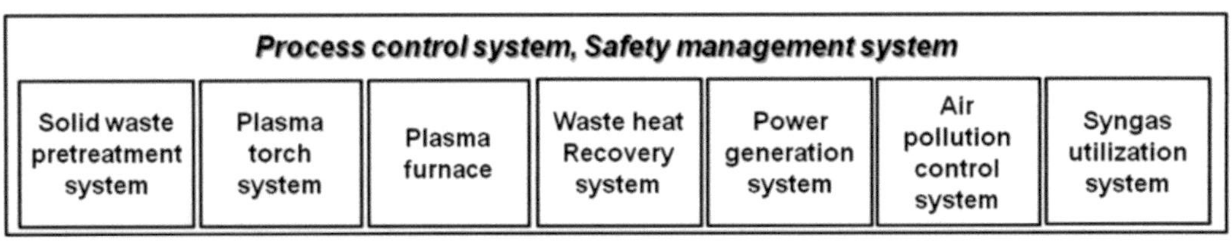

Figure 1. Technologies boundary for the thermal plasma gasification plant for solid wastes treatment

The most important part for the mentioned specific systems is the gasification furnace equipped with a thermal plasma generator. Direct current (DC) arc plasma has been mainly used for the treatment of MSW. It has generated through torch-shaped plasma generator. The plasma torch generates and maintains a gaseous electrical conducting element (the plasma) and uses the resistance of the plasma to convert electricity into heat energy. The use of plasma torches is not new. Westinghouse (now a subsidiary of AlterNRG) reportedly began building plasma torches for the National Aeronautics and Space Administration (NASA) in conjunction with the Apollo Space Program as long ago as the 1960s for the purpose of testing heat shields for spacecraft [33]. In DC arc plasma, the plasma state is maintained between two electrodes of the plasma torch by electrical and mechanical stabilization that are built into the plasma torch hardware. Two arc attachment points are required to generate a plasma column: one attachment point at the solid-gas interface at the cathode electrode and another at the gas-solid interface at the anode electrode [34]. The electrodes are separated by an insulator to preserve the potential difference between them. Very high temperatures are encountered at the attachment points of the plasma that exceed the melting temperature of any electrode material. Therefore, the vaporization of electrode materials at the attachment points is accepted and water cooling is used to minimize the rate of vaporization of electrode materials to increase the lifetime of electrode. Arc plasma torches can be classified as rod type and well type cathodes according to electrode geometry [35]. Thermal plasma torches can be also divided into transferred and non-transferred types depending on whether or not arc attaches onto a substrate directly. Tailored thermal plasma characteristics such as input power level, plasma flame volume, temperature field, velocity distribution, and chemical composition can be achieved for each application.

Generally, the plasma gasification furnace is a type of vertical shaft conventionally used in the foundry industry for the re-melting of scrap iron and steel. Solid wastes have been injected into the top of furnace. The furnace is internally lined with the appropriate refractory to withstand high internal temperatures and the corrosive operating conditions within the furnace. The plasma torches were installed in the bottom of the furnace to enhance the melting of inorganic materials contained in solid wastes. The preliminary size of the standard plasma gasification furnace, for example AlterNRG is 9.7 m outer diameter at its widest point and 19 m overall height [36]. Recently, Solenagroup designed a new furnace concept with a plenum zone (residence time ~ 2 sec) [37], which is a secondary combustion chamber for allowing sufficient residence time at elevated temperatures to assure complete reactions and gasification of soot. Solid wastes are injected into the sides of furnace. In both AlterNRG and Solenagroup's furnaces, coke is added with the solid wastes, which is consumed in the furnace at a much lower rate than the waste material due to its low reactivity, and forms a bed onto which the MSW falls and is quickly gasified. The coke bed also provides voids for molten flux, slag, and metal to flow downward as the gas flows upward. The coke also reacts with the incoming O_2 to provide heat for the gasification of the feed materials. Its role is similar with that of coke in a blast furnace of a steel-making plant.

The components of the other processes for the thermal plasma gasification are shown schematically in Figure 2. To gasify solid wastes, they must be properly treated before adding them into the thermal plasma furnace. The pretreatment process is typically composed of sorting and crushing units like a conventional incineration facility. The pretreated MSW is injected into a gasification furnace equipped with thermal plasma torches. Sometimes, LPG burner is installed to raise the initial temperature and to add heat when the heat value of solid waste is not enough. Coke is also a good assistant heating material as mentioned above. The gas temperature is very high (>1,200 ^oC) in the thermal plasma furnace, so the temperature of the gas emitted from the thermal plasma furnace must be decreased. Generally, a heat exchanger is installed behind the thermal plasma furnace to recover the heat from the gas. The recovered heat can also be utilized as an energy source using a steam turbine. The cooled gases passed through the heat exchanger must be purified to generate clean syngas.

There are many options to treat gases containing syngas. Generally, bag filters and wet scrubbers are used to remove fly ash and acidic gases. One advantage in using thermal plasma gasification is that NO_x removal process such as selective non-catalytic reduction and selective catalytic reduction is not necessary because thermal plasma gasification does not emit NO_x since the inside of the plasma furnace is O_2 starved.

After air pollution control the purified syngas (H_2 and CO) can be used as an energy source. First, electricity can be generated by steam and/or gas turbines. For electricity generation by steam turbine, the syngas is just combusted, generating steam which is injected into the steam turbine. The gas turbine can be used for the generation of electricity from the syngas even though additional equipment such as gas purifiers and syngas compressors are necessary. Second, high value chemicals can be produced from the syngas by the combination of chemical processes. CH_4 can be generated with a methanation process, and chemical wax can also be generated using a Fischer-Tropsch process. It also can generate high purity H_2 which can be used as raw material in fuel cells to generate electricity. However, currently, the combination of thermal plasma process with methanation, Fischer-Tropsch, and fuel cell with high purity H_2 processes has not been implemented. We have believed that, if thermal plasma gasification process will combine with them, its applicability will be also widen. As shown in Figure 2, thermal plasma technology for the gasification of solid wastes is comprised of multiple combined element technologies.

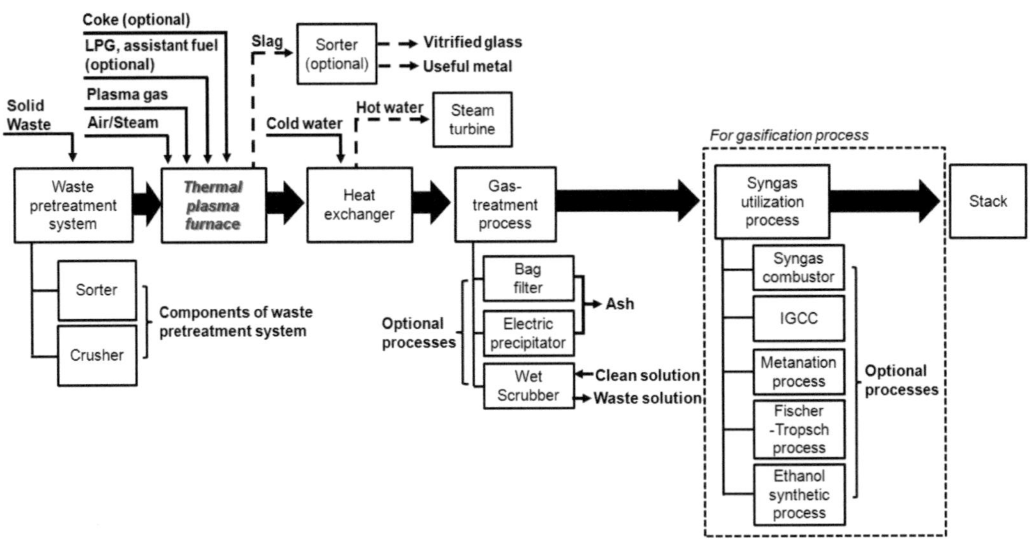

Figure 2. Schematic diagram of the overall process for the gasification of solid waste using thermal plasma

Locations	Population	Materials	Capacity (TPD)	Start date
Europe				
Landskrona, Sweden	27,889	Fly ash	200	1983
Bordeaux, France	1.01 million	Ash from MSW	10	1998
Morcenx, France	4,993	Asbestos	22	2001
Bergen, Norway	213,000	Tannery waste	15	2001
North America				
Anniston, Albama	24,276	Catalytic converters	24	1985
Jonquiere, Canada	54,872	Aluminum dross	50	1991
Honolulu, Hawaii	374,676	Medical waste	1	2001
Richland, Weshington	46,155	Hazardous waste	4	2002
Alpoca, West Virginia	613	Ammunition	10	2003
USA Navy	-	Shipboard waste	7	2004
USA Army	-	Chemical agents	10	2004
Hawwthorne, Nevada	3,311	Munitions	10	2006
Ottawa, Canada	1.1 million	MSW	85	2007
Madison, Pennsylvania	510	Biomass, Const. waste	18	2009
Asia				
Kinura, Japan	40,806	MSW Ash	50	1995
Mihama-Mikata, Japan	28,817	MSW/Sewage sludge	28	2002
Utashinai, Japan	5,221	MSW/ASR	300	2002
Shimonoseki, Japan	1.5 million	MSW Ash	41	2002
Imizu, Japan	94,313	MSW Ash	12	2002
Kakogawa, Japan	268,565	MSW Ash	31	2003
Maizuru, Japan	89,626	MSW Ash	6	2003
Lizuka, Japan	78,201	Industrial waste	10	2004
Taipei, Taiwan	22.2 million	Medical and battery waste	4	2005
Osaka, Japan	2.6 million	PCBs (Poly chlorinated Biphenyl)	4	2006
Cheongsong, Korea	150,000	MSW	10	2008

Table 1. Commercial thermal plasma plants of solid waste treatment [39]

There are a number of applications of commercial thermal plasma facilities for various solid wastes treatment in the EU, the USA, and Asia (Table 1). Especially, Japan and the EU have constructed many thermal plasma processing plants; the largest of which is located in Utashinai, Japan has a 300 TPD capacity for the treatment of MSW and ASR (auto shredder residue). Several major companies manufacture thermal plasma torches (Westinghouse, Europlasma, Phoenix, and Tetronics). Westinghouse has supplied a maximum 2.4 MW thermal plasma torch having an approximately 1,500 hr lifetime [38]. Europlasma has

Supplier	Nation	Materials
AlterNRG	Canada	MSW, RDF (refuse derived fuel), ASR, tire, coal and wood, hazardous waste, petcoke
Advanced Plasma Power (APP)	UK	RDF
Bellwether Gasification Technologies	Germany	MSW, RDF
Bio Arc	USA	Agricultural waste, medical waste
Blue Vista Technologies	Canada	MSW, hazardous liquids and gaseous wastes
Environmental Energy Resources (EER)	Israel	MSW
Encore Environmental Solutions	USA	Hazardous waste
Enersol Technologies	USA	LLR (low level radioactive), munitions
Enviroarc Technologies	Norway	Tannery waste, other hazardous waste, ash
Europlasma	France	Hazardous waste, ash, MSW, tires, syngas cleaning
GS Platech	Korea	MSW, biomass, ASR, industrial waste, hazardous waste, sludge, radioactive waste
Hera Plasco	Spain	MSW
Hitachi Metals	Japan	MSW and ASR, MSW and sewage sludge
Hitachi Zosen	Japan	Ash
Hungaroplazma Services	Hungary	MSW
InEnTec	USA	Medical waste, hazardous waste
International Scientific Center of Thermophysics and Energetics (ISCTE)	Russia	Transformer oil, pesticide, medical wastes, waste oil and coal slimes
Kawasaki Heavy Industries	Japan	PCBs and asbestos
Kinectrics	Canada	MSW, waste plastics
Mitsubishi Heavy Industries	Japan	Ash
MPM Technologies	USA	ASR, sewage sludge, waste tires and petcoke, biomass
MSE Technology Applications	USA	Military, hazardous waste
Plasma Energy Applied Technology (PEAT) International	USA	Hazardous waste, medical, industrial process and pharmacy waste
Phoenix Solutions	USA	Ash
Plasco Energy	Canada	MSW
Pyrogenesis	Canada	Shipboard waste, industrial waste

Supplier	Nation	Materials
Radon	Russia	LLR and hazardous waste
Retech Systems	USA	Hazardous wastes, LLR wastes
SRL Plasma	Australia	Solvent, waste chemicals and CFC's (chloro fluoro carbon)
Startech Environmental	USA	MSW
Tetronics	UK	Ash, APC residues and hazardous waste, catalyst waste, steel plant wastes, hazardous waste, RDF

Table 2. Suppliers and treated materials for the treatment of solid wastes in the world [43]

developed a maximum 4.0 MW transferred torch, also with 1,500 hr lifetimes [40]. Phoenix has developed transferred, non-transferred, and convertible thermal plasma torches with a maximum power of 3 MW and lifetimes of about 2,300 hr [41]. Tetronics has developed transferred, non-transferred, and twin torches having approximately 1,000 hr lifetimes [42]. In addition, many suppliers have also widely distributed (especially North America and EU) for various material treatments using thermal plasma (Table 2) [43]. These findings lead us to believe that thermal plasma technology for the treatment of solid waste is well-established technology and is immediately usable for solving problems for waste treatment.

3. Characteristics of thermal plasma process for the gasification of MSW

Combustion can play a number of important roles in an integrated MSW management system as follows: it can (1) reduce the volume of waste, therefore preserving landfill space, (2) allow for the recovery of energy from the MSW, (3) permit the recovery of minerals from the solid waste which can then be reused or recycled, (4) destroy a number of contaminants that may be present in the waste stream, and (5) reduce the need for the "long-hauling" of waste.

The recovery of energy from MSW combustion typically involves the conversion of solid waste to energy resulting in the generation of electricity from the recovered heat, and/or the generation of hot water or steam to use for community-based industrial, commercial or residential heating applications. Conventional combustion technologies include mass burn incineration. On the basis of chemical analysis, the average composition of combustible materials in MSW can be expressed by the formula $C_6H_{10}O_4$ [44]. When this hypothetical compound is combusted with air, the reaction is [44]:

$$C_6H_{10}O_4 + 6.5O_2 + (24.5N_2) \rightarrow 6CO_2 + 5H_2O + (24.5N_2) \qquad \Delta H=-6.5\ \text{MWh/ton} \qquad (7)$$

Although, incineration technology has been widely utilized to reduce the total volume of waste and recover the energy from MSW, the emissions of pollutants such as NO_x, SO_x, HCl, harmful organic compounds, and heavy metals are high. Another problem is the serious corrosion of the incineration system by alkali metals contained in solid residues and fly ash

[45]. Thermal plasma technology has been applied for the treatment of MSW as an alternative to solve these problems [46-48].

Thermal plasma technology can make extremely high temperatures in the absence of or near-absence of O_2, with MSW containing organics and other materials. Organics are converted into syngas and other materials dissociated into constituent chemical elements that are then collected and vitrified to produce an inert glass-like slag; most of the heavy and alkali metals (with the exception of mercury, zinc and lead, which can vaporize at high temperatures and be retained in fly ash and syngas) are retained in the vitrified slag. The vitrified slag obtained after cooling can be used as construction materials. The simple gasification reaction of MSW using thermal plasma can be expressed as follows [44]:

$$C_6H_{10}O_4 + 3O_2 \rightarrow 3CO + 3CO_2 + 4H_2 + H_2O \qquad \Delta H=-1.3\ \text{MWh/ton} \tag{8}$$

The principal product of plasma gasification of MSW is a low to medium calorific value syngas composed of CO and H_2 as shown in equation (8). This gas can be burned to produce heat and steam, or chemically scrubbed and filtered to remove impurities before conversion to various liquid fuels or industrial chemicals. Syngas combusts according to the following equations [44]:

$$3CO + 4H_2 + 3.5O_2 \rightarrow 3CO_2 + 4H_2O \qquad \Delta H=-1.5\ \text{MWh/ton} \tag{9}$$

Occasionally, steam has been injected with MSW into plasma gasification furnaces to increase the energy efficiency and syngas yield according to equations (1)-(4). Nishikawa et al. reported that steam enhanced the reduction of the weight of charcoal and production of hydrogen through laboratory experiments [49]. Qinglin et al. also investigated the effect of steam injection in pilot scale thermal plasma gasification plant of MSW [45], showing that the cold gas efficiency and syngas yield are improved with the increase of steam injected.

Table 3 shows the important differences mentioned above between incineration and thermal plasma gasification. Main differential factors between them are amount of added O_2 and temperature inside a furnace. Incinerators have designed to maximize CO_2 and H_2O, indicating complete combustion, however thermal plasma treatment system is designed to maximize CO and H_2, indicating incomplete combustion. These complete and incomplete combustions have been controlled using added O_2 amounts. Incinerators add a large quantity of excess air, but thermal plasma treatment systems add a limited quantity of O_2. Therefore, inside of incineration furnace is an oxidizing environment, causing the generation of NO_x and SO_x, but inside of thermal plasma process is a reducing environment, prohibiting the generation of NO_x and SO_x. Temperature of incineration furnaces is around 800 °C which is below an ash melting point. This makes inorganic materials contained in MSW to convert to bottom and fly ash. However, temperature of thermal plasma processes is around 1,400 °C, which is above an ash melting point. This makes inorganic materials contained in MSW to convert to vitrified slag which can be utilized as a source of construction materials.

Differential factors	Incineration process	Thermal plasma process
Definition	- Mass burning process	Gasification process
Amount of O_2	- Designed to maximize CO_2 and H_2O - Added large quantity of excess air - Oxidizing environment - Generating NO_x and SO_x	- Designed to maximize CO and H_2 - Added limited quantity of O_2 - Reducing environment - Prohibiting the generation of NO_x and SO_x
Temperature	- Operating at temperature below ash melting point - Inorganic materials are converted to bottom ash and fly ash - Bottom ash and fly ash are collected, treated, and disposed as hazardous wastes.	- Operating at temperature above ash melting point - Inorganic materials are converted to glassy slag and fine particulate matter - Slag is non-leachable, non-hazardous and suitable for use in construction materials

Table 3. Comparison between the incineration and thermal plasma gasification processes for MSW treatment

4. Operation status of 10 TPD scale thermal plasma gasification plant for MSW treatment

A 10 TPD capacity thermal plasma plant for MSW treatment is located in Cheongsong, Korea. The local population of 30,000 generates 15 TPD MSW. The characteristics of the MSW are shown in Table 4 [27]. The thermal plasma plant was constructed in early 2008 and optimized for 6 months beginning in September, 2008. The plant continues to operate normally for over 3.5 years without any problems.

During the optimization period, several attempts were made to improve the performance of the process for normal waste feeding system, control of the hot air flow rate added into the furnace, removal of bag-filter and other considerations were made. Operating data of the thermal plasma gasification plant have been obtained during the normal operation periods (3.5 years). An exterior image and interior 3D design scheme of the thermal plasma plant are shown in Figures 3 and 4, respectively. The thermal plasma gasification plant mainly consists of a waste feeding system, integrated thermal plasma furnace, heat exchanger, bag filter, water quencher and scrubber, and secondary combustion chamber. Currently we are not using bag filter because we reduced the generated fly ash by employing a centrifugal force using flow jet of the thermal plasma torch inside the furnace, which melted the fly ash and unburned MSW onto the wall of the integrated furnace.

Items		Values
Heating value (kcal/kg)	Higher heating value	4491.09
	Lower heating value	2999.90
Proximate analysis (wt%)	Moisture	24.83
	Combustible	67.54
	Non-combustible	7.64
Ultimate analysis (Dry basis, wt%)	C	45.21
	H	6.37
	N	0.87
	S	0.18
	Cl	0.88

Table 4. Heating values and chemical composition of MSW gathered in Cheongsong, Korea [27]

Figure 3. Exterior image of 10 TPD thermal plasma gasification plant located in Cheongsong, Korea

Detailed specifications of 10 TPD thermal plasma gasification plant can be found in reference [27]; in this chapter, we summarize specifications and performances compactly. The integrated furnace, equipped with two non-transferred thermal plasma torches, is the central apparatus where the gasification takes place. The operating pressure of the integrated furnace is maintained at -10 mmAq gauge pressure (Figure 5(a)); the increase of pressure after November was caused by the modification of operation conditions for the combination of thermal plasma process with fuel cells. The integrated furnace is composed of the furnace, two non-transferred thermal plasma torches, a preheating burner that uses liquefied petroleum gas (LPG), a MSW feeding system, an outlet for the melted slag, and a hot air injection port. The role of the preheating burner is to preheat the integrated furnace (up to 600 °C for the initial 12 hr). The temperatures inside the integrated furnace and syngas combustor are maintained at 1,400 and 880 °C, respectively (Figure 5(b)). The MSW is initially partially oxidized by the injection of hot air (600 °C, 700 Nm^3/hr), which is taken from the air preheater/gas cooler, and then melted by the plasma torches. This partial oxidation of MSW can reduce the electrical energy used for the thermal plasma torches. The melted slag is tapped out into a water tank from a hole located at the bottom of the furnace. This molten slag is quenched with water to produce granulated slag, which is removed using the slag conveyer belt. Two non-transferred thermal plasma torches are installed into the integrated furnace (Figure 6) at a 30° angle to induce a centrifugal force in the furnace. An image of the thermal torches and an interior view of the plasma torch are shown in Figure 7. The power capacity of each plasma torch is 200 kW, with an operational voltage and current of 571 ± 30 V and 293 ± 10 A, respectively (Figure 5(c)). The efficiency of the plasma torches is approximately 70% and the lifetime of the electrode is about 500 hr. Air (500 L/min, at 490 kPa) is supplied to the thermal plasma torches using a compressor.

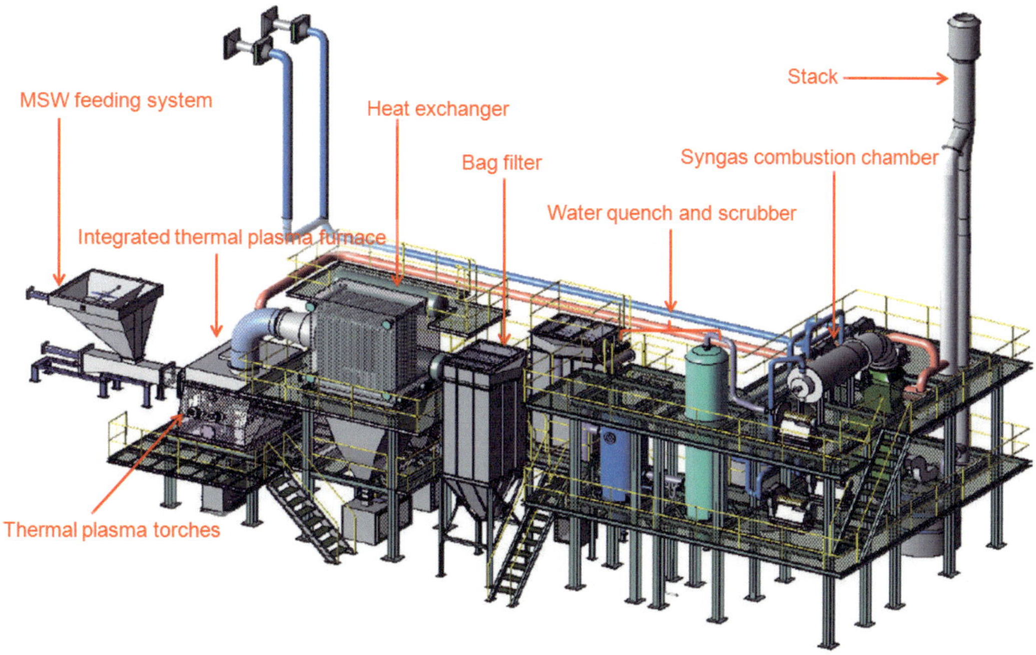

Figure 4. Interior 3D scheme of 10 TPD thermal plasma gasification plant located in Cheongsong, Korea

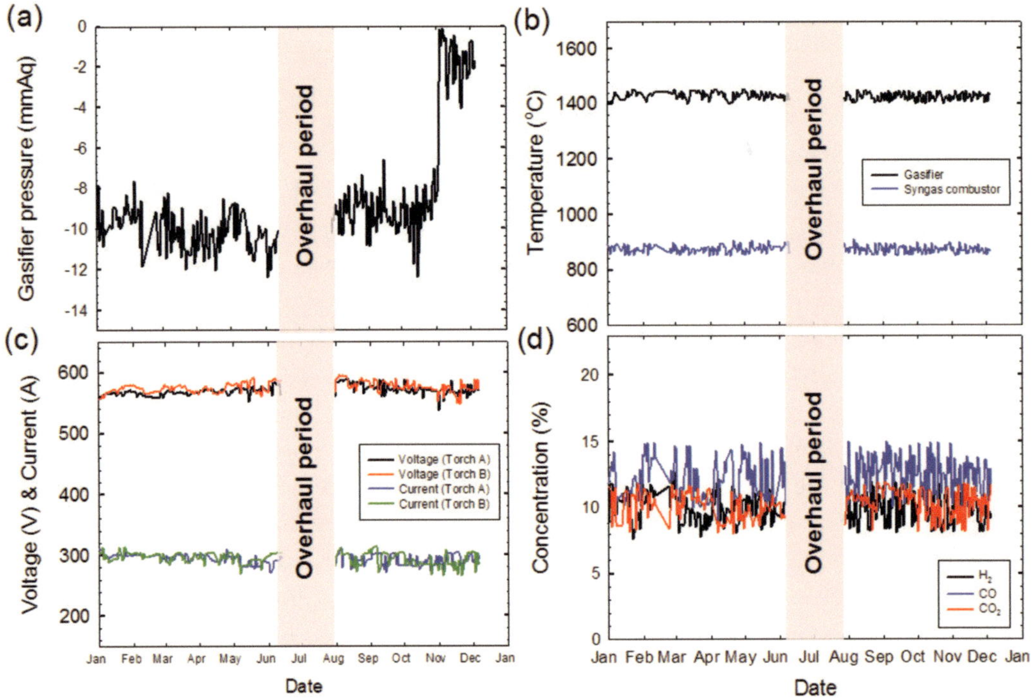

Figure 5. Measured characteristics of thermal plasma gasification plant for 1 year. (a) Pressure inside the integrated thermal plasma furnace and syngas combustion chamber. (b) Temperature inside the integrated thermal plasma furnace. (c) Applied voltage and current at two thermal plasma torches. (d) H_2, CO, and CO_2 concentration of integrated thermal plasma furnace

The gas produced in the integrated furnace (1,400 °C) enters the steam generator, where the temperature is cooled to 180 °C. Steam is generated at 1.2 ton/hr, and then injected into the steam condenser and recirculated to the steam generator. The gas cleaning systems eliminate acidic gases prior to the gas entering syngas combustion chamber. For this purpose water quencher and scrubber are installed in series. The water quencher and scrubber are located at the outlet of the heat exchanger. The role of the water quencher is to cool the gas to 30 °C with NaOH solution (40%). Although the remaining acidic gases are also partially removed by the water quencher, almost all acidic gases are removed in the scrubber. The scrubbing solution is controlled at pH 9.0, and recirculated. A syngas combustion chamber is installed to combust the H_2 and CO gases emitted. Air taken out from the MSW storage is added to the chamber to completely combust the H_2 and CO with an LPG burner, which utilizes 4.76 Nm^3/ton of LPG. The temperature of the syngas combustion chamber is maintained close to 900 °C (Figure 5(b)).

The purpose of the air preheater and gas cooler is to increase the temperature of the air taken from the MSW storage to 600 °C, and decrease the temperature of the gas emitted from the syngas combustion chamber to 200 °C. The heated air is injected into the integrated furnace to partially oxidize the MSW (700 Nm^3/hr).

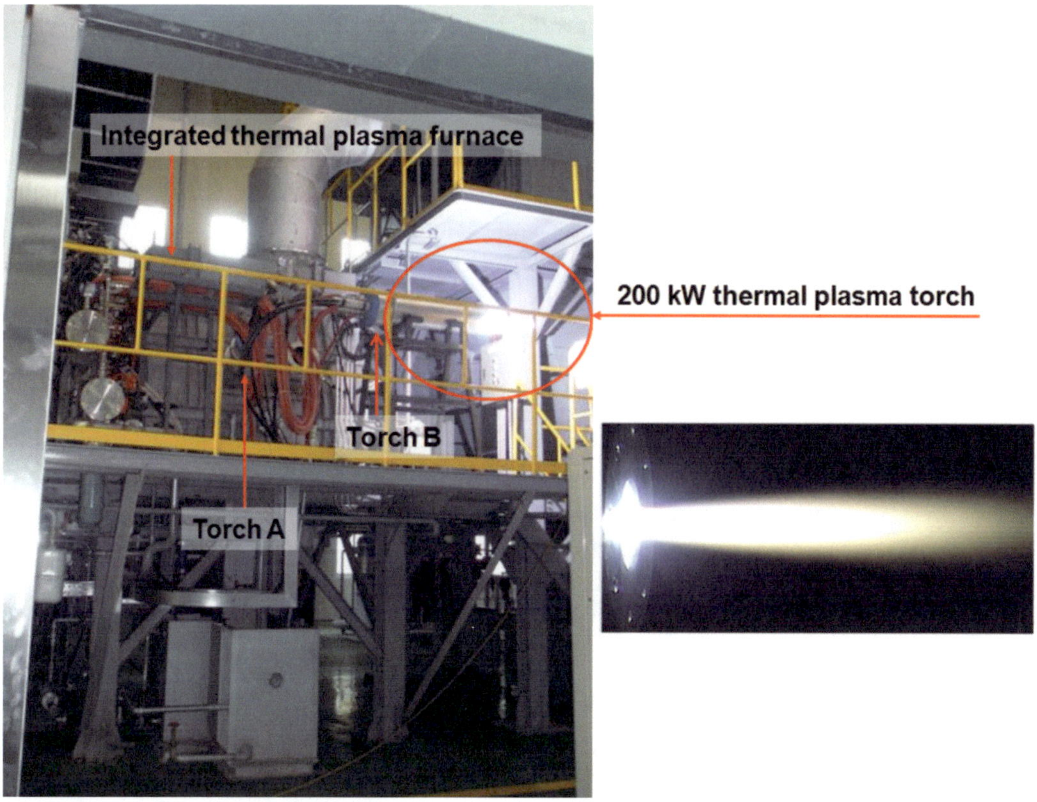

Figure 6. Image (left) of the generated thermal plasma using 200 kW thermal plasma torch beside the integrated furnace. Detailed image of the generated thermal plasma (right)

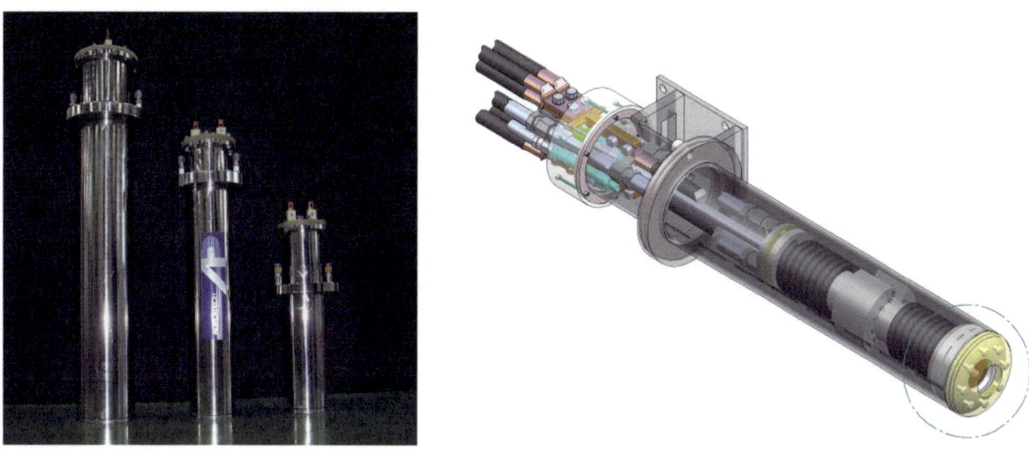

Figure 7. Images of thermal plasma torches manufactured from GS Platech (left) and interior view of thermal plasma torch (right) installed at the integrated furnace

Solid, liquid, and gaseous byproducts are generated from the thermal plasma gasification plant. Solid byproducts originate from the integrated furnace as slag (75.8 kg/ton (7.8% to the inlet MSW)) (Figure 8), the liquid byproducts are from the wastewater treatment system (0.43 Nm^3/ton), and the gaseous byproducts are from the gasification of MSW and the combustion of syngas.

The composition of the slag was analyzed and shown to have SiO_2, Al_2O_3, CaO, Fe_2O_3, Na_2O, and MgO as the major constituents, with a total percentage of 97%. The weight percentage of the inlet MSW that becomes slag is 7.58%. In addition, the rate of MSW volume reduction to slag was up to 99% (the density of MSW: 0.09 ton/m^3, slag: 2.6 ton/m^3). This result indicates that the treatment of MSW using thermal plasma processes can greatly reduce the volume of MSW. A toxicity characteristics leaching procedure was performed on three slag samples. No heavy metals were eluted from the slag (Table 5). This result indicated that the slag produced in the thermal plasma process from the treatment of MSW is non toxic.

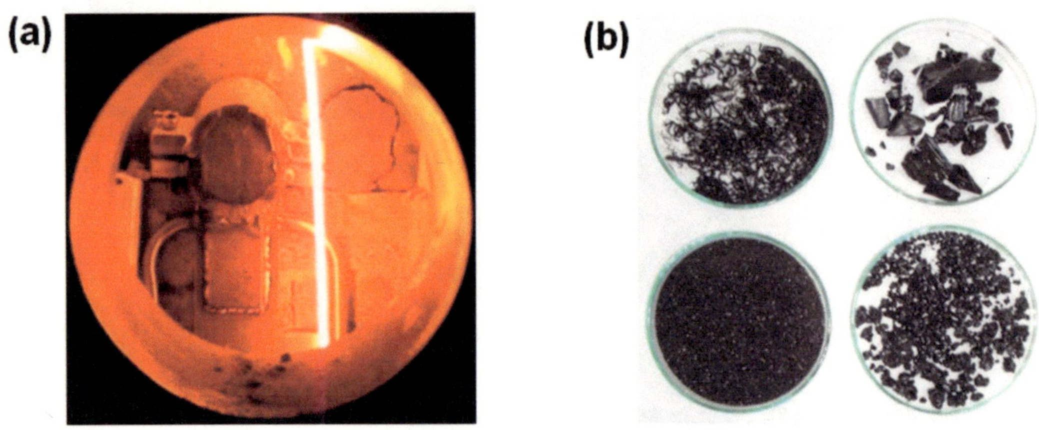

Figure 8. Images of (a) molten slag tapped from integrated furnace and (b) vitrified slag

Measurement times	Pb (mg/L)	Cd (mg/L)	As (mg/L)	Cu (mg/L)	Hg (mg/L)	Cr^{6+} (mg/L)
1	N.D.	N.D.	N.D.	N.D.	N.D.	N.D.
2	N.D.	N.D.	N.D.	N.D.	N.D.	N.D.
3	N.D.	N.D.	N.D.	N.D.	N.D.	N.D.
4	N.D.	N.D.	N.D.	N.D.	N.D.	N.D.
5	N.D.	N.D.	N.D.	0.05	N.D.	N.D.
Average	N.D.	N.D.	N.D.	N.D.	N.D.	N.D.

N.D.: not detected

Table 5. Results of toxicity characteristics leaching procedure for vitrified slag

The discharged liquid originating from the water quenching and scrubbing, which was used to decrease the gas temperature and remove the acidic gas, amounted to 0.43 Nm^3/ton. This wastewater was treated via the wastewater treatment system and recycled into the water quencher and scrubber.

The most crucial point in the operation of a thermal plasma process is the nature and amount of the final air emissions. The concentrations of air pollutants were measured at two ports: one at the outlet of the integrated furnace, and the other at the stack. We also measured the concentration of gaseous emission at two ports for 1 year (Figure 5(d) and Table 6); we measured syngas continuously at the outlet of the integrated furnace (Figure 5(d)) and air pollutants periodically at the stack (Table 6). The flow rates at the outlet of the integrated furnace and stack were 1,161 and 2,654 Nm^3/hr, respectively. The higher flow rate at the stack was due to the syngas combustion chamber that used air and LPG. The concentrations of O_2 at the outlet of the integrated furnace and scrubber were 0.4 ± 0.2 and 1.1 ± 0.4%, respectively. The small concentration of O_2 in the integrated furnace indicates that the inside was under O_2 starved conditions. The average concentrations measured at the outlet of the integrated furnace were 10.4% for H_2 and 14.2% for CO with 10% CO_2 (Figure 5(d)); extra gases are mostly N_2. The fluctuation of concentrations was caused by the variation of waste composition and water content. The syngas combustion chamber was employed for the combustion of H_2 and CO at the present 10 TPD scale without the reuse of the syngas. It is worth noting that no NO_x and SO_x were detected at the outlet of the integrated furnace due to the O_2-starved conditions inside the integrated furnace. The concentrations of dioxin were 1.04 ng-TEQ/Nm^3 at the outlet of the integrated furnace and 0.05 ng-TEQ/Nm^3 at the stack, which were much lower than those of conventional incineration plants. This result suggests that negligible amounts of PCDD/DFs were produced in the thermal plasma gasification plant due to the high temperature of the integrated furnace. The concentrations of NO_x and SO_x were 10 and 4 ppm, respectively, which is increased somewhat at the stack. This is because of the syngas combustion chamber. The concentrations of CO, HCl, and dust are 5 ppm, 1.92 ppm, and 4.15 mg/Sm^3, respectively, which satisfied the requirements of current legislation. These results indicated that the thermal plasma process for the treatment of MSW is an environmentally friendly process.

As mentioned above, we don't reuse the generated syngas for the recovery of energy at 10 TPD thermal plasma gasification plant; we have just combusted syngas in the syngas combustion chamber. However, recently, we have tried to utilize syngas generated from MSW as an energy source. We combined the thermal plasma gasification plant with 50 kW proton exchange membrane fuel cell (PEMFC) from November, 2010 to October, 2011. We installed WGS and PSA to make high purity H_2 (> 99.999%); we already demonstrated to make high-purity H_2 (>99.99%) using WGS and PSA in 3 TPD thermal plasma gasification plant using paper mill waste [26]. Finally, we succeed to make high-purity H_2 (>99.999%) and generate electricity from 50 kW PEMFC. We will report those results in time. We have believed strongly that these trials also can widen the applicability of thermal plasma process for MSW.

Date	CO (ppm)	HCl (ppm)	Dust (mg/Sm^3)	NO_x (ppm)	SO_x (ppm)
14/01/2010	2	1.29	3.7	5	3
29/01/2010	2	2.03	3.4	10	3
12/02/2010	7	1.67	3.5	13	7
03/03/2010	4	1.2	4.1	8	5
12/03/2010	3	2.09	3.9	11	6
26/03/2010	5	1.11	5.2	22	3
16/04/2010	4	1.19	3.4	8	5
30/04/2010	2	3.01	4.7	7	7
15/05/2010	6	1.74	4.9	8	5
28/05/2010	3	2.40	4.2	9	3
14/06/2010	5	1.78	4.0	8	3
04/08/2010	2	1.56	4.40	8	3
13/08/2010	6	2.64	5.20	8	7
27/08/2010	9	2.13	4.60	19	4
10/09/2010	10	1.79	4.20	6	4
02/10/2010	1	2.75	3.90	6	3
08/10/2010	2	2.62	2.50	5	3
22/10/2010	9	2.08	3.80	13	4
05/11/2010	5	1.62	4.60	16	3
19/11/2010	2	2.23	3.60	16	3
14/12/2010	6	1.43	5.40	13	5
Average	5	1.92	4.15	10	4

Table 6. Gas composition measured at the stack of the thermal plasma gasification plant for 1 year

5. Design parameters for a 100 TPD scale thermal plasma gasification plant

Based on the obtained data from the 10 TPD thermal plasma plant, we could obtain design parameters for a 100 TPD plant. It is considered that the MSW has 3,300 kcal/kg of heating value. Figure 9 shows the schematic of overall process of 100 TPD thermal plasma plant for MSW treatment. A 100 TPD thermal plasma plant consists of six main sections for the gasification of MSW: (1) An MSW storage unit and feeding system, (2) an integrated furnace equipped with two non-transferred thermal plasma torches, (3) effluent gas treatment systems, including water quencher and scrubber, (4) a syngas combustion chamber, (5) an air preheater/gas cooler, and (6) a steam turbine (which was not included in the 10 TPD plant). An energy balance for the overall process is presented in Figure 10. The third line of the table inserted in Figure 10 shows the latent heat of the produced syngas. The specific different characteristics between the 10 and 100 TPD scales are also tabulated in Table 7. At 10 TPD capacity, the power consumption of the plasma torch used for the treatment of 1 ton of MSW was 0.817 MWh/ton. At 100 TPD, use of 0.447 MWh/ton of thermal plasma power is planned. At 10 TPD, the heat loss of the overall process through the wall was 14% and the energy contained in the effluent gases of the stack was 16%. However, we considered, at 100 TPD, the heat loss of the overall process through the wall would be 7% and the energy contained in the effluent gases of the stack would be 10%. In addition, at a 10 TPD scale, syngas and the heat generated from heat exchanger have not been reused, however, at 100 TPD, the energy generated from syngas and heat exchanger through steam generators would be used. The energy reused by the two steam generators would be 73% of the input energy (a ratio of 12 plus 13 (16,679 Mcal/hr) to 1 plus 2 (22,858 Mcal/hr) in Figure 10).

Items	10 TPD scale	100 TPD scale
Thermal plasma consumption power	0.817 MWh/ton	0.447 MWh/ton
Heat loss from effluent gases of stack	16%	10%
Heat loss through system walls	14%	7%
Energy recovery	Not used	Used through steam turbine

Table 7. Comparison of the characteristics between 10 and 100 TPD thermal plasma plants for MSW treatment

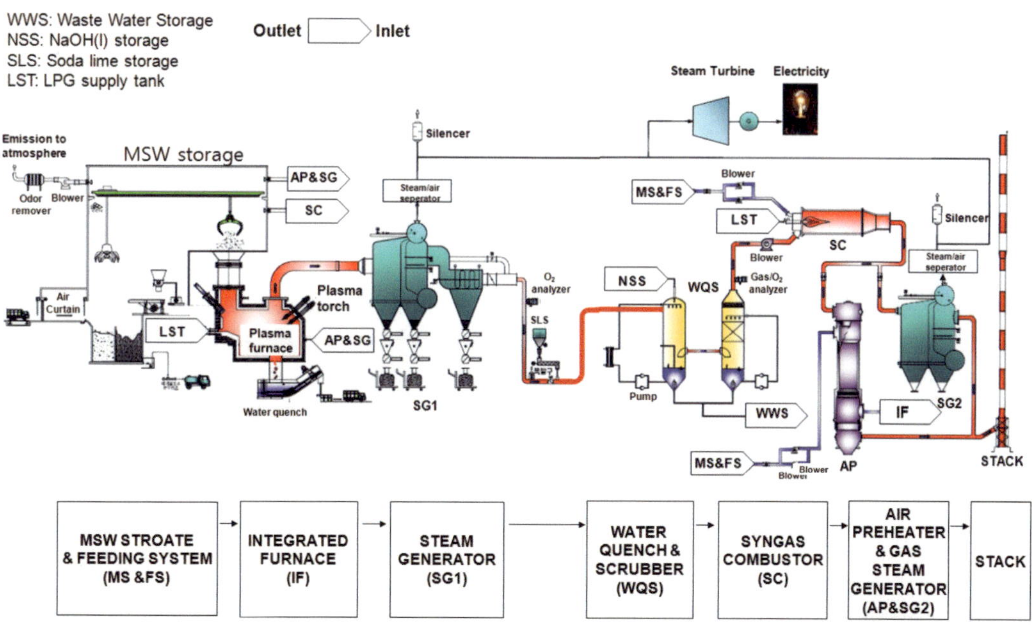

Figure 9. Schematic of the overall process for 100 TPD thermal plasma gasification plant

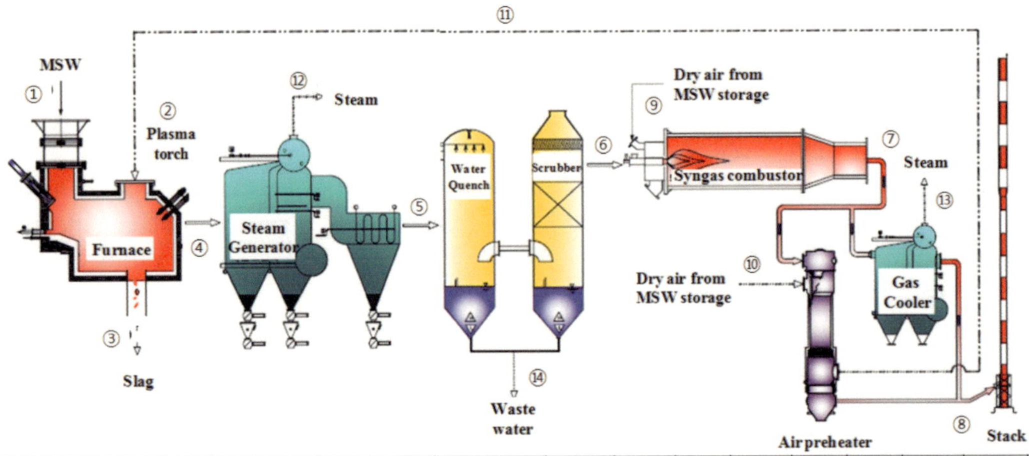

NO.	1	2	3	4 Sensible Heat	5 Sensible Heat	6 Sensible heat	7	8	9	10	11	12	13	14
UNIT	Mcal/hr													
Value	21,740	1,118	251	8,182	937	198	14,967	3,919	178	81	1,695	7,245	9,434	19.2
Latent heat				15,341	15,341	15,341								

Figure 10. Energy balance for 100 TPD thermal plasma gasification plant

6. Economic evaluation of the thermal plasma gasification plant

The major disadvantage of thermal plasma gasification processes mentioned by many scientists and engineers is the use of electricity, which is an expensive energy source [32]. The economics of thermal plasma gasification processes have many variable parameters such as regional characteristics, types of solid wastes to be processed, capacity, and others. In the USA, the cost of a landfill is approximately 30-80 US$/ton and the average incineration cost is 69 US$/ton [50]. However, the average cost of landfills and incinerators in small countries such as Japan and European countries is approximately 200-300 US$/ton since land is more scarce [50], meaning that the economics of thermal plasma gasification for MSW is improved in these regions. Presently, the average construction cost of thermal plasma plants is estimated to approximately 0.13-0.39 million US$/TPD. Dodge estimated that the construction cost of a 750 TPD is 150 million US$, which is equivalent to 0.2 million US$/TPD [51]. The construction cost of the 300 TPD plant in Utashinai, Japan was approximately 0.17 million US$/TPD. A 600 TPD thermal plasma plant in St. Lucie, Canada planned by Geoplasma using Alter NRG's thermal plasma torch is also 0.17 million US$/TPD. The initial project planning to construct a 2,700 TPD by Geoplasma in St. Lucie had a 0.13 million US$/TPD construction cost. Figure 11 shows the trend of construction cost according to capacity; cases of GS Platech (10 and 100 TPD scales) will be discussed detailed in below. Although the prices of each country are different and data are not enough fully, the trend of construction cost according to capacity could be identified. 0.39 million US$/TPD applies to the 10 TPD plant constructed by GS Platech in Korea. For capacities between 250 and 750 TPD, around 0.17-0.22 million US$/TPD is applicable. Above 2,000

TPD, 0.13 million US$/TPD is applicable. These results indicate that thermal plasma gasification processes are more economical if the treatment capacity is increased. Presently, detailed operational costs of each case are not available other than GS Platech. In addition, there are many methods to utilize byproducts generated during MSW gasification. For example, syngas, which could be used for the generation of high value products such as fuel, chemical compounds, and high purity hydrogen, would work to this effect. This means that, although thermal plasma technology is well-established, there are still many fields to investigate for enhancing the economics of the process.

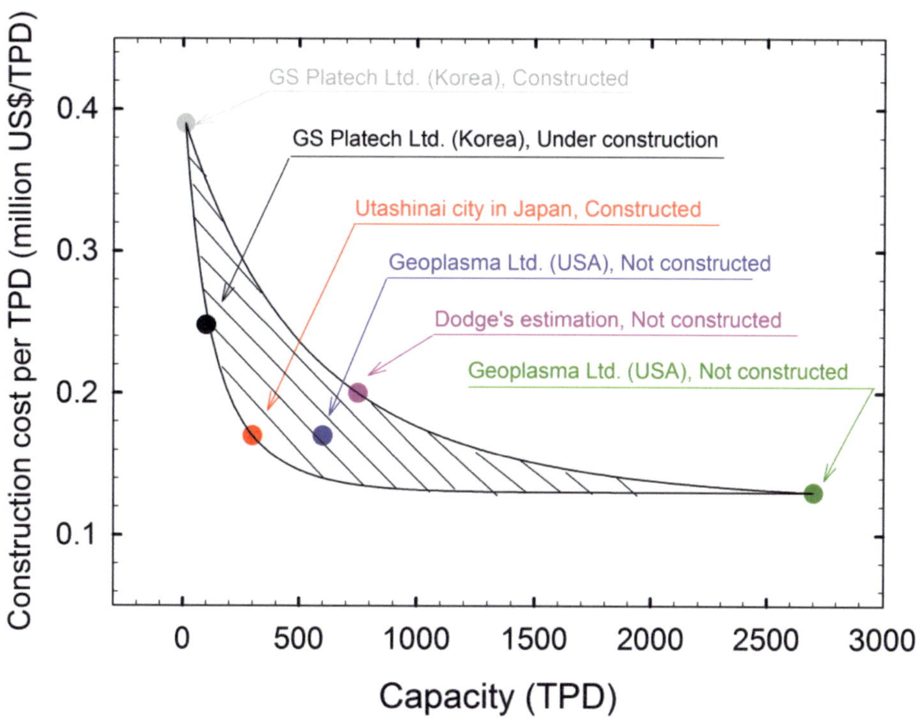

Figure 11. Construction cost (million US$) of thermal plasma treatment plants according to treatment capacity (TPD)

We can obtain detailed economic evaluations for a 10 TPD plant, including construction and operation costs (Table 8). 3.9 million US$ was the total construction cost of a 10 TPD or 0.39 million US$/TPD. Operation costs include labor costs, depreciation cost, overhead charges, and insurance. Labor cost for 12 labors and overhead charges are 0.49 and 0.24 million US$/year, respectively. Depreciation cost and insurance are 0.26 and 0.02 million US$/year, respectively. Total operation costs are 0.99 million US$/year. This is equivalent to 300 US$/ton without VAT. 110 US$/ton is received from local government for treating MSW in Cheongsong, Korea, which would vary by region. Therefore, total profit is negative (-190 US$/ton). However, economics will be improved if treatment scale is increased because of the following three reasons. First, the construction cost will be decreased as the capacity is increased, as mentioned above. This will cause a decrease in depreciation cost. Second, syngas can generate profit as an energy source. Presently, we are abandoning generated syngas because the amounts generated are not sufficient to use as an energy source. Lastly, the operation of a plant is an economy of scale. As the capacity increases, labor costs, overhead charge, and etc will decrease. Although these numerical economics were obtained for a 10 TPD plant, these experiences indicate that the thermal plasma gasification process is a viable alternative economically if the scale increases.

Items			Costs
Construction cost			3.9 million US$
Operation cost per year	Labor costs	12 labors	0.49 million US$/year
	Depreciation cost	Depreciation period = 15 years	0.26 million US$/year
	Variable costs	Maintenance cost Electricity cost Chemical cost Wetted cost Etc	0.24 million US$/year
	Insurance	0.5% of construction cost	0.02 million US$/year
	Total		0.99 million US$/year
	Operation cost per ton of MSW	Total operation cost/330 day x 0.01 day/ton	330 US$/ton (with V.A.T.)
			300 US$/ton (without V.A.T.)

Table 8. Economic evaluation of a 10 TPD thermal plasma gasification plant for MSW treatment (These data based on the operation for 3.5 years.)

Based on this information, total construction cost for a 100 TPD scale plant would be 24.8 million US$, or 0.25 million US$/TPD. Operation cost consists of fixed cost, variable cost, and insurance. In fixed cost, labor cost, depreciation cost, and overhead charges such as fringe benefits, safe maintenance costs, training expense, and per diem and travel expenses are included; total fixed cost would be 2.39 million US$/year. Variable cost including maintenance, electricity, chemical, water costs would be 0.82 million US$/year. All of the variable costs with insurance is 0.94 million US$/year. Based on the energy balance and operational costs (Figure 10 and Table 9), profit from selling electricity generated from steam turbines would also be generated (Table 10). The recovery heat values from two steam generators are 16,679 Mcal/hr (12 plus 13 in Figure 10). Considering the total efficiency of a steam supply and power generation using a steam turbine as 26%, 4,286 Mcal/hr of electricity could be generated, which is equivalent to 5,000 kW of electricity. 2,000 kW of electricity is necessary to generate thermal plasma torches and utilities meaning that 3,000 kW of electricity could be sold to grid and is equivalent to 23.8 million kWh/year.

Considering the selling price of electricity as 10.9 cent/kWh, total profit per year from selling electricity would be around 2.6 million US$/year; the selling price of electricity recovered from MSW is relatively high compared to other electricity prices due to the government's renewable portfolio standards (RPS) policy promoting the use of renewable energy in Korea. In addition, profit could be obtained from treating MSW. 110 US$/ton is paid by the local government for treating MSW in Cheongsong, Korea, which means that, 100 TPD MSW is treated, profit for treating MSW would be 3.6 million US$/year. Therefore, total profits are 6.2 million US$/year (2.6 million US$/ year plus 3.6 million US$/ year). Considering the operation cost (3.34 million US$/year), it can be concluded that total margin for a 100 TPD MSW treatment plant using thermal plasma gasification would be about 2.86 million US$/year (6.2 million US$/year minus 3.34 million US$/year), which is equivalent to 86 US$/ton.

Based on these design parameters, energy balance, and economic evaluation, a 100 TPD thermal plasma plant for RPF (refused plastic fuel) gasification is now under construction in Yeoncheon, Korea. As soon as construction and initial operation is finished, those results will be reported.

<table>
<tr><td colspan="4">Items</td><td>Costs</td></tr>
<tr><td colspan="4">Construction cost</td><td>24.8 million US$</td></tr>
<tr><td rowspan="10">Operation cost per year</td><td rowspan="4">Fixed costs</td><td>Labor costs</td><td>14 labors</td><td>0.57 million US$/year</td></tr>
<tr><td>Overhead charges</td><td>Fring benefits
Safe maintenance cost
Train expense
Per diem and travel expenses
Etc</td><td>0.17 million US$/year</td></tr>
<tr><td>Depreciation cost</td><td>Depreciation period = 15 years</td><td>1.65 million US$/year</td></tr>
<tr><td colspan="2">Sub total</td><td>2.39 million US$/year</td></tr>
<tr><td colspan="2">Variable costs</td><td>Maintenance cost
Electricity cost
Chemical cost
Wetted cost
Etc</td><td>0.82 million US$/year</td></tr>
<tr><td colspan="2">Insurance</td><td>0.5% of construction cost</td><td>0.12 million US$/year</td></tr>
<tr><td colspan="3">Total</td><td>3.34 million US$/year</td></tr>
<tr><td colspan="2" rowspan="2">Operation cost per ton of MSW</td><td rowspan="2">Total operation cost/330 day × 0.01 day/ton</td><td>111 US$/ton (with V.A.T.)</td></tr>
<tr><td>101 US$/ton (without V.A.T.)</td></tr>
</table>

Table 9. Economic evaluations of a 100 TPD thermal plasma gasification plant for MSW treatment. These data are obtained based on experiences obtained from a 10 TPD thermal plasma gasification plant. All costs are based on Korean price. Exchange rate between USA and Korea is 1,130 won/US$.

Items	Values	Note
Power generation	5,000 kW	steam supply and power generation system
Consumed electric power	2,000 kW	
Sold electric power	3,000 kW	
Operation day per year	330 day	
Operation hour per day	24 hr	
Amount of electricity sales	23.8 million kWh/year	
Unit cost of electricity sales	10.9 cent/kWh	10.6 cent/kWh (SMP, system marginal prices) 4.4 cent/kWh (RPS, Renewable Portfolio Standards)
Profit from selling electricity	2.6 million US$/year	23,760,000 kWh/year × 10.9 cent/kWh
Profit from treating MSW	3.6 million US$/year	100 TPD × 330 day/year × 110 US$/ton (MSW treatment cost)
Total profit per ton of MSW	187 US$/ton	Total profit per year (6.2 million US$/year) /330 day × 0.01 day/ton

Table 10. Calculation of profits on the basis of used electricity, selling electricity, and treating MSW costs. These data are obtained based on experiences from a 10 TPD thermal plasma gasification plant. All costs are based on Korean price. Exchange rate between USA and Korea is 1,130 won/US$.

7. Conclusions

Thermal plasma technology is a mature, reliable, and proven method for generating high temperatures at atmospheric pressure, which is not achievable by burning fuels. Recently, thermal plasma technology has been applied for the treatment of MSW directly from trucks in pilot and commercial plants. Thermal plasma gasification processes convert organics contained in MSW into syngas, and dissociate other materials into constituent chemical elements that are then collected and vitrified to produce an inert glass-like slag retaining most of the heavy and alkali metals from the waste. The vitrified slag can be used as construction materials. In addition, NO_x and SO_x are not emitted due to O_2-starved conditions inside the thermal plasma furnace. The concentrations of dioxins are also very low compared to conventional incinerators for MSW treatment due to the high temperature of the integrated furnace. Therefore, thermal plasma processes are an environmentally friendly alternative for the gasification of MSW.

A commercial thermal plasma gasification plant for MSW was constructed at a 10 TPD scale using an integrated furnace equipped with two non-transferred thermal plasma torches, and has operated for 3.5 years without any problems. It was successfully demonstrated that the thermal plasma process converted MSW into innocuous slag, with much lower levels of environmental air pollutant emissions and producing syngas as a potential energy source (287 Nm^3/ton for H_2 and 395 Nm^3/ton for CO), using 1.14 MWh/ton of electricity (thermal plasma torch (0.817 MWh/ton) + utilities (0.322 MWh/ton)) and 7.37 Nm^3/ton of liquefied petroleum gas (LPG). Data obtained for 3.5 years of operation provided many insights into plant operation such as economic factors and design parameters to extend capacity.

We obtained a detailed evaluation of economics for a 10 TPD scale including construction cost and operation cost. Total operation costs are 0.99 million US$/year (300 US$/ton without VAT). In addition, 110 US$/ton is paid by the local government for treating MSW. This means that total margin is negative 190 US$/ton at a 10 TPD scale. However, based on this experience, we are absolutely convinced that economics of the process will be improved if treatment scale is increased due to decrease of construction cost with increased capacity, profits from the utilization of syngas as an energy source, and the decrease of total operation costs such as labors cost and overhead charges. We also evaluated the economics for a 100 TPD thermal plasma gasification process for MSW. As a result, we calculated that total operation costs are 3.34 million US$/year (101 US$/ton without VAT) and total profits from selling electricity and treating MSW would be about 6.2 million US$/year for a 100 TPD plant, which is equivalent to 187 US$/ton. This means that total margin is positive 86 US$/ton at a 100 TPD scale.

Although the technical feasibility of thermal plasma gasification of MSW has been well demonstrated, it is not presently clear that the process is economically viable on the global market because regional variation of the costs of MSW treatment. However, it is clear that the reuse of vitrified slag and energy production from syngas will improve the commercial viability of this process, and there have been continued advances towards further development of the process.

Author details

Youngchul Byun
School of Chemical Engineering and Analytical Science, The University of Manchester, Manchester, UK

Moohyun Cho
Department of Physics, Pohang University of Science and Technology (POSTECH), Pohang, Republic of Korea

Soon-Mo Hwang
Research Center, GS Platech, Daejeon, Republic of Korea

Jaewoo Chung*
Department of Environmental Engineering, Gyeongnam National University of Science and Technology (GNTECH), Jinju, Republic of Korea

* Corresponding Author

8. References

[1] U.S. Environmental Protection Agency (2010) Municipal solid waste in the United States: 2009 Facts and Figures. Washington, DC.

[2] N. Okafor (2011) The disposal of municipal solid wastes in environmental microbiology of aquatic and waste systems, 1 edition. Springer Science+Business Media BV.

[3] F.F. Chen (1984) Introduction to plasma physics and controlled fusion, Volume 1: Plasma physics, 2 edition. New York: Plenum Press.

[4] C. Tendero, C. Tixier, P. Tristant, J. Desmaison, P. Leprince (2006) Atmospheric pressure plasmas: A review. Spectrochim. Acta Part B 61: 2-30.

[5] A. Fridman (2008) Plasma Chemistry. New York: Chambridge University Press.

[6] J.R. Roth (1995) Industrial Plasma Engineering: Principles. Institute of Physics Publishing.

[7] H. Zhang, G. Yue, J. Lu, Z. Jia, J. Mao, T. Fujimori, T. Suko, T. Kiga (2007) Development of high temperature air combustion technology in pulverized fossil fuel fired boilers. Proc. Combust. Inst. 31: 2779-2785.

[8] U. Kogelschatz (2004) Atmospheric-pressure plasma technology. Plasma Phys. Controlled Fusion 46: B63-B75.

[9] S.-W. Kim, H.-S. Park, H.-J. Kim (2003) 100 kW steam plasma process for treatment of PCBs (polychlorinated biphenyls) waste. Vacuum 70: 59-66.

[10] Y. Cheng, M. Shigeta, S. Choi, T. Watanabe (2012) Formation Mechanism of Titanium Boride Nanoparticles by RF Induction Thermal Plasma. Chem. Eng. J. 183: 483-491.

[11] M.S. Choi, D.U. Kim, S. Choi, B.-H. Chung, S.J. Noh (2011) Iron reduction process using transferred plasma. Curr. Appl. Phys. 11: S82-S86.

[12] P. Fauchais, A. Vardelle (1997) Thermal Plasmas. IEEE Trans. Plasma Sci. 25: 1258-1280.

[13] P. Fauchais, A. Vardelle, A. Denoirhean (1997) Reactive thermal plasmas: ultrafine particle synthesis and coating deposition. Surf. Coat. Technol. 97: 66-78.

[14] P. Fauchais, M. Vardelle (1994) Plasma spraying: present and future. Pure Appl. Chem. 66: 1247-1258.

[15] T. Iwao, M. Yumoto (2006) Portable application of thermal plasma and arc discharge for waste treatment, thermal spraying and surface treatment. IEEJ T. Electr. Electr. 1: 163-170.

[16] J.P. Chu, I.J. Hwang, C.C. Tzeng, Y.Y. Kuo, Y.J. Yu (1998) Characterization of vitrified slag from mixed medical waste surrogates treated by a thermal plasma system. J. Hazard. Mater. 58: 179-194.

[17] G. Rutberg, A.N. Bratsev, A.A. Safronov, A.V. Surov, V.V. Schegolev (2002) The technology and execution of plasma chemical disnfection of hazardous medical waste. IEEE Trans. Plasma Sci. 30: 1445-1448.

[18] S.K. Nema, K.S. Ganeshprasad (2002) Plasma pyrolysis of medical waste. Curr. Sci. India 83: 271-278.

[19] T. Inaba, M. Nagano, M. Endo (1999) Investigation of plasma treatment for hazardous wastes such as fly ash and asbestos. Electr. Eng. Jpn. 126: 73-82.

[20] R. Poiroux, M. Rollin (1996) High temperature treatment of waste: From laboratories to the industrial stage. Pure Appl. Chem. 68: 1035-1040.

[21] H. Jimbo (1996) Plasma melting and useful application of molten slag. Waste Manage. 16: 417-422.

[22] K. Katou, T. Asou, Y. Kurauchi, R. Sameshima (2001) Melting municipal solid waste incineration residue by plasma melting furnace with a graphite electrode. Thin Solid Films 386: 183-188.

[23] C.-C. Tzeng, Y.-Y. Kuo, T.-F. Huang, D.-L. Lin, Y.-J. Yu (1998) Treatment of radioactive wastes by plasma incineration and vitrification for final disposal. J. Hazard. Mater. 58: 207-220.

[24] L.I. Krasovskaya, A.L. Mossé (1997) Use of electric-arc plasma for radioactive waste immobilization. J. Eng. Phys. Thermophys. 70: 631-638.

[25] International Atomic Energy Agency (IAEA) (2006) Application of thermal technologies for processing of radioactive waste.

[26] Y. Byun, M. Cho, J.W. Chung, W. Namkung, H.D. Lee, S.D. Jang, Y.-S. Kim, J.H. Lee, C.R. Lee, S.M. Hwang (2011) Hydrogen recovery from the thermal plasma gasification of solid waste. J. Hazard. Mater. 190: 317-323.

[27] Y. Byun, W. Namkung, M. Cho, J.W. Chung, Y.S. Kim, J.H. Lee, C.R. Lee, S.M. Hwang (2010) Demonstration of thermal plasma gasification/vitrification for municipal solid waste treatment. Environ. Sci. Technol. 44: 6680-6684.

[28] A. Mountouris, E. Voutsas, D. Tassios (2006) Solid waste plasma gasification: Equilibrium model development and exergy analysis. Energy Convers. Manage. 47: 1723-1737.

[29] A.S. An'shakov, V.A. Faleev, A.A. Danilenko, E.K. Urbakh, A.E. Urbakh (2007) Investigation of plasma gasification of carbonaceous technogeneous wastes. Thermophys. Aeromech. 14: 607-616.

[30] G. Galeno, M. Minutillo, A. Perna (2011) From waste to electricity through integrated plasma gasification/fuel cell (IPGFC) system. Int. J. Hydrogen Energy 36: 1692-1701.

[31] M. Minutillo, A. Perna, D.D. Bona (2009) Modelling and performance analysis of an integrated plasma gasification combined cycle (IPGCC) power plant. Energy Convers. Manage. 50: 2837-2842.

[32] J. Heberlein, A.B. Murphy (2008) Thermal plasma waste treatment. J. Phys. D: Appl. Phys. 41: 053001.

[33] S.L. Camacho (1988) Industrial worthy plasma torches: State-of-the-art. Pure Appl. Chem. 60: 619-632.

[34] J.M. Park, K.S. Kim, T.H. Hwang, S.H. Hong (2004) Three-dimensional modeling of arc root rotation by external magnetic field in non-transferred thermal plasma torches. IEEE Trans. Plasma Sci. 32: 479-487.

[35] M. Hur, S.H. Hong (2002) Comparative analysis of turbulent effects on thermal plasma characteristics inside the plasma torches with rod- and well-type cathodes. J. Phys. D: Appl. Phys. 35: 1946-1954.

[36] Juniper.com (2008) The alter NRG/Westinghouse plasma gasification process: Independent waste technology report, in, Juniper.com, Bisley, England.

[37] Solenagroup Company. Available: http://www.solenagroup.com. Accessed 2011 May 15.
[38] J. Bowyer, K. Fernholz (2010) Plasma gasification: An examination of the health, safety, and environmental records of established facilities, in, Dovetail Partners, Inc..
[39] Westinghouse-plasma Company. Available: http://www.westinghouse-plasma.com. Accessed 2011 May 20.
[40] Europlasma Company. Available: http://www.europlasma.com. Accessed 2011 May 17.
[41] Phoenixsolutions Company. Available: http://www.phoenixsolutionsco.com/psctorches.html. Accessed 2011 May 16.
[42] Tetronics Company. Available: http://www.tetronics.com. Accessed 2011 May 19.
[43] Juniper.com (2007) Plasma technologies for waste processing applications: Juniper ratings report. Bisley, England.
[44] M.J. Castaldi, N.J. Themelis (2010) The case for increasing the global capacity for waste to energy (WTE). Waste Biomass Valor. 1: 91-105.
[45] Q. Zhang, L. Dor, K. Fenigshtein, W. Yang, W. Blasiak (2012) Gasification of municipal solid waste in the plasma gasification melting process. Appl. Energy 90: 106-112.
[46] E. Leal-Quirós (2004) Plasma Processing of Municipal Solid Waste. Braz. J. Phys. 34: 1587-1593.
[47] H. Cheng, Y. Hu (2010) Municipal solid waste (MSW) as a renewable source of energy: Current and future practices in China. Bioresour. Technol. 101: 3816-3824.
[48] U. Arena (2012) Process and technological aspects of municipal solid waste gasification. A review. Waste Manage. 32: 625-639.
[49] H. Nishikawa, M. Ibe, M. Tanaka, M. Ushio, T. Takemoto, K. Tanaka, N. Tanahashi, T. Ito (2004) A treatment of carbonaceous wastes using thermal plasma with steam. Vacuum 73: 589-593.
[50] D. Cyranoski (2006) One man's trash. Nature 444: 262-263.
[51] E. Dodge (2008) Plasma-gasification of waste. Clean production of renewable fuels through the vaporization of garbage. Cornell University-Johnson Graduate School of Management.

Chapter 8

Environmental Impacts of Hydrogen Production by Hydrothermal Gasification of a Real Biowaste

Sevgihan Yildiz Bircan, Kozo Matsumoto and Kuniyuki Kitagawa

Additional information is available at the end of the chapter

http://dx.doi.org/10.5772/50329

1. Introduction

Energy consumption is increasing regularly with increasing human population [1]. Finite resources of fossil fuels [2], security of other energy sources (especially nuclear energy), and concerns over greenhouse gases produced by combustion of fossil fuels have all motivated the search for renewable energy sources [3]. Energy from biomass could reduce the increase of carbon dioxide in the atmosphere and provide 14% of the world's energy needs [4, 5]. Also biomass gasification through the hydrothermal process has the added advantage of disposing of wastes [6]. Therefore, biomass has been selected for generation of energy by using hydrothermal gasification.

Hydrogen gas is anticipated as a fuel for clean power systems such as fuel cells. Many techniques have been reported for producing hydrogen gas [7, 8]. Hydrothermal gasification in sub or supercritical water has also been studied as a promising process for hydrogen production. The fluid can dissolve and decompose organic compounds [9]. Hydrothermal gasification is carried out at a relatively low temperature (about 400 °C) and occurs rapidly, compared with fermentation processes [10, 11]. Furthermore, hydrothermal gasification is carried out in supercritical fluid water, so this method is applicable to wet biomass samples without the necessity for a drying process, while the conventional thermal gasification needs excessive energy to dry wet biomass before it is gasified [4, 9, 12]. This process is therefore more suitable for biowastes with high water content, such as food wastes and animal dungs, than the conventional thermal gasification process that requires additional energy to overcome the latent heat of water.

There have been numerous studies related to the hydrothermal gasification process, and conducted for wide range of materials. Morimoto *et al.* [13] of Kyoto University studied hydrothermal gasification process of brown coal. Antal *et al.* [14] reported the gasification of cornstarch and wood dust. Yoshida *et al.* [15] studied supercritical water gasification of cellulose, hemi-cellulose, and lignin. However, this process has not been studied for animal waste, because animal wastes were thought to have the potential for environmental pollution [16, 17, 18].

Toxic compounds might be produced through the hydrothermal gasification of real biomass. Some chlorinated organic compounds are very toxic and can cause serious damage to the human body even with exposures of trace amounts. This study has also made a determination of resulting dioxins as these are among the most toxic substances.

This method would not be an optimum solution for disposing biowaste. However, hydrogen production by hydrothermal gasification of biowaste appears to be a promising source for the predicted hydrogen fuel production needs [19].

1.1. Hydrothermal gasification

Hydrothermal processing describes the thermal treatment of wet biomass at elevated pressures to produce carbohydrate, liquid hydrocarbons, or gaseous products depending upon the reaction conditions [20].

The processing pressure must be increased as the reaction temperature increases to prevent boiling of water in the wet biomass. At temperatures around 100 °C, extraction of high-value plant chemicals such as reins, fats, phenolics, and phytosterols is possible. At 200 °C and 2 MPa, fibrous biomass undergoes a fractionation process to yield cellulose, lignin, and hemicellulose degradation products such as furfural. Further hydrothermal processing can hydrolyze the cellulose to glucose. At 300-350 °C and 12.2-18.2 MPa, biomass undergoes more extensive chemical reactions, yielding a hydrocarbon-rich liquid known as biocrude. At 600-650 °C and 30.4 MPa the main products are gases, including a significant fraction of methane [20].

Hydrothermal pyrolysis is also known as hydrothermal liquefaction. Hydrothermal pyrolysis is a feasible method for waste treatment and conversion of wastes into liquid bio-products such as bio-oil. Hydrothermal liquefaction of biomass is a depolymerization process to break the solid organic compounds into smaller fragments [21].

In hydrothermal liquefaction, water simultaneously acts as a reactant and so this process is significantly different from pyrolysis [22].

Biomass can be thermally processed through either gasification or pyrolysis to produce hydrogen and other fuels. In general, the main gaseous products from the pyrolysis of biomass are H_2, CO_2, CO, and hydrocarbon gases, whereas the main gaseous products from the gasification of biomass are H_2, CO_2, CO, and N_2 [23].

Hydrothermal biomass gasification benefits from the special properties of near- and supercritical water as the solvent and its presence as the reaction partner. Relatively fast hydrolysis of biomass in sub and supercritical water leads to a rapid degradation of the polymeric structure of biomass [9].

1.2. Super critical water

A supercritical fluid (SCF) is any substances at a temperature and pressure above the critical point. Above the critical temperature of a substance, the pure, gaseous component cannot be liquefied regardless of the pressure applied. The critical pressure is the vapor pressure of the gas at the critical temperature. In the supercritical environment only one phase exists. The fluid, as it is termed, is neither a gas nor a liquid and is best described as intermediate to the two extremes. This phase retains solvent power approximating liquids as well as the transport properties common to gases.

At conditions around the critical point water has several valuable properties. Among them are low viscosity and high solubility of organic substances, making subcritical water an excellent medium for fast, homogeneous and efficient reactions. Supercritical water gasification is a promising technology for gasifying biomass with high moisture content [24]. Use of water as a reaction medium obviates the need to dry the feedstock and allows a fast reaction rate [25]. However corrosion in the subcritical water is a key issue [22].

There are two approaches to biomass gasification in supercritical water. The first: low-temperature catalytic gasification employs a reaction temperature ranging from 350 to 600 ºC (above 22.05 MPa) and gasifies the reaction material with the aid of metal catalysts. The second: high-temperature supercritical water gasification employs reaction temperatures ranging from 500 to 750 ºC (above 25 MPa), either without a catalyst or with non-metallic catalysts [10].

For the disposal of chicken manure, the advantages of hydrothermal gasification method are summarized in the Figure 1 below, which also shows some disadvantages of other methods.

1.3. Experimental equipment

The experimental setup was developed in this work for hydrothermal gasification. A stainless steel tube of SUS 316 of 1/2 inch in O.D., 12 cm in length is used as the reactor. One side of the reactor was sealed with a connector (Swagelok Co.) and the other side was connected with a 1/2 to 1/8 inch reducing union to which the Tee was connected. The strain amplifier for pressure measurement (Kyowa-Dengyo, Co., Japan) was connected to the one side of the Tee, and the stop valve was to the other side. A gas chromatograph oven (Hewlett Packard, 5890 GC) was used for heating the reactor at a programmed temperature [26, 27].

1.4. Reagents

Chicken manure (G.I. Ltd., Japan) containing 9% phosphorus was selected as a real biomass waste.

As a model sample containing phosphorus element, O-Phospho-DL-serine (Wako Chemical Co. Ltd, Japan) was used. O-Phospho-DL-serine, as the name implies, has a serine, which is an amino acid with the formula $HO_2CCH(NH_2)CH_2OH$. It is one of the proteinogenic amino acids. By virtue of the hydroxyl group, serine is classified as a polar amino acid. O-Phospho-DL-serine consists with phosphorylation of serine. Aspartate, glutamate, proline and serine are abundant amino acids in chicken manure [28]. Some of the constituent amino acids were found in a range from 24.7% (for valine) to 76.4% (for serine) in poultry manure [29]. O-phospho-DL-serine also contains the P atom in the molecule. Therefore it was chosen as the test sample.

$Ca(OH)_2$ used as an additive was purchased from Wako Chemical Co. Ltd, Japan.

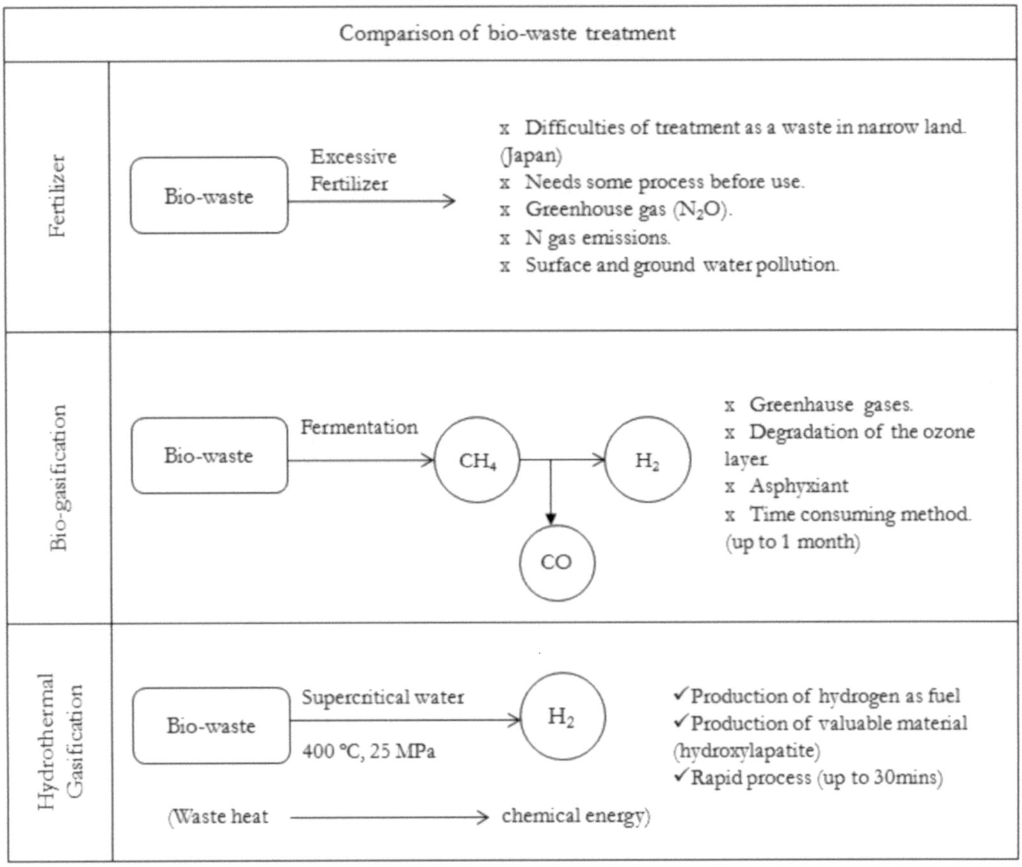

Figure 1. Diagram of comparison for bio-waste treatment.

1.5. Procedure for hydrothermal gasification

The biomass sample (chicken manure or O-Phospho-DL-serine) was weighed (about 100 mg) and put into the reactor. Additionally, the alkaline additive $Ca(OH)_2$ was weighed and added into the reactor (without $Ca(OH)_2$, with 2 mmol and 3 mmol $Ca(OH)_2$). Then 5 ml water was added. N_2 gas was introduced to purge the residual O_2 gas in the reactor. After the reactor was connected to the reducing unit with the pressure gauge and the stop valve, the reactor was placed in the oven. Then the oven was heated to 400°C at 0.025°C min^{-1}. The reactor was kept at 400°C for 40 minutes to complete the hydrothermal reaction under a pressure of 26~27 MPa. Subsequently, the oven was cooled down to room temperature and the components generated were analyzed [26, 27].

The experimental procedures are illustrated in Fig. 2 and comprised three main stages; sample preparation, hydrothermal gasification, and analysis of the compounds produced.

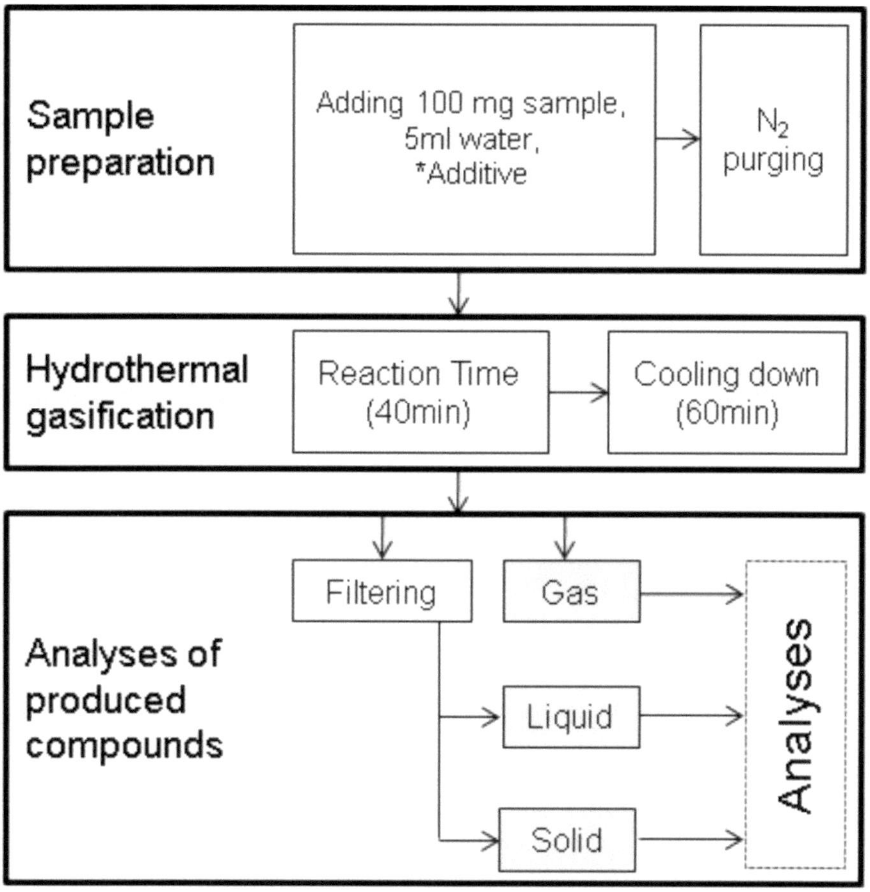

Figure 2. Experimental procedures of hydrothermal gasification

1.6. Analytical equipment

GC-TCD (Gas Chromatography - Thermal Conductivity Detector)

A 5A Shimadzu Gas Chromatograph (GC) of equipped with a thermal conductivity detector (TCD) was used for the analysis of chemical species in the gas phase.

IC (Ion Chromatography)

A Shimadzu (HIC-SP) Ion Chromatograph (IC) was used for the analysis of ionic species in the liquid phase.

GC/MS (Gas Chromatography/Mass Spectrometry)

More than 100 ml of the liquid sample was required for the determination of dioxins by GC/MS. The reaction procedure was repeated 27 to 30 times for each sample.

Dioxin analysis was performed on the resulting liquid and solid samples using a gas chromatograph combined with a mass spectrometer. An Agilent model 6890-GC interfaced with a JMS-700D double focus MS (JEOL, Japan) was used for the analysis [27].

2. Generation of hydrogen gas

For effective production of hydrogen gas and reduction of the formation of pollutants, optimum conditions for hydrothermal gasification of biowaste were examined under various experimental conditions by using O-phospho-DL-serine as a test sample. Next, chicken manure was used as a real biomass waste sample for the production of hydrogen gas by the hydrothermal gasification and for the suppression of the pollutants.

Additives were used to enhance the reaction rate of the hydrothermal gasification in sub or supercritical water at low reaction temperature [10]. The study also looked at whether the addition of catalysts could also enhance the hydrogen yield [30].

Several additives were used in earlier studies. The effects of the various alkaline metals on the amounts of generated gases have been reported [31]. When $Ca(OH)_2$ was used, only hydrogen gas was produced without production of other gases. This would be explained by the following Equations 1 and 2.

$$H_2O + CO \dashrightarrow CO_2 + H_2 \tag{1}$$

$$Ca(OH)_2 + CO_2 \rightarrow CaCO_3 + H_2O \tag{2}$$

The effects of the addition of various kinds of alkaline metals on the amounts of phosphate ion were also studied. The addition of Na_2CO_3 or K_2CO_3 was found to have no suppression effect on the production of phosphate ions in the liquid phase. However, when $Ca(OH)_2$ was added, no phosphate ions were detected. From these experimental results [31], it can be concluded that reasonable alkaline element compound, $Ca(OH)_2$ was a suitable additive because it could suppress the production of heteroatom pollutants in the gas phase and enhance the hydrogen yield [26].

2.1. Gas phase

The effects of the amounts of additive and temperature on the yield of gases generated were studied.

Without the additive, the main produced gas is CO, while hydrogen gas is also generated. 0.1943 mmol H_2, 0.2617 mmol CO, 0.0244 mmol CO_2, 0.0024 mmol CH_4, and 0.0088 mmol C_2H_4, 0.0010 C_2H_6 were detected [26].

With the addition of 2 mmol $Ca(OH)_2$, the yield of CO, CO_2 and C_2H_4 gases were suppressed. However, the generation of hydrogen gas was decreased in yield. 0.1459 mmol H_2, 0.0019 mmol CO, 0.0009 mmol CO_2, 0.0039 mmol CH_4, and 0.0019 mmol C_2H_4, 0.0003 C_2H_6 were detected [26].

With the addition of 3 mmol $Ca(OH)_2$, the main gas is hydrogen gas, while other gases were hardly detected. With addition of 3 mmol $Ca(OH)_2$, the main gas is hydrogen gas. Generation of hydrogen gas increases with an increase of gasification temperature. 0.2007 mmol H_2, 0.0002 mmol CO, 0.0009 mmol CO_2, 0.0017 mmol CH_4, and 0.0012 mmol C_2H_4, 0.0016 C_2H_6 were detected [26].

The enhancement of H_2 yield by adding alkali was due to water-gas shift reactions. These results indicate that the most suitable conditions for obtaining pure hydrogen gas from the hydrothermal reaction of the model sample, O-Phospho-DL-serine, are as follows: 3 mmol of additive $Ca(OH)_2$, reaction temperature at 400°C, and pressure of 22 MPa (super critical state).

2.2. Liquid phase

The effects of the added amount of $Ca(OH)_2$ on the yield of phosphate ion dissolved in the liquid phase through the hydrothermal reaction under the supercritical conditions at 400°C were also studied. When no additive was used, the yield of phosphate ion in the liquid phase found was 93.3% of the P in the original sample. However, the addition of 2mmol $Ca(OH)_2$ resulted in the suppression of the formation of phosphate ion in the liquid phase. When 3mmol of $Ca(OH)_2$ was added, the generation of phosphate ion was further decreased to 5.6%. Phosphorus containing compounds were barely detectable in the liquid phase. Phosphorus in the sample would be converted and precipitated as solid compounds (Figure 3).

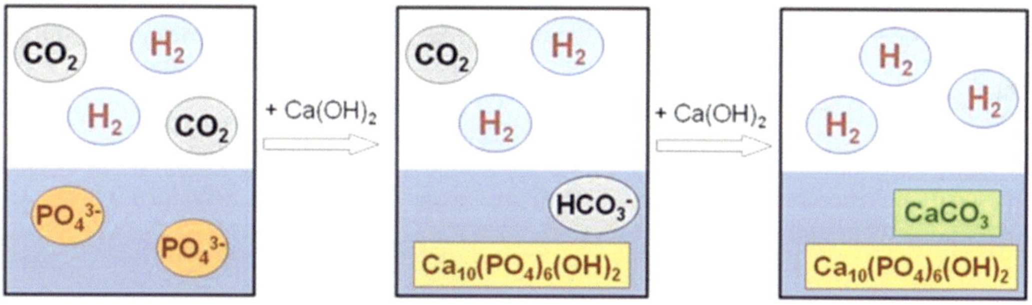

Figure 3. Estimation of phosphorus conversion.

2.3. Hydrothermal gasification of chicken manure

Chicken manure, which contains phosphorus, was selected as a real biowaste for the production of hydrogen gas and suppressing formation of pollutants by the hydrothermal reaction. Various reaction conditions were investigated for suitable conditions. The same optimum conditions were obtained as those of the hydrothermal reaction of the model test compound, O-Phospho-DL-serine. With the same conditions of 3mmol $Ca(OH)_2$ and 400°C, hydrogen gas was mainly produced in the gas phase. 0.1122 mmol H_2, 0.0044 mmol CO, 0.2088 mmol CO_2, 0.0025 mmol CH_4, and 0.0114 mmol C_2H_4, 0.0014 C_2H_6 were detected. H_2 yields were increased and other gasses were suppressed by using the additive, especially in the case of CO_2, which was suppressed very effectively. It was concluded that the enhancement of H_2 yield by adding the alkali was due to the water-gas shift reactions (Equations 3 and 4). Equation (5) shows the production of $CaCO_3$ after hydrothermal gasification by adding $Ca(OH)_2$ [26].

$$\text{Org. C} + H_2O \rightarrow CO + H_2 \tag{3}$$

$$H_2O + CO \rightarrow CO_2 + H_2 \tag{4}$$

$$CO_2 + Ca(OH)_2 \rightarrow CaCO_3 + H_2O \tag{5}$$

Additionally, phosphate ion was hardly detected in the liquid phase as in case of the model sample. The phosphorus compounds in the real sample are decomposed and new compounds would be produced and precipitated in the solid phase by the hydrothermal reaction. From these results the following equation is obtained (Equation 6). When the sample includes phosphorus, the P element would be converted into PO_4^{3-} by the hydrothermal reaction [26]. The ion, PO_4^{3-}, would react with Ca^{2-} ion and some insoluble compound would be produced.

$$P(\text{ in a Sample}) + H_2O \rightarrow PO_4^{3-} \tag{6}$$

When $Ca(OH)_2$ was used as the additive, the main produced gas was hydrogen gas, and the generation of CO_2 gas was suppressed efficiently. Additionally, calcium ion easily reacts with heteroatoms, and would form insoluble solid material in water. The cost of $Ca(OH)_2$ is less expensive than other additives. To treat a large amount of bio-wastes, reasonable reagents are more preferable. $Ca(OH)_2$ was decided to use as the additive for understanding the reaction mechanisms for disposal of hetero-atom containing compounds under the hydrothermal process.

In the hydrothermal reactions with the use of $Ca(OH)_2$ as the additive, the suppression of CO_2 and the promotion of H_2 generation are expected from the reactions which are expressed on Equation 3 and 4.

3. Dioxins analysis

3.1. Dioxins

The name "dioxins" is often used for the family of structurally and chemically related polychlorinated dibenzo para dioxins (PCDDs) and polychlorinated dibenzofurans (PCDFs). Certain dioxin-like polychlorinated biphenyls (PCBs) with similar toxic properties are also included under the term "dioxins". Some 419 types of dioxin-related compounds have been identified but only about 30 of these are considered to have significant toxicity, in which TCDD (2,3,7,8- tetrachlorodibenzo para dioxin) is the most toxic [32]. The formation mechanisms for them are not yet completely understood because of their complex production mechanisms [33]. Dioxins do have a damaging effect on human health and the environment [32, 34], and 30 dioxins are known to have significant toxicity [32]. When biomass-containing chlorine is gasified in supercritical water, PCDDs, PCDFs and PCBs might be formed. In this study, dioxins in the liquid and solid phases produced through the hydrothermal reaction of chicken manure were determined.

TEF and TEQ

TEQ (toxic equivalent quantity) is total toxicity of dioxins contained in a sample and calculated by the Equation (7),

$$TEQ = \sum f_i g_i \tag{7}$$

f_i : toxic equivalency factor for i_{th} dioxin (TEF, WHO 2006 [35])

g_i : the abundance of i_{th} dioxin in the sample.

3.2. Experimental procedure

In order to examine the effect of the additive and the effect of temperature on dioxin formation in the chicken manure, the experiments were performed under six different conditions (Figure 4).

Figure 4. Photographs of solid (a) and liquid (b) samples from six different conditions. R1; without additive, 200 ºC, R2; 3 mmol $Ca(OH)_2$, 200 ºC,R3; without additive, 300 ºC, R4; 3 mmol $Ca(OH)_2$, 300 ºC, R5; without additive, 400 ºC, R6; 3 mmol $Ca(OH)_2$, 400 ºC.

The samples produced under the various experimental conditions were separated into liquid and solid phases by filtration.

Solid samples

For determination of toxic equivalent quantity (TEQ) of each dioxin for the solid phase, the hydrothermal gasification experiment was carried out under the various conditions for the chicken manure. PCDDs and PCDFs were not detected. Three kinds of PCBs were only detected. These were T4CB#77 (Fig. 5) (TEF=0.0001), P5CB#118 (Fig. 6) (TEF=0.00003), and P5CB#105 (Fig. 7) (TEF=0.00003).

The total TEQ values for solid samples were 0.00237, 0.00357, 0.00647, 0.00196, 0.00172, and 0.00148 pgTEQg^{-1} for Run 1, 3, 5, 2, 4, and 6, respectively.

The highest total TEQ of 0.00647 pgTEQg^{-1} was observed for the reaction temperature of 400°C without additive (Run 5). This level is well below the permitted Japanese level for solid residue (3000 pgTEQg^{-1}) [36].

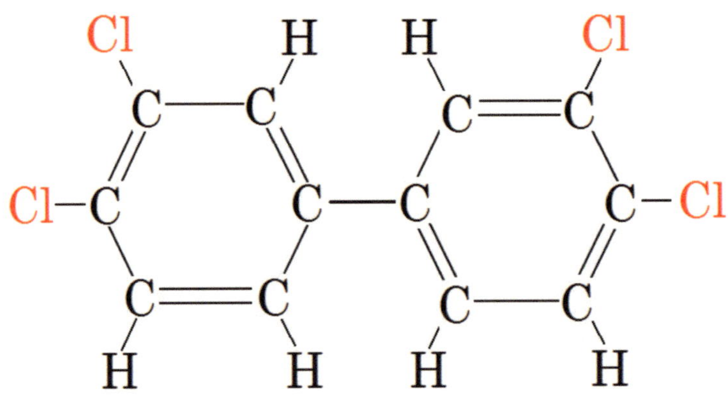

Figure 5. Chemical structure of T4CB#77.

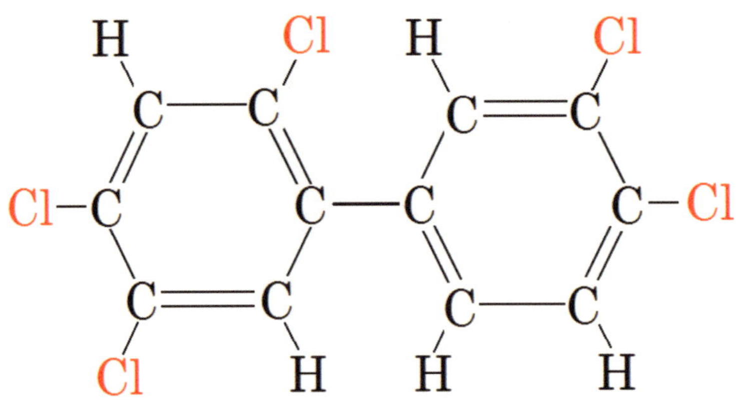

Figure 6. Chemical structure of P5CB#118.

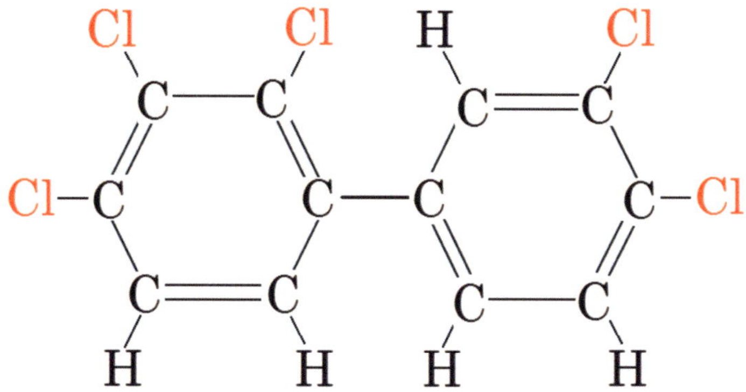

Figure 7. Chemical structure of P5CB#105.

Liquid samples

In the case of the liquid phase products, PCDDs and PCDFs were not detected as they were in the case of the solid phase products. Two kinds of PCBs were detected (vs. three in the solid phase material). These were P5CB#118 (TEF=0.00003) and P5CB#105 (TEF=0.00003).

The total TEQ values were 0.00026, 0.00054, 0.00029, 0.00023, 0.00028 and 0.00042 $pgTEQL^{-1}$ for Run 1, 3, 5, 2, 4, and 6, respectively.

With and without the additive, the total TEQs are nearly equal to the level of tap water. The results show that reaction temperature has little effect on the formation of dioxins. However, the addition of $Ca(OH)_2$ increases the value of the TEQ at reaction temperatures of 300 °C and 400 °C. The highest total TEQ measured was 0.00054 $pgTEQL^{-1}$, observed at the reaction temperature of 200 °C without the additive (Run 1). This total TEQ was well below the permitted Japanese limit for liquid residue (10 $pgTEQL^{-1}$) [36].

4. Conclusions

Increase in energy consumption, limited energy capacity, environmental concerns related to fossil fuels, and security/safety concerns of some energy sources have all motivated the search for renewable energy sources.

A real biowaste, chicken manure, was used as an energy source and $Ca(OH)_2$ was the most effective additive among the tested additive candidates for producing hydrogen in this study by the hydrothermal gasification process. Almost pure hydrogen gas could be obtained by adding $Ca(OH)_2$ under supercritical conditions. It was found that the generation of hydrogen gas through hydrothermal gasification could be conducted without considering the toxicity of dioxins. Dioxins were detected, but they were far below the environmental regulation values. An added benefit found was that this process solves the problem of treatment of chicken manure while producing hydrogen.

This newly developed method of hydrothermal gasification of chicken manure is a promising method for producing hydrogen as a fuel and for disposing of the biowaste.

Author details

Sevgihan Yildiz Bircan*
Department of Mechanical Science and Engineering, Graduate School of Engineering, Nagoya University, Furo-cho, Chikusa-ku, Nagoya, Japan

Kozo Matsumoto
EcoTopia Science Institute, Nagoya University, Furo-cho, Chikusa-ku, Nagoya, Japan

Kuniyuki Kitagawa
EcoTopia Science Institute, Nagoya University, Furo-cho, Chikusa-ku, Nagoya, Japan

5. References

[1] US Census Bureau, BP (2002)

[2] M. King Hubbert, Nuclear energy and fossil fuels, Shell Development Company, Exploration and Production Research Division, Houston, Texas (1956)

[3] D. J. K. MacKay, Sustainable Energy-Without the hot air, UIT Cambridge, England (2009)

[4] L. Kong, G. Li, B. Zhang, W. He, H. Wang, Hydrogen production from biomass wastes by hydrothermal gasification, Energy Sources Part A, 30 (2008), Pages: 1166-1178

[5] D. B. Levin, H. Zhu, M. Beland, N. Cicek, B. E. Holbein, Potential for hydrogen and methane production from biomass residues in Canada, Bioresource Technology 98 (2007), Pages: 654-660

[6] Y. Calzavara, C. Joussot-Dubien, G. Boissonnet, S. Sarrade, Evaluation of biomass gasification in supercritical water process for hydrogen production, Energy Conversion and Management 46 (2005), Pages: 615-631

[7] W. Lijun, C.L. Weller, D.D. Jones and M.A. Hanna, 2008. Review: Contemporary issues in thermal gasification of biomass and its application to electricity and fuel production. Biomass Bioenergy 32, Issue 7, (2008), Pages: 573–581

[8] Y. Yurum, Hydrogen Energy System Production and Utilization of Hydrogen and Future Aspects : Proceedings of the NATO Advanced Study Institute, Akcay, Turkey, August 21-September 3 (1994)

[9] A. Kruse, Hydrothermal biomass gasification, J. of Supercritical Fluids 47 (2009), Pages: 391–399

[10] Y. Matsumura, T. Minowa, B. Potic, S. R. A. Kersten, W. Prins, W. P. MV. Swaaij, B. V. D. Beld, D. C. Elliott, G. G. Neuenschwander, A. Kruse, Jr. M. J. Antal, Biomass gasification in near- and super-critical water: Status and prospects (Review), Biomass and Bioenergy 29 (2005), Pages: 269–292

[11] A. Chu, D. S. Mavinic, H. G. Kelly, and C. Guarnaschelli, The influence of aeration and solids retention time on volatile fatty acid accumulation in thermophilic aerobic digestion of sludge. Environmental Technology 18 (1997), Pages: 731–738

* Corresponding Author

[12] M. Momirlana, T. N. Veziroglu, The properties of hydrogen as fuel tomorrow in sustainable energy system for a cleaner planet, International Journal of Hydrogen Energy 30 (2005), Pages: 795–802
[13] M. Morimoto, H. Nakagawa, K. Miura, Hydrothermal extraction and hydrothermal gasification process for brown coal conversion, Fuel 87 (2008), Pages: 546–551
[14] M.J. Antal, S.G. Allen, D. Schulman, X. Xu, R.J. Divilio, Biomass gasification in supercritical water. Industrial & Engineering Chemistry Research 39 (2000), Pages: 4040–53
[15] T. Yoshida, Y. Matsumura, Gasification of cellulose, xylan and lignin mixtures in supercritical water, Industrial & Engineering Chemistry Research 40 (2001), Pages: 40:5469
[16] D. R. Edwards, T. C. Daniel, Environmental impacts of on-farm poultry waste disposal: A review Bioresource Technology 41 (1992), Pages: 9-33
[17] M. G. M. Berges, P. J. Crutzen, Estimates of global N_2O emissions from cattle, pig and chicken manure, including a discussion of CH4 emissions, Journal of Atmospheric Chemistry 24 (1996), Pages: 241-269
[18] R. Khaleel, K. R. Reddy, M. R. Overcash, Transport of potential pollutants in runoff water from land areas receiving animal wastes: A review, Water Research 14 (1980), Pages: 421-436.
[19] J. A. Onwudili and P. T. Williams. Role of sodium hydroxide in the production of hydrogen gas from the hydrothermal gasification of biomass, International Journal of Hydrogen Energy 34 (2009), Issue 14, Pages: 5645-5656
[20] R. C. Brown, Thermochemical Processing of Biomass: Conversion Into Fuels, Chemicals and Power (2011)
[21] S. Xiu, A. Shahbazi, V. Shirley, D. Cheng, Hydrothermal pyrolysis of swine manure to bio-oil: Effects of operating parameters on products yield and characterization of bio-oil, Journal of Analytical and Applied Pyrolysis 88 (2010), Pages: 73-79
[22] S. S. Toor, L. Rosendahl, A. Rudolf, Hydrothermal liquefaction of biomass: A review of subcritical water technologies, Energy 36 (2011), Pages: 2328-2342
[23] C.A.C. Sequeira, P.S.D. Brito, A.F. Mota, J.L. Carvalho, L.F.F.T.T.G. Rodrigues, D.M.F. Santos, D.B. Barrio, D.M. Justo, Fermentation, gasification and pyrolysis of carbonaceous residues towards usage in fuel cells, Energy Conversion and Management 48 (2007), Pages: 2203–2220
[24] A. Demirbas, Biorefineries: Current activities and future developments, Energy Conversion and Management 50 (2009), Pages: 2782–2801
[25] T. Yoshida, Y. Oshima, Y. Matsumura, Gasification of biomass model compounds and real biomass in supercritical water, Biomass and Bioenergy (2004), Volume: 26, Issue: 1, Pages: 71-78
[26] S. Yildiz Bircan, H. Kamoshita, R. Kanamori, Y. Ishida, K. Matsumoto, Y. Hasegawa, K. Kitagawa, Behavior of heteroatom compounds in hydrothermal gasification of biowaste for hydrogen production, Applied Energy Volume 88, Issue 12, December 2011, Pages: 4874-4878

[27] S. Yildiz Bircan, R. Kanamori, Y. Hasegawa, K. Ohba, K. Matsumoto, K. Kitagawa GC-MS ultra trace analysis of dioxins produced through hydrothermal gasification of biowastes, Microchemical Journal (2011) Volume 99, Issue 2, November 2011, Pages 556-560

[28] I. Nachamkin, C. M. Szymanski, M. J. Blaser, Campylobacter (2008), Page: 47

[29] R. Prabakaran, Good practices in planning and management of integrated commercial poultry production in South Asia (2003), Page: 88

[30] Z. Fanga, T. Minowa, C. Fang, Jr. R. L. Smith, H. Inomata, J. A. Kozinski, Catalytic hydrothermal gasification of cellulose and glucose, International Journal of Hydrogen energy 3, (2008), Pages: 981–990

[31] R. Kanamori, K. Matsumoto, K, Kitagawa, Behavior of phosphorus compounds in hydrothermal decomposition of biomass, World Renewable Energy Congress, Bangkok, Thailand (2009)

[32] World Health Organization (WHO), http://www.who.int/mediacentre/factsheets/fs225/en/

[33] T. Katami, A. Yasuhara, T. Okuda, T. Shibamoto, Formation of PCDDs, PCDFs, and Coplanar PCBs from Polyvinyl Chloride during Combustion in an Incinerator, Environmental Science & Technology 36 (2002), Pages: 1320-1324.

[34] F. Coulston, F. Pocchiari, Accidental Exposure to Dioxins: Human Health Aspects, Academic Press, New York (1983)

[35] M. Van den Berg, L.S. Birnbaum, M. Denison, M. De Vito, W. Farland, M. Feeley, H. Fiedler, H. Håkansson, A. Hanberg, L. Haws, M. Rose, S. Safe, D. Schrenk, C. Tohyama, A. Tritscher, J. Tuomisto, M. Tysklind, N. Walker, R.E. Peterson, The 2005 World Health Organization reevaluation of human and mammalian toxic equivalency factors for dioxins and dioxin-like compounds, Toxicological Sciences 93 (2006), Pages: 223-241.

[36] Japan Ministry of the Environment, Overview on the treatment of POPs in Japan (2006)

Section 4

Cost and Economics of Gasification

Chapter 9

Gasification a Driver to Stranded Resource Development

Yuli Grigoryev

Additional information is available at the end of the chapter

http://dx.doi.org/10.5772/53565

1. Introduction

When evaluating natural gas projects, stakeholders and decision-makers have been traditionally limited by the requirement for large reserves to be recoverable before any investment can be committed. Technological advancement and commercial know-how has unlocked many potential reserves that would technically be considered stranded or entirely overlooked by IOCs (International Oil Companies) and NOCs (National Oil Companies). This has important implications for policymakers as it can affect the way natural gas is utilised as an indigenous supply, an export resource or substitution fuel.

There have been some structural changes in the way the gas sector operates over the last decade. This has been an exciting time for market observers, as political, economic, financial and technical inputs have driven major changes in the industry.

One of the ways that the industry has changed is the attention to using gas in less traditional methods. To understand how this has developed, this chapter analyses the major recent trends, the traditional market characteristics and discusses the outlook of potential future changed in the field of natural gas development, production and processing.

When looking at gas field development, stakeholders evaluate a number of options to monetise gas. These options are limited by a host of factors, each with a unique position with respect to geography, government, market and political dynamics.

This chapter acts as an overview of the monetisation routes and the options that are becoming available to decision-makers.

2. Current and future trends of oil and gas market

Over the past decade, there has been a paradigm shift in the behaviour of the oil and natural gas markets. Compared to the traditional model, where gas production was secondary to the production and marketing of oil, and prices of gas were naturally linked to the price of oil (or a basket of oil products), we have seen gas emerge as in increasingly important fuel with a decoupling of prices. This has been particularly evident in the North American market, where competitive forces and regulation of the midstream sector allowed for an emergence of a separate gas market, marked by consistently high liquidity. There has also been a discrepancy in the regional gas price, which lends to arbitrage activity by spot traders, and some LNG (Liquefied Natural Gas) cargoes have been redirected from initial destinations to other markets, even whereby destination clauses in contracts have been broken.

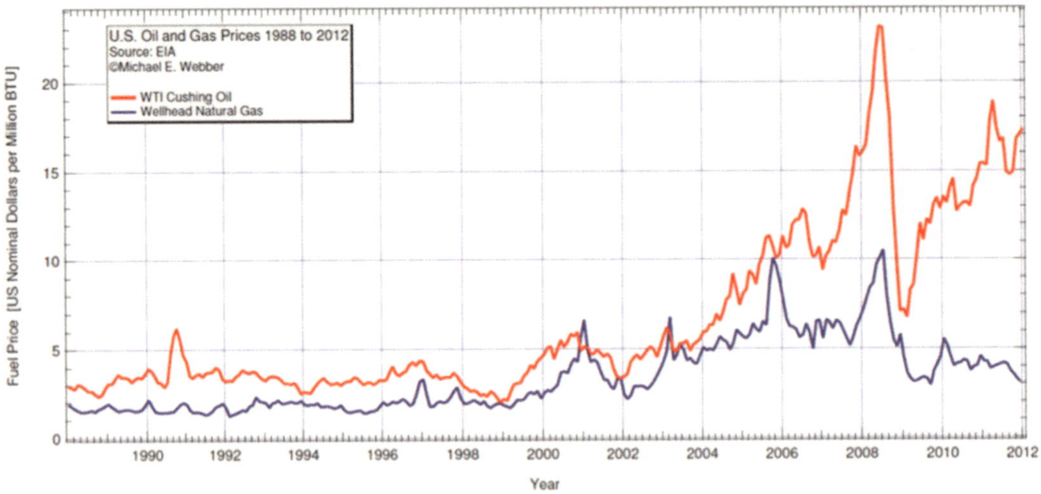

Source: IEA, Michael E Webber, 2012Figure 1

Figure 1.

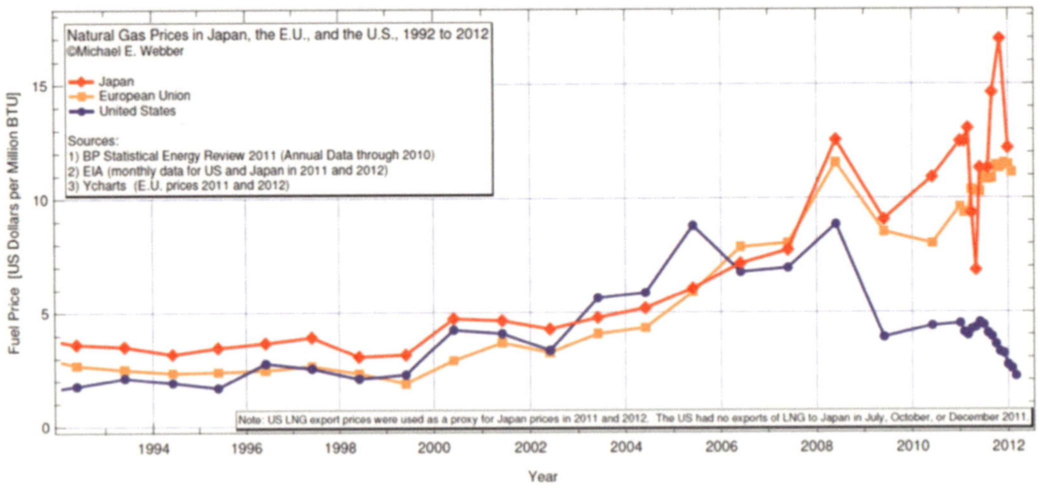

Source: IEA, BP Stats Review, Michael E Webber, 2012

Figure 2.

There is some expectation that natural gas consumption may over take oil consumption by 2030, which will result from a number of pressures on oil consumption, ranging from economic and environmental, to issues relating to security of supply.

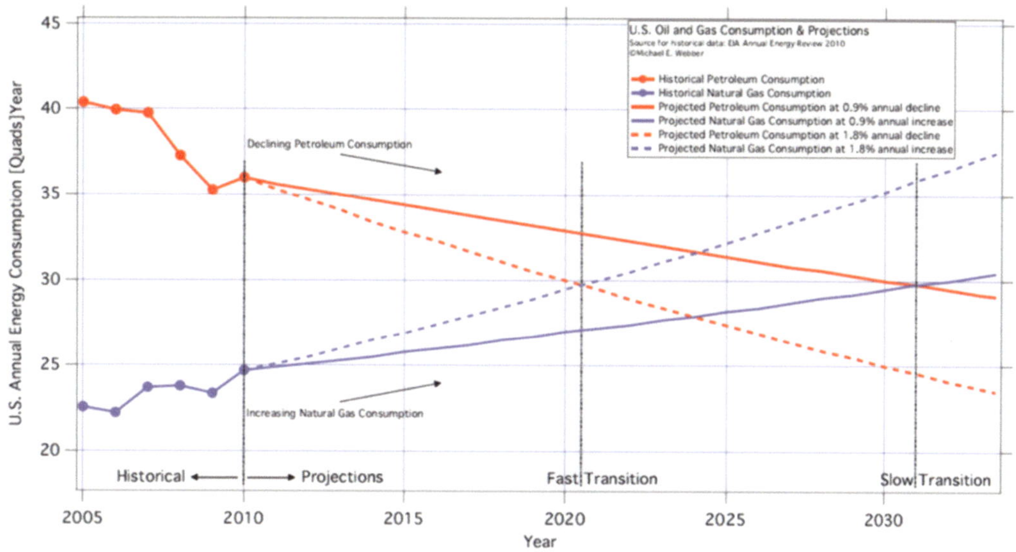

Source: EIA, M Webber, 2012

Figure 3.

In Europe, this shift has seen a slower uptake. A primary reason for this is the separation of supply and demand centres, with the EU zone increasingly relying on imported gas. The biggest supplier to this market is Russia, also the world's largest producer of natural gas, which as been supplying gas to Europe via soviet built high pressure interregional trunk pipelines.

The majority of contracts are long term take-or-pay contracts which have a price formula as an index linked to a basket of refined products (the "substitution fuels"). Historically, long-term contracts have played an important role in the development of the European gas market by providing a risk sharing arrangement between producers and buyers, enabling important new investment into production and infrastructure projects to be undertaken. The Eurozone realised that their growing gas needs, the bulk of which are met with Russian gas, can only be adequately supplied if Russia is able to invest in new gas fields and pipeline construction. They took a position that if gas is supplied exclusively through spot transactions, gas suppliers, Gazprom included, will not be willing to shoulder the risks associated with multi-billion dollar investments and substantial quantity risks. Corporate strategy aside, it would be impossible to access the international capital markets without guaranteed offtake contracts being in place. [1] Thus, contracts of 20 years or more have been a normal occurrence in the European continent.

[1] The dynamics of funding such projects are very complex and are out of the scope of this article; however it is worth mentioning that domestic markets of major producers lack the hard currency to finance national champions whilst international capital markets generally shun away from risks associated with emerging market domestic consumption.

Source: IEA, 2010

Figure 4.

country	contract term	volume of annual supplies	description	additional agreements
Austria	2012–2027	7 billion m^3	contract signed by Gazprom with OMV, EconGas as well as with GWH and Centrex Europe Energy & Gas AG, affiliated to Gazprom	The rules for gas reception have changed: - change of gas recipient from OMV to EconGas ; - GWH and Centrex, which are affiliated with Gazprom, were granted the right to sell gas (nearly 1.75 billion m^3 annually) to Austrian end-users
Germany (VNG)	2014–2031	5,25 billion m^3 (possibility of incre-asing supplies)	contract signed by WIEH (a joint venture of Gazprom and Wintershall) and Verbundnetzgas (VNG). Gazprom sells gas via WIEH in the markets of other EU member states	
Germany (E.ON)	2011–2036	4 billion m^3 (Nord Stream)	contract signed by Gazprom and E.ON on supplies via NS gas pipeline	
Germany (E.ON)	2020–2035	20 billion m^3	contract signed by Gazprom and E.ON, being an extention of the currently binding contracts (to expire in 2020)	
Italy	2017–2035	22 billion m^3	contract signed by Gazprom and ENI which constitutes a part of the new strategic partnership agreement; extention of the terms of the currently binding contracts	- companies affiliated with Gazprom (GMT Italia, Centrex) have gained access since 2007 to the Italian domestic market, and sales of up to 3 billion m^3 annually in 2010 is allowed - ENI has been promised the right to buy assets in Russia
France	2012–2030	12 billion m^3 + 2,5 billion m^3 (Nord Stream)	contract signed by Gazprom and Gaz de France ; the currently binding contract is prolonged, and an agreement for supplies via the Nord Stream gas pipeline is concluded	- Gazprom's subsidiary, GMT France, has been granted access to the French domestic market, and is allowed to sell up to a maximum of 1.5 billion m^3 of gas annually from 2007
Czech Republic (RWE)	2014–2035	9 billion m^3	contract signed by Gazprom and RWE Transgas; prolongation of the previous agreement on gas supplies to, and transit through, the Czech Republic	
Czech Republic (Vemex)	2008–2012	0,5 billion m^3	contract concerns supplies of gas to the Gazprom-controlled company which supplies gas to some Czech industrial customers	
Bulgaria	2011–2030	3 billion m^3	Gazprom and Bulgargaz have signed a contract prolonging the previously binding contract	
Romania (WIEH)	2012–2030	4,5 billion m^3	the previously binding contract has been prolonged	The contract's prolongation is made dependent on its renegotiation, and switching to cash payments for gas transit. The agreement provides for the increase of the gas transit through Bulgaria in exchange for co-operation in the implementation of Russian gas pipeline projects.
Romania (Conef)	2010–2030	2,5 billion m^3	Gazprom and Conef Energy SRL have signed a contract; supplies for Alro Stalina aluminium plant	
Denmark	2011–2031	1 billion m^3	Gazprom and DONG Energy have signed a contract for supplies from NS	

Based on *Gazprom in Europe: Faster expansion in 2006*, Ewa Paszyc, Centre for Eastern Studies, Warsaw, February 2007; data quoted from the companies' websites and news agencies

Source: Ewa Paszyc, Centre for Eastern Studies, 2008

Figure 5.

A number of drivers have begun to put significant pressure on the traditional model. Working in tandem, the economic growth on the continent, together with significant global environmental concerns and directives, has delivered a growing demand for natural gas. At the same time, as indigenous supplies begin to plateau and decline[2] and governments become more reliant on imports, Security of Supply issues begin to make their way up national policy agendas. From a security of supply standpoint, there have traditionally been three pillars of national strategy for policymakers – development of indigenous supplies, diversification of suppliers and reduction in consumption.[3] From the three pillars, the fastest route is evidently the diversification of suppliers as consumption reduction and indigenous suppliers requires significant lead-times. Certainly it is difficult to diversify in a timely manner if transit is to take place via pipelines, however, as the market of liquefied natural gas became more mature, it allowed for an efficient way to introduce new suppliers. Regasification terminals are significantly less complex than liquefaction terminals, and began to appear in a host of European coastlines.

---

[2] From the major producers, the UK North Sea production is in decline, Dutch production is capped, and Norwegian fields are in plateau (although there is heavy E&P activity).

[3] Temporary relief was seen during the 2007-2011 financial crisis, as demand destruction allowed for a temporary shift from "sellers market" to "buyers market" and attention of the ministers was diverted to dealing with the financial economy and the failing UN Framework Convention on Climate Change.

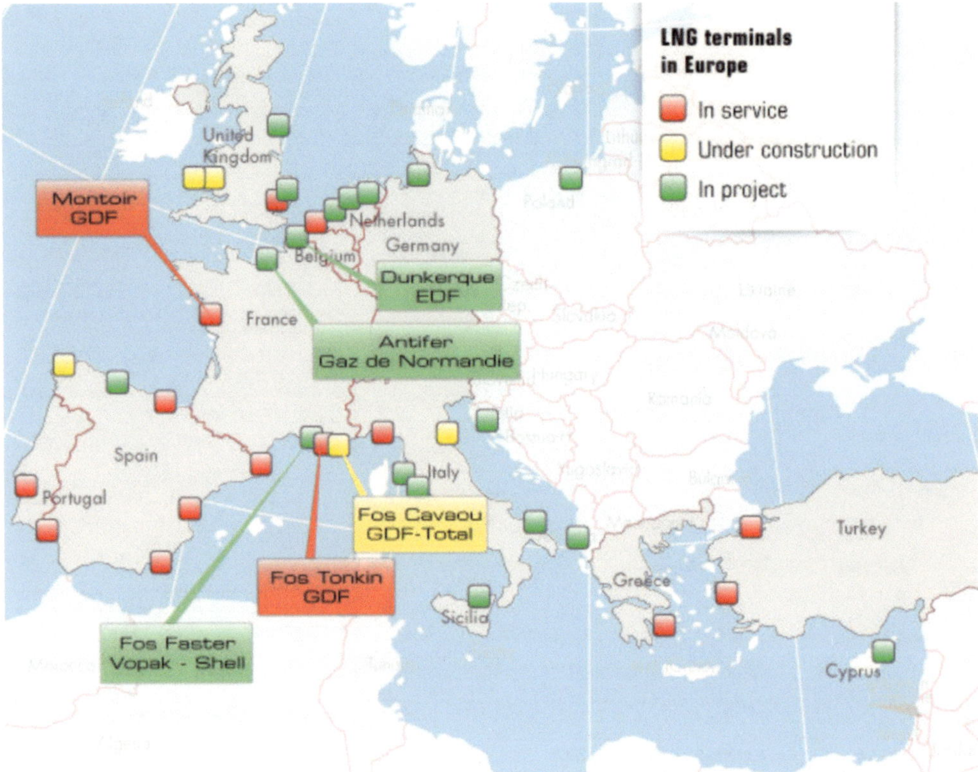

Source: GdF, 2011

Figure 6.

European competition rules have created somewhat of a stumbling block for these initially, but investment arrived in sufficient quantity to allow for an emerging spot market in the European gas hubs. The net effect of this has been an evolution of long-term contracts with certain traditional terms being re-examined and renegotiated. Some of the centrally important clauses such as duration/period are seeing a decrease from the frequently encountered fifteen to twenty-five years to perhaps eight to twelve years in length. This is, in part, due to the contract volumes also decreasing with new project supplying between three and ten BCM (Billion Cubic Metres) annually as opposed to the traditional ten to twenty BCM. Take-or-pay obligations are also become less stringent, with increasing "carry-forward" and "make-up" rights. Index pricing is being replaced in highly competitive markets by daily pricing derived from a liquid short term market, such as the UK National Balancing Point. Certainly this trend will apply to some of the new export contracts yet others, which intend to supply large volumes and require substantial infrastructure investment, will be done under traditional terms.[4]

What cannot go unmentioned is the shale gas development. The flurry of exploration activity has seen significant results in adding major volumes to reserves in the US, and has

[4] Nord Stream and South Stream, for example.

become a game-changer in the US domestic market. The effect on global markets has not yet been so dramatic, although exploration activity for shale gas in the Eurozone has excited many a journalist and energy observer. Thus far, however, the UK has enforced a temporary ban on shale gas fracking and Poland's estimate of reserves has so far been cut by a factor of ten. How this develops could have a profound effect on the industry.

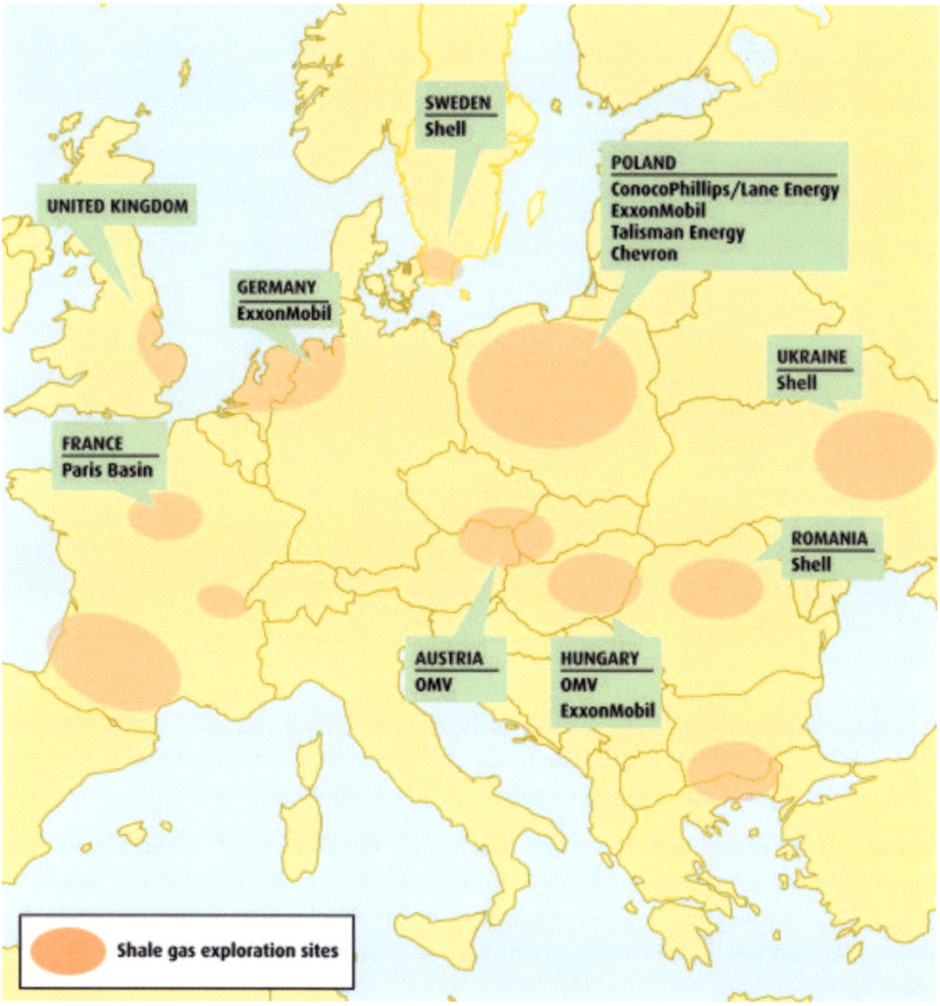

Source: Energy Tribute, 2011

Figure 7.

3. Traditional gasification uses

Where gas has been discovered in abandoned supply, the stakeholders had a very clear picture of how this asset can be monetised. The most straightforward solution has been the construction of a pipeline from the supply centre to the consumption centre, where gas would be used for heat and power generation, industry and grid supply. Pipelines can run for many thousands of kilometres, over different and difficult terrain, across borders, through mountains and under water.

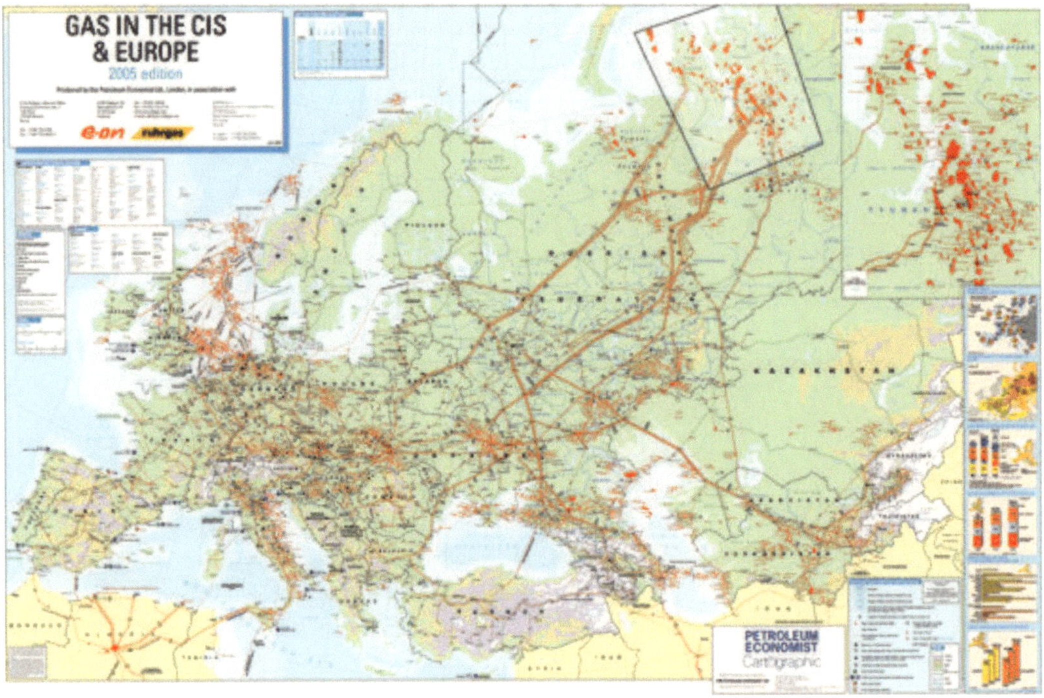

Source: Petroleum Economist, 2006

Figure 8.

If the consumption centres were satisfactorily supplied, or if the cost of the pipeline was prohibitively expensive, gas was either left in the ground, or converted into a final product that could be transported as a liquid or solid to other distant markets.

These conversion routes are what are known as the "Gas-to- " technologies and are specifically Gas-to-Chemicals, Gas-to-Liquids and Gas-to-Power. Recent advances in technology have allowed these processes to become available as economic methods not only to utilise stranded gas but to take advantage of pipeline gas that may be limited in its transport options.[5]

Gas-to-Liquids (GTL) is a process that was initially discovered by Fisher and Tropsch during the World War II and has seen various applications thereafter. In essence, it is a petrochemical process that converts methane (major component of natural gas) into a synthetic diesel fuel that is environmentally clean as it contains no sulphur and is aromatics free. The first major commercial GTL facility was built in South Africa by Sasol, using coal gasification to produce the feedstock and manufacturing diesel oil. It is generally accepted

[5] Russian independent gas producers are prohibited from exporting natural gas by law (Federal Law "On Gas Export", 2006).

that due to economies of scale, GTL facilities become economic with large output capacities and extremely low feedstock cost. As such, new plants are expected to be in excess of 100,000 barrels per day (bpd) of product, and located in the Middle East or Africa[6]. Multi-billion dollar projects such as Shell's Pearl GTL in Qatar and Oryx GTL (Qatar Petroleum and Sasol) are leading examples of this technology in application. Another 200,000 bpd plant has been proposed in Australia.

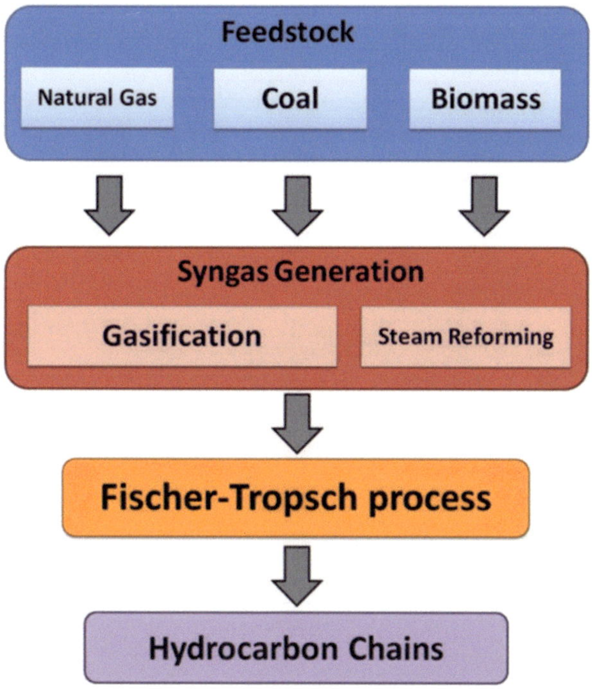

Source: Stamford University, 2010

Figure 9.

The Gas-to-Chemicals (GTC) process is a very mature process. This involves the conversion of methane to a chemical product, either an intermediate or final stage. Indeed, more value is captured the further down the process chain that one is able to proceed. The most common product is Ammonia which is used in the production of fertilizers. The high oil price has been somewhat of a double-edged sword for the price of fertilizers since the increase in the feedstock (where natural gas is still tied to the oil price) and also the increase in demand driven by the biofuels surge as a means to find alternative energy solutions. Because natural gas makes up about 70% of the cost of production, European based producers can no longer compete with producers with a low cost base such as Russia, or even Ukraine (due to special relationships with Russia[7]). There is a clear link between

[6] SassolChevron is in the process of building a 34,000 bpd plant in Nigeria.

[7] This does not refer to inter-governemental price agreements but to private agreements between Gazpromexport and Ukrainian fertiliser producers.

financially stable fertiliser producers and a low gas source. As more producers are forced to shut down or relocate, and food scarcity continues to haunt developing countries, fertiliser production will remain a highly lucrative option for GTC processing. Another common product is methanol, and whilst a very price volatile product, it can itself be used as an intermediate to produce more valuable products. The methanol to olefins (MTO) process chain is a lucrative way to capture added value. Given the recent worldwide rise in the use of polymers, this particular process has spurred a myriad of activity. The process can be tuned to produce polypropylene and polyethylene. Given the issues outlined above, it makes sense to commission boutique-plants with capacities not exceeding 150,000 tonnes per year. Certainly scale economies are also achieved in this process, but given factors such as political risk and competitive pressures from new producers, it seems prudent to seek a short project pay-back period. Thus, given that the project is Capex sensitive, it is advisable to seek new, low cost technology that has become available in China and has half the cost of similar European technology.

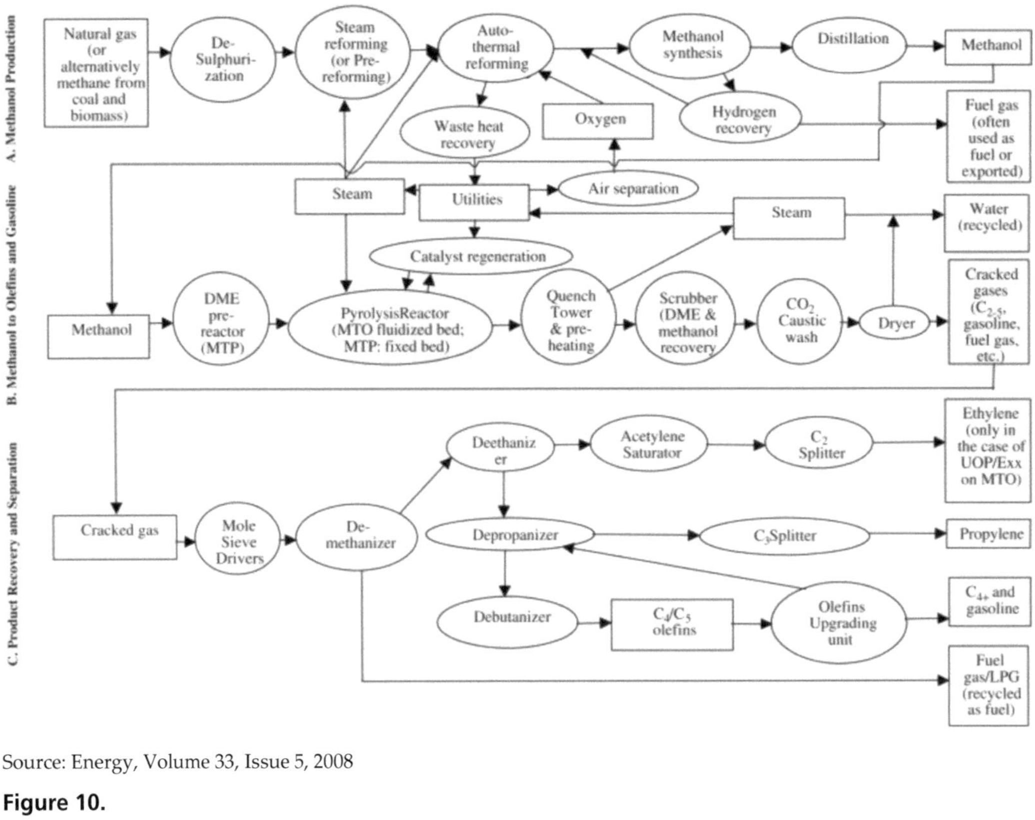

Source: Energy, Volume 33, Issue 5, 2008

Figure 10.

Other polymers that may be of interest in the GTC segment are PBT (Polybutylene terephthalate) and PET (Polyethylene terephthalate). These are thermoplastic polymers that are used in the production of electrical insulators or plastic bottles and synthetic fibres respectively. PBT is a less widespread product, however its versatile nature sees the market grow at 7% annually. One route to its production would be via 1,4-Butanediol which can be produced from propylene, itself produced from methanol. Like many of these processes, the rights are protected by international patents and it is necessary to approach the patent holders to implement them. The manufacturing process of PBT via 1,4-Butanediol is patented by Zimmer, now part of Lurgi AG. In most cases, patent holders are willing to grant user licenses to return research and development costs. This is not the case with PET however, as the necessary intermediate is Acetic Acid, produced from methanol via only two economical routes. These routes are patented by BP Chemicals and Celanese, which between them control the acetic acid market. These companies do not grant licenses to third parties and as such, gas owners would need to yield a majority stake to the licensors. Nevertheless, acetic acid, and subsequent PET production is an extremely lucrative method to monetise stranded or semi-stranded gas. These projects are indeed capital intensive and require considerable upfront investment. A 500,000 tonne acetic acid plant with a PET production line could cost in excess of $800+ million.

Gas-to-Power (GTP) involves the conversion of natural gas to electricity and normally implies in-situ generation. GTP has become a viable option since the introduction of the combined cycle gas turbine (CCGT), a system where the gas turbine is driven to produce electricity whilst the heat is used to manufacture steam to generate additional electricity through a steam turbine. These plants are much more efficient than their traditional counterparts[8] and are more compact is size. Furthermore, the construction lead times usually do not exceed two years, which is a significant improvement on traditional power plants. In situ GTP is particularly applicable when gas is found in undeveloped urban centres with a low degree of residential and commercial gasification. Gasification refers specifically to the level of development and infiltration of the low pressure distribution networks that supply gas to local residents or small commercial users. Africa and India, both of which have discovered gas near populated areas, would see great benefit from such technology. Nevertheless, Nigeria, which holds Africa's largest natural gas resources, flares more gas than any other country, after Russia.

In fact, there has been some discussion about applying old jet engines as temporary gas turbines for local power production. Because gas is considered a clean fuel, and due to the CHP Directive[9] in the EU urging the construction of such plants, CCGTs are likely to take a dominant role in the addition of new generating capacities, on the demand centre side. The major draw-back is the necessity to be located next to high-voltage electrical infrastructure, which makes it highly likely that gas transport pipelines will be found in the vicinity, in such a case yielding preference to the GTC process. In Russia's case, if a CCGT plant may be located near a European border, then, receiving access to the grid, it may be possible to export electricity to Russia's neighbours. However, as most of Russia's gas is located thousands of kilometres from the borders and large distances from major residential or industrial areas, CCGT is not a viable option. Instead, it becomes as viable option only at the

[8] A CCGT plant shows to have a conversion efficiency of 65%, as compared to a traditional gas turbine of 33%.

[9] Directive on the promotion of cogeneration based on a useful heat demand in the internal energy market and amending Directive 92/62/EEC

receiving end of a gas chain. As there is significant delay and uncertainty surrounding the nuclear power route in Europe and the UK, CCGT will play an ever increasing role.

Who Are the Top Gas Flarers?				
	OFFICIAL DATA for 2004		WHAT IMAGERY SHOWS for 2004	
	Country	**Billion Cu. M.'s**	**Country**	**Billion Cu. M's**
1	Nigeria	24.1	Russia	50.7
2	Russia	14.9	Nigeria	23.0
3	Iran	13.3	Iran	11.4
4	Iraq	8.6	Iraq	8.1
5	Angola	6.8	Kazakhstan	5.8
6	Venezuela	5.4	Algeria	5.5
7	Qatar	4.5	Angola	5.2
8	Algeria	4.3	Libya	4.2
9	Indonesia	3.7	Qatar	3.2
10	Eq. Guinea	3.6	Saudi Arabia	3.0
11	USA	2.8	China	2.9
12	Kuwait	2.7	Indonesia	2.9
13	Kazakhstan	2.7	Kuwait	2.6
14	Libya	2.5	Gabón	2.5
15	Azerbaijan	2.5	Oman	2.5
16	Mexico	1.5	North Sea	2.4
17	U.K.	1.6	Venezuela	2.1
18	Brazil	1.5	Uzbekistan	2.1
19	Gabon	1.4	Malaysia	1.7
20	Congo	1.2	Egypt	1.7

Source: World Bank, 2005

Figure 11.

The status quo of the industry has thus far been a dominating position of major conglomerates and IOCs that have been controlling the entire value chain. Although one of the main barriers to entry for new players has been the extremely capital intensive nature of such projects, the technical complexity of these large scale undertakings has also been limiting the ability of niche operators to enter the market. Nevertheless, even if these issues were to be overcome, the proprietary technology required to efficiently run these processes sit with a handful of licence holders. As such, companies like Shell, Sasol, ExxonMobil and Statoil control the GTL process, for example. Independent producers that have access to natural gas, have engineering expertise and access to capital (such as in Russia or Latin America) must work with these license holders to implement GTL projects. This often adds a difficult commercial angle to an otherwise difficult technological process.

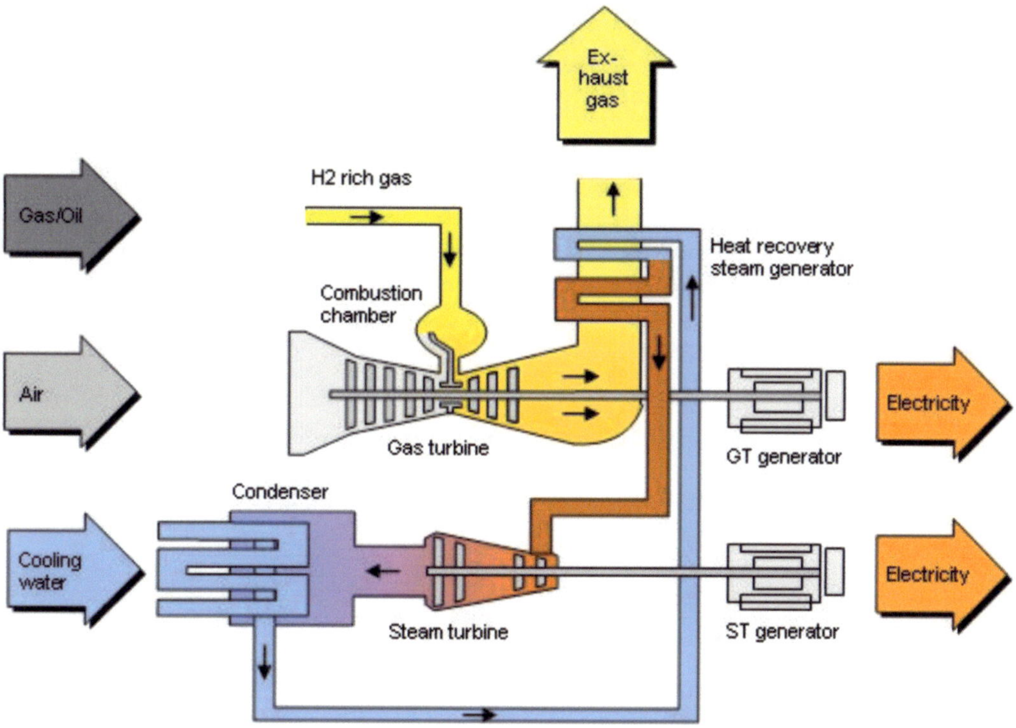

Source: Siemens, 2012

Figure 12.

4. New gasification markets

One of the key ways in which the traditional model outlined above is changing is in the shift to boutique production – small and medium sized projects in which natural gas is used in situ to produce final, value added products. In the last five years, there has been a significant amount of research in reducing the size of gas conversion technology, from micro LNG developed in Australia to micro GTL being developed in America and Asia. This has been, in part, as a result of technological advancement in the field of materials, processing, catalyst and engineering. Two American firms have made significant process in showing the commercial viability of GTL processes without recourse to the proprietary technology of the majors. Rentech and Syntroleum have both developed technology which has seen application outside of the laboratory conditions. As a specific example, Rentech, a medium sized US listed technology company, originally developed GTL technology as part of Texaco, and after a successful spin-out, remained as an independent developer. Whereas traditional GTL technology processes employ the use of cobalt catalysts with fixed bed reactors, Rentech has developed a way to use an iron based catalyst, which is seen as cheaper, and more efficient with a slurry bed reactor. Once capex costs are reduced, the economy of scale element becomes a secondary metric to reach required project rate of returns. Ultra Low Sulphur Diesel (ULSD) manufactured from Rentech production facilities in America has been used in vehicles and aircraft over the last ten years. The US Air Force has used Rentech GTL derived A1 Jet fuel in its aircraft, as part of its security of supply policy.

Boutique application means that stranded or semi-stranded gas reserves of a much smaller size can be successfully monetised. By decreasing required output from 100,000 bpd to 10,000 bpd or less for commercial production, fields of 5 BCM of recoverable reserves open up opportunity for GTL production. A huge market can be identified as CBM (Coal Bed Methane), where large coal deposits in areas such as China, Australia, Indonesia, Mongolia, Ukraine, can begin to utilise gas otherwise unable to reach a value generating market. The United States have significant CBM potential.

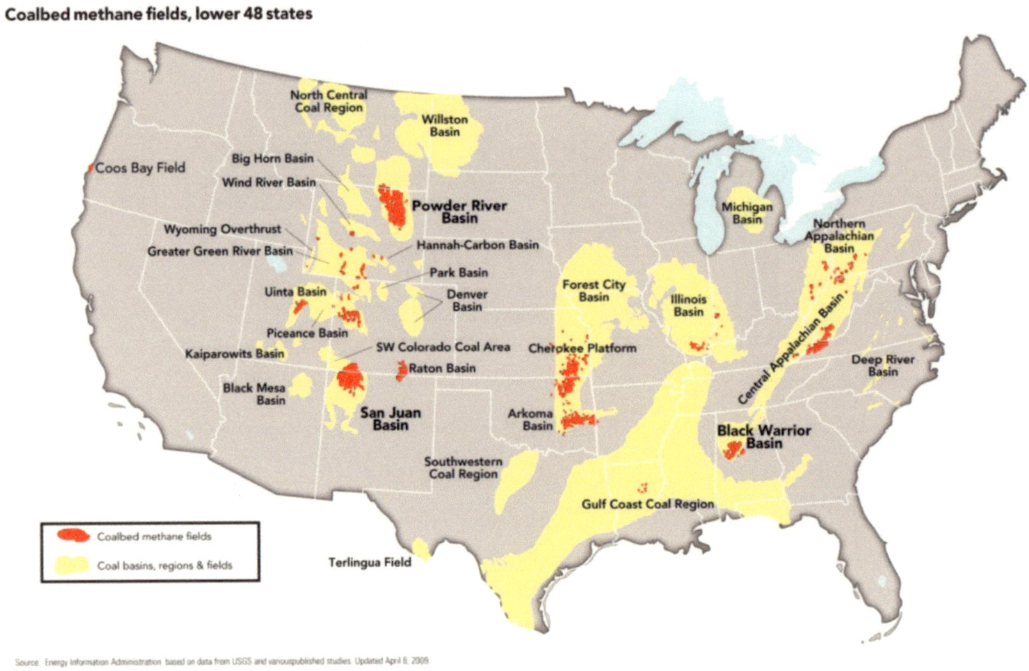

Source: Oil and Gas Journal, 2012

Figure 13.

In normal circumstances, small sized fields that are away from gas infrastructure means that they are stranded, meaning that investment required into pipeline construction render field development uneconomical. Small reactors allow for the production of liquid products that can be stored in canisters and transported by road or rail. For solid products, such as urea, it is possible to build up volumes in any port storage areas and then loaded on to larger vessels (typically 50,000+ tonnes).[10] Urea market is fairly liquid, however, the price volatility means that feedstock costs must be fairly low in order to avoid risks of prolonged loss making.

[10] This operation must be carried out in a fairly timely fashion, as urea has a tendency to degrade over time.

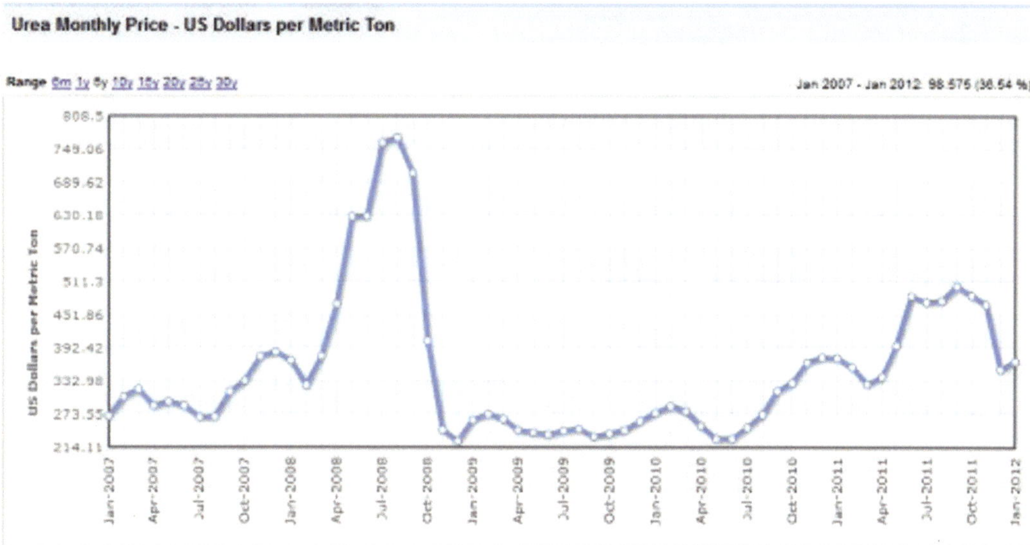

Source: Bloomberg, 2012

Figure 14.

One area which has seen significant attention is the gasification of biomass. This can be done on a micro-scale, meaning the most obvious applications are those of the municipal waste bodies and utilities or operators with large volumes of biomass waste products such as sawmills or sunflower oil producers. Some companies are claiming that they are able to achieve ULSD production in the volumes of 2,200 litres per day from a 10 tonne per day feedstock requirement. Woody biomass is gasified in a two-stage gasifier to produce Syngas with a 2:1 ration of hydrogen to carbon monoxide. This is then processed in a Fischer – Tropsch reactor to produce synthetic diesel fuel. Such small scale, modular application can reroute waste resources traditionally used to produce solid biomass fuels (pellets, briquettes, torrefied biomass etc) that can only be used in power or heat generation to liquid fuels that can be used in the transportation sector, either as blended additives, or for direct internal combustion. The key for a quick uptake of this technology is to reduce capex costs to a level where a 3 – 4 year simple payback can be achieved.

The advantage of the Biomass gasification is that it does not compete with food sources for feedstock, unlike traditional bio-fuels and hence is not party to significant pressure from political commentators and various pressure groups. By-products of agricultural production cycles, or forestry operations can increase efficiency and reduce transport / operations fleet carbon output.

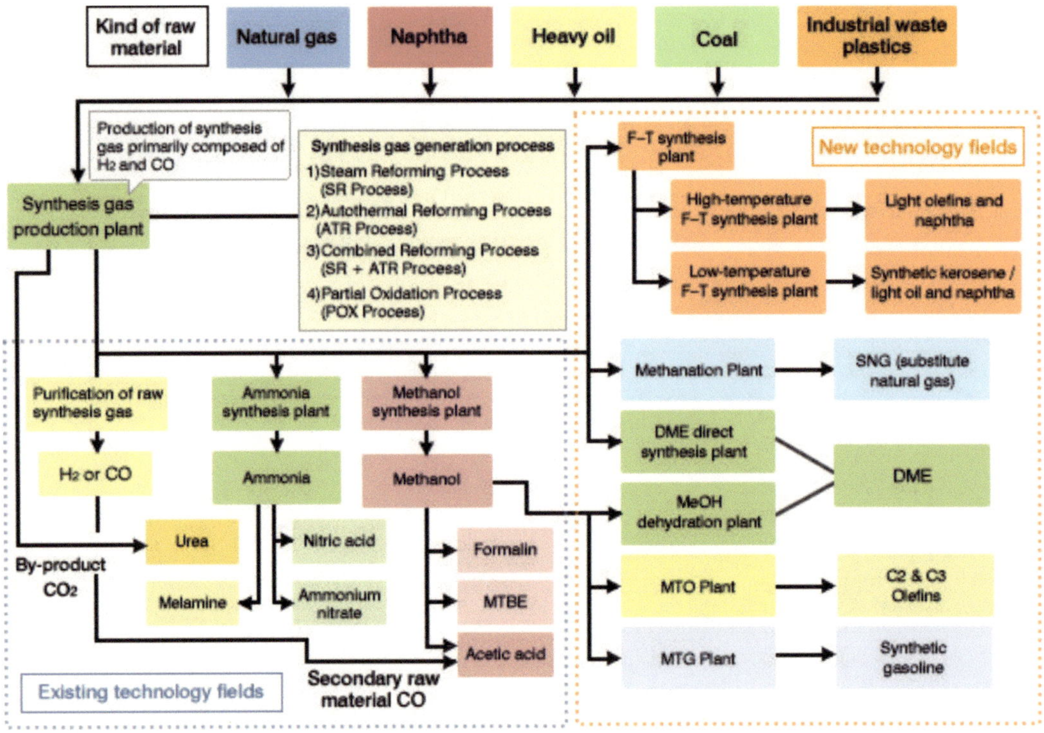

Source: Chiyoda, 2012

Figure 15.

5. Economics of small and mid-sized gasification

The economics of boutique synthetic fuels production have recently shown similar parameters to those of major projects undertaken by IOCs / NOCs. The author was directly involved in a feasibility study undertaken for a medium-sized GTL project, based on non-stranded gas in the Republic of Kazakhstan. The project economics were accepted by the contracted engineers and major, global investment banks focusing on natural resources.

The project entailed a facility with production of 120,000 metric tonnes per year of synthetic fuel (70% ULSD, 20% Naphtha, 10% kerosene), with a feedstock requirement of only 200 MMSCM of natural gas (dry, pipeline quality, high pressure) annually. This can be gas that is received from the gathering system of flared gas collections system, and directed to a processing facility or direct production or even pipeline gas. The price of gas was taken as US $2 per MMBTU.

Capital costs were considered at $150 million (which equates to c. 50,000 /bbl /day), with operational costs estimated at $7 / bbl. Although major operators are able to achieve a lower throughput costs, due to the super premium nature of ULSD and high conversion ratio, project profitability is more sensitive to capital costs. In this case, the project had a 4 year pay back and a 38% IRR.

Investment Analysis												
Discount Rate												15.00%
	Years											
Transaction	0	1	2	3	4	5	6	7	8	9	10	11
Cashflow from Operations	8,571	2,143	57,337	60,509	60,598	57,789	57,856	57,571	74,872	74,765	74,996	74,889
Investments	-75,000											
Net Cashflow	-66,429	2,143	57,337	60,509	60,598	57,789	57,856	57,571	74,872	74,765	74,996	74,889
Cumulative Cashflow	-66,429	-57,072	265	60,774	121,373	179,162	237,017	294,588	369,460	444,225	519,221	594,110
Discounted Cashflow	-66,429	1,863	43,355	39,786	34,647	28,732	25,013	21,643	24,476	21,253	18,538	16,097
Net Present Value	-66,429	-57,320	-13,965	25,821	60,468	89,199	114,212	135,855	160,331	181,584	200,122	216,218
Internal Rate of Return												37%

Source: Author, Midstream Energy Ltd, 2010

Figure 16.

The key for success of this project, and indeed any boutique application of gas conversion, is the ability to avoid "green-field" development. By placing new facilities on existing infrastructure, such as working refineries or old and abandoned heavy facilities, the capex figure can be kept to a manageable level to give satisfactory project returns. The Former Soviet Union (FSU), for example, has a large number of old chemical facilities that have ceased to operate and with a low cost of domestic gas, become good candidates for boutique GTL or GTC processing. One major advantage is the existence of transport infrastructure, both for the gas via pipeline and product via rail.

6. Impact on policy

Ever since the major discoveries in the US of shale gas, new opportunities have arisen for application of "gas to" technologies. Observers have predicted that the US will become self-sufficient with respect to natural gas, and may become an exporter in the next decade. This has been further compounded by the recent permissions granted to Cheniere Energy for an LNG export terminal. It is incorrect to say that the US will become a net exporter, as it likely there will be imports of natural gas from Canada and some volumes of LNG from further afield. However, Shell has already announced that it is evaluating a large GTL project in the US. The key for such projects is the differential in price between natural gas and high-end products, a situation which reflects the current market in the US very well. Recent prices in the US (Henry Hub Futures) have been hovering at around the US $2 / MMBtu, whilst low sulphur diesel is currently trading at between USD $800 and $1000. If the price of crude oil continues to stay at or about $90+ per bbl, GTL projects become economically viable. This will also have a positive effect on Supply Security concerns, as the more transport fuels can be derived from domestic natural gas, the less dependence there is on oil imports.

When looking at other regions, there are similar advantages for China, as there are large opportunities in the near term for CBM gasification, and in the mid-term for shale gas development. China has announced significant finds of shale gas, and this can help to reduce dependence on oil imports. In fact, China is aware of the strategic disadvantage of having the bulk of the oil imports from the Middle East being shipped via the Malacca Strait. A well planned military operation can block this channel, effectively cutting China off from its oil flow.

African states, especially mature oil development areas such as Nigeria, have been unable to capitalise on the associated gas production, with various methods being undertaken to reduce gas flaring. In situ gas conversion, certainly in the first instance to power, and subsequently to fertiliser production, would be a coherent road map to develop the country's resources.

In Europe, there is less scope for this application, simply because due to liberalised markets, gas prices do not allow for economic production of other products, except for power generation and commercial and residential sectors. Furthermore, there is simply no spare capacity in the system to divert supplies from power and other sectors to gas processing. Economically, it makes more sense to produce in areas of low cost feedstock and deliver final products to the EU market.

7. Future applications

One of the most advantages characteristics of synthetic fuels or more traditional gas processing products is the ability to utilise these in existing infrastructure without the need for a stock change. The biggest future growth will come from GTL, BTL and CTL processes and environmental concerns will play a role to increase the uptake of these fuels. As more stringent regulation places greater standards on reduced sulphur content in transportation fuels, more ULSD will be used as a blending fuel. Once the technological costs come down the cost curve, and producers will be incentivised to invest in direct GTL technology versus traditional deep refining, pressure will applied to the aviation industry to use synthetic fuels. Aviation is responsible for a major share of Green House Gas (GHG) emissions, and as such is a great potential consumer of synthetic fuels will come from this sector.

8. Conclusion

Natural gas is a versatile raw material that has traditionally been characterised by large complex infrastructure products, requiring full value chain integration. When not used as a fuel for power generation, natural gas has been an invaluable element in many household items and industrial chemicals. Due to the fact that supply and consumption centres have traditionally been separated by large distances, most natural gas projects required capital intensive pipeline construction. The financing of these required the mitigation of risks via long term offtake contracts. This was not the case in the Former Soviet Union, as government central planning directed investment and energy flows according to internal economic planning.

As a result, only large gas bearing basins were developed, with small fields either ignored or considered uneconomic for development. Oil reservoirs that contained a high gas-oil ratio were considered cumbersome in production areas where flaring was unacceptable, and in others where flaring was acceptable, natural gas remained as a nuisance.

With various advancements in technology, reduction in costs and improvements in technical knowhow, as well as economic and environmental conditions, there has been a focus on natural gas as the fuel of choice, ahead of crude oil, in most of the applications. This is likely to drive a trend where the growth in the consumption of gas will overtake oil in the long run, and perhaps become a major contributor to power, transportation and chemical sectors.

Author details

Yuli Grigoryev *
CEPMLP, University of Dundee, UK

* Corresponding Author

9. References

Grigoryev Y. (2007) The Russian Gas Industry, Its Legal Structure, And Its Influence On World Markets. Energy Law Journal. Vol 28:125.

Grigoryev Y. (2007) Today or not today: Deregulating the Russian gas sector. Energy Policy 35: 3036–3045.

Kang Z and Yundong T. (2009) The Status of World Shale Gas Resources Potential and Production Status as well as Development Prospect of China's Shale Gas. Petroleum & Petrochemical Today. 2009-3.

Grigoryev Y. (2012) Future Of European Energy Security And Its Impact On Global Demand And Supply Centres. Proceedings World Gas Conference 2012, Kuala Lumpur.

Grigoryev Y. (2012) Challenges of the Russian regulatory framework and the monetisation of gas resources, Proceedings 2012 Russia & CIS Executive Summit, Downstream Oil and Gas, Dubai.

Chapter 10

Cost Estimates of Coal Gasification for Chemicals and Motor Fuels

Marek Sciazko and Tomasz Chmielniak

Additional information is available at the end of the chapter

http://dx.doi.org/10.5772/48556

1. Introduction

Solid fuels gasification technology has been understood and applied for a long time. The current directions in developing coal gasification technology are primarily related to power generation in combined systems involving steam and gas turbine implementation, which considerably increases fuel use efficiency. Compared to the first gasifying installations, the current solutions have a much higher conversion intensity and are more reliable. Integrated power generation-related gasification technology developments have created increased interest in chemical products, such as liquid motor fuels, methanol and hydrogen. At the present time, the basic reason for the increase of coal use as a raw material for chemical production is the dynamic industrial growth in countries with high economic potential that do not have their own natural gas and oil resources and have limited access to international sources of the above minerals. China is a good example of a country in this situation, and it constitutes the largest coal gasifying economy in the world. In China alone, more than 100 million tonnes of coal is gasified yearly. We expect that countries such as the USA and India will follow China in coal gasification-based production growth.

The crucial driver of gasification technology development is the necessity of a drastic reduction in CO_2 emission from anthropogenic sources, which is considered to be one of the main contributors to the greenhouse effect. Among fossil fuels, the most important CO_2 emitter is coal, which is characterised as having the highest concentration of carbon element compared to its caloric value. In the coal gasification process, carbon dioxide is removed from the processed gas by the absorption of acid components, which constitutes an inherent part of the technology. In case of chemical plant the acid gases, i.e. H_2S and CO_2 must be removed from the processed gas, regardless of the chemical facility's production profile because H_2S can damage the catalysts used during chemical synthesis, and the content of CO_2 is corrected to the expected composition of a synthesis gas. This removal step can alleviate the need for additional CO_2 separation so that the costs associated with dehydration and compression are the primary costs remaining. These two processes are critically important to system, as they ensure safe transport of the CO_2 to the storage (sequestration) area.

In the case of integrated gasification combined cycle (IGCC, power generation), the removal of sulphur compounds (H_2S, COS) is required to protect the gas turbine, and CO_2 removal is conducted only to reduce atmospheric emissions. However, because of the high concentrations of carbon dioxide and the high-pressure of the treated gas, the removal of CO_2 from syngas (i.e., pre-combustion removal) is less expensive than if the CO_2 were separated from the flue gases (post-combustion removal). Pre-combustion CO_2 removal results in better process and economic efficiency of IGCC systems (in case of CO_2 sequestration) compared to conventional power plants based on coal combustion.

The development state of coal gasification technology

A review of the global development state of gasification technologies has been performed based on a 2010 database developed by the U.S. Department of Energy (US Department of Energy & National Energy Technology Laboratory [US DOE & NETL], 2010a). The results of this analysis have been compared in three categories characterising the current status of technology development: plants that are operational, plants that are under construction (or start-up) and, plants that are in the development phase (this category includes plants in varying degrees of implementation, including plants at the stages of planning, conceptual work and designing). When analysing the data for the various systems, plants that use natural gas as a fuel have been omitted as these plants are not considered to be gasification systems but rather are plants for the partial oxidation of natural gas. The total power of the aforementioned systems (the thermal capacity of syngas output) amounts to 15,281 MW_{th}, of which 72 % (10,936 MW_{th}) is attributed to a plant using a Shell pressure reactor that is under construction in Qatar.

The published data show that there are 116 gasification plants equipped with 342 reactors with a total power of 50,104 MW_{th} are currently operating worldwide. The total power represents the chemical energy in the gas that is produced but does not include the systems for the partial oxidation of natural gas. Seventeen systems are under construction (28 gasification reactors, 16,289 MW_{th}; coal), and 37 plants (76 reactors, 40,432 MW_{th}) are at the planning stage with systems to be implemented in the years 2011 – 2016. Since the last review in 2007, the installed power increased by 7 %, resulting in the largest recorded increase for coal gasification at 18 %. For other fuels, there was a clear decrease in the amount of gas produced, particularly for biomass and petroleum coke (-68 % and -37 %, respectively) (Table 1). The implementation of all current and planned investment projects will contribute to more than a doubling of gas production (106,825 MW_{th}).

The largest percentage of gasification systems is operating in the Asia and Oceania region (39 % of total global gas production), primarily because of extremely dynamic technology developments in China (78 % of this region). In this region, which includes China, Australia,

South Korea and Vietnam, the majority of systems are now under construction and planned for implementation in the next few years. Long-term plans exist for technology development in North America (primarily in the U.S.), the implementation of which would move this region into second place in the global production of gas from gasification (30.4 % of global gas production).

Coal, the basic feedstock for gasification, is used in gasification plants that are currently operating and accounts for 61.6 % of global gas production (Fig. 1). Petrochemical industry by-products rank second (35.8 %), and the remaining 2.6 % of gas production is attributed to petroleum coke and biomass. For plants that are under construction and planned for start-up by 2016, the role of coal as the basic fuel will be maintained, and the share of gas produced from this raw material will increase to 79 %.

The basic products of operational systems using gasification processes comprise chemicals such as ammonia, hydrogen and oxy-chemicals (46 % of world gas productions), products of Fischer-Tropsch synthesis (30 %), power (16 %) and gaseous fuels (8 %) (Fig. 2). Chemicals will also be the main products of the plants that are under construction (72 %). In the case of plants planned for implementation, the largest share will be power-generating plants (37.5 %), which is probably related to the attractiveness of power systems that are integrated with gasification, particularly in the context of the necessity for CO_2 emission reduction (Fig. 2).

Feedstock		Operating 2010 (operating, construction, start-up)	Operating 2007 (operating, construction, start-up)	Difference	%
Coal	MW_{th}	36,315	30,825	5,490	18
	Gasifiers	201	212	-	-
	Plants	53	45	-	-
Petroleum	MW_{th}	17,938	18,454	-516	-3
	Gasifiers	138	145	-	-
	Plants	56	59	-	-
Petcoke	MW_{th}	911	1,441	-530	-37
	Gasifiers	5	8	-	-
	Plants	3	5	-	-
Biomass/ waste	MW_{th}	373	1,174	-801	-68
	Gasifiers	9	21	-	-
	Plants	9	13	-	-
Total	MW_{th}	55,537	51,894	3,643	7
	Gasifiers	353	386	-	-
	Plants	121	122	-	-

Table 1. Comparison of the state of worldwide existing gasification technologies in the years 2007 and 2010 (US DOE & NETL, 2007, 2010a).

In the case of coal use, the most popular gasification plants are now fixed bed gasification technology, which is practically no longer developed but still accounts for 57 % of gas production due to the high manufacturing potential of the Sasol plant in South Africa.

Processes using entrained flow reactors are the most intensively developed technologies (operating plants, 43 % of gas production) as confirmed by the projects that are under construction and planned for start-up by 2016, which are nearly all related to this reactor design. Fig. 3 shows the structure of the operational plants and the coal gasifiers planned for start-up in terms of the technological solutions used.

Of the technologies used for coal gasification in entrained flow reactors (operating plants), the Shell (dry feeding) and GE/Texaco (slurry feeding) have the dominant share in gas production (77 %), followed by the ECUST (15.3 %) technology. The third place position of the use of ECUST (East China University of Science and Technology) technology in developing plants is noteworthy because of the rapid pace of the ECUST technology development. Beginning with a pilot plant (22 t/d of fuel) in 1996, the technology led to operational demonstration plants in the years 2001 – 2005 (750 and 1,150 t/d of coal) and 17 commercial gasifiers that were implemented by 2010 (capacity of up to 2,000 t/d of coal) (Liu, 2010).

The highest percentage of plants planned for start-up that are under construction and in the development phase will use the Shell gasification technology (26.7 %; 11,913.2 MW_{th}) followed by ECUST (20.8 %), Udhe PRENFLO (16.8 %), Siemens (14.7 %), ConocoPhillips E-Gas (11.3 %), GE Energy/Texaco (5.3 %) and MHI (3.7 %) (Fig. 3). The fluidised bed coal gasification reactor technologies GTI U-GAS and TRIG (KBR Transport Gasifier) will be developed apart from the entrained flow technologies.

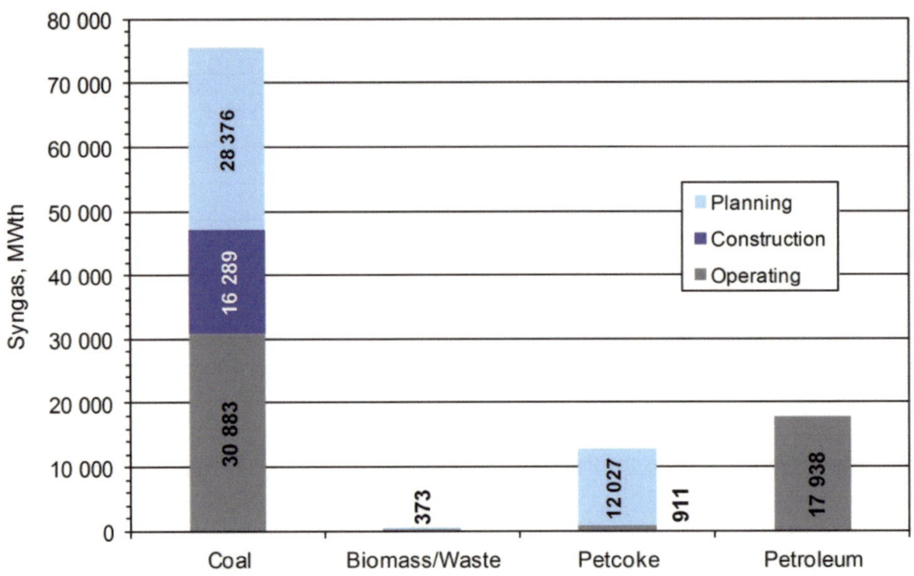

Figure 1. Total capacity of gasifiers versus fuel used (current and forecast by 2016).

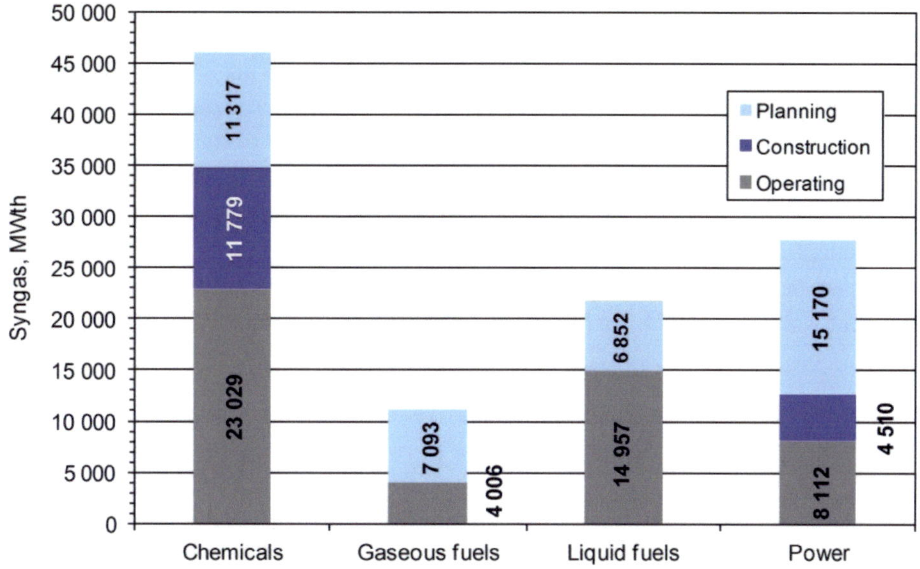

Figure 2. Total capacity of gasifiers versus product manufactured (current and forecast by 2016).

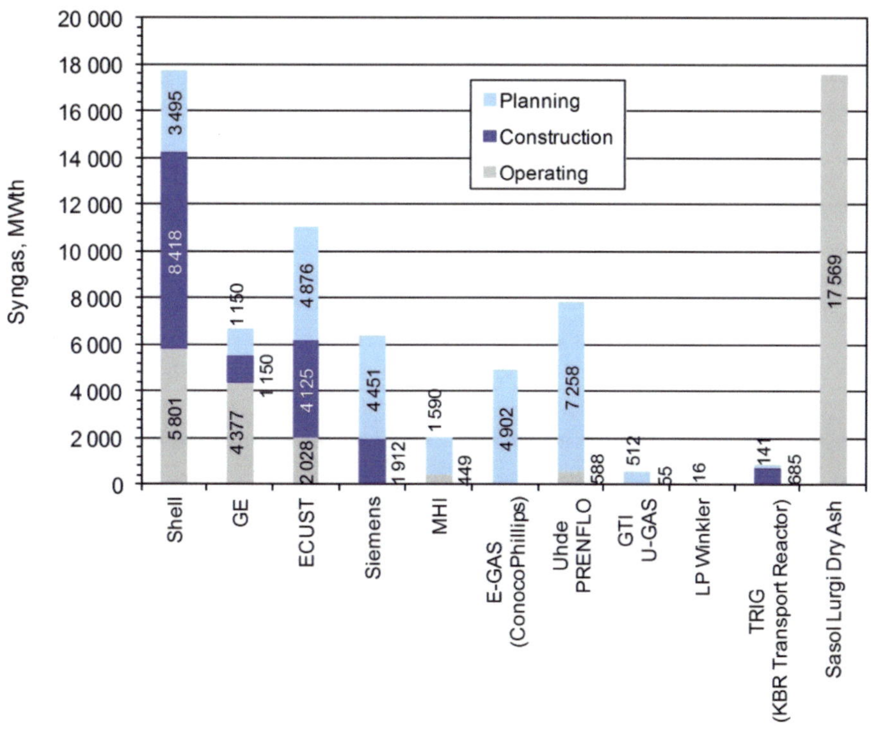

Figure 3. Total capacity of reactors using coal as the main fuel, breakdown by technological groups (current and forecast by 2016).

2. Technological option

The review of the global development state of gasification technologies shows that gasification systems will be used for syngas production in power generation systems (IGCC) and particularly in chemical synthesis to obtain liquid and gaseous fuels including methanol and hydrogen. The analysis of the above processes is the subject of this study. Four cases for coal gasification applications involving chemical synthesis and electricity generation have been analysed and discussed in detail. The options include: liquid fuel production, hydrogen generation, methanol production (options I-III) (Dreszer & Mikulska, 2009), and power and syngas production (Polygeneration Plant, option IV) (Chmielniak et al., 2008; Energoprojekt-Katowice S.A & Institute for Chemical Processing of Coal [EPK & IChPW], 2008).

Option I – A system of six gasifiers, which requires an annual coal consumption of approximately 5,600,000 t/y. The adopted scale of coal processing results from preliminary cost-effectiveness studies for liquid fuel production from coal, which have shown that the operation of a production plant starts to be profitable only at a production level exceeding 1 million tonnes of liquid fuels, which corresponds to the adopted scale of coal consumption. The plant products have been defined as technical propane-butane (LPG), diesel oil and a semi-product for the diesel oil that is not further processed into final commercial products.

Option II and III – one gasifier system. The adopted scale makes it possible to accomplish the following:

- cover the demand for hydrogen on the scale of a single standard chemical plant fertiliser production train (no network for high-volume hydrogen distribution was assumed) (option II).
- produce methanol from the gas originating from coal gasification on the scale of 500,000 t/y (option III).

Option IV – a system of two gasifiers operating in parallel technological trains to produce syngas (for methanol synthesis) and power (IGCC). Due to their identical capacity, gasifiers operating in an integrated system can provide mutual back-up functionality for each other, increasing the annual availability of syngas or electricity production units, depending on the adopted production programme. The scale of production allows to manufacture of approximately 500,000 t/y of methanol what ensures the profitability of its production.

Each of the analysed options consists of a syngas generation unit, i.e., a coal gasification system including units for converting and cleaning syngas.

Gasification technologies in the entrained flow reactors play an essential role in the production of syngas from coal and are offered by a number of providers. The final choice of gasification technology must therefore be made using a separate analysis based on detailed data from the technology providers, including investment and operational cost and the assessment of coal suitability for processing.

GE/Texaco technology has been selected for the analysis of the considered cases for the following factors:

- mature technology / solution used for the longest recorded period,
- one of the largest shares in the coal gasification sector (33 %, operational plants),
- absence of inert gases in syngas, which constitute a redundant ballast in chemical synthesis and result in an increase in equipment size needed due to the increased gas volume in the circuit.

The disadvantage of this technology is the lower energy efficiency of the gasification process compared to technologies using dry coal feeding. However, it has been assumed that to assess different fuel production systems based on coal gasification using conceptual studies, it will be less risky to assume process guidelines for GE/Texaco technology with a coal-water slurry feeding system.

3. Description of considered technological systems

3.1. Coal gasification – GE/Texaco technology

The coal-water slurry (62-68 % coal) and oxygen from the air separation system are fed through a system of valves to the feedstock injector in the top part of the reactor where gasification proceeds at a temperature of 1,260 – 1,480 °C. The hot processed gas with molten ash flows to the bottom part of the reactor, the radiant cooler, where it is cooled down to approximately 730 °C and then is taken off of the reactor to a convective cooler and a scrubber. After being cooled down to approximately 230 °C, the raw gas is directed to the gas conversion and/or cleaning systems. High-pressure (HP) steam is produced in the radiant and convective coolers. The molten ash flows down to the water bath in the bottom part of the reactor where, after solidification and cooling, it is taken off the system through a lockhoppers. Fly ash that is separated from the gas is also taken off together with slag (the ash separation from the gas occurs through a sudden change of its flow direction before leaving the reactor). After water separation, the slag is directed to a waste landfill. Separated fly ash with a carbon fraction of approximately 30 % is delivered to the coal-water slurry preparation system for recirculation to the reactor. The spraying water from the scrubber and the water from slag dewatering is returned to the scrubber after the removal of solid particles (fine slag/fly ash), and its excess is fed to a water treatment plant (US DOE & NETL, 2002).

In addition to the technology option described above, General Electric commercially offers two other configurations of gasification plants (US DOE & NETL, 2010b):

- a reactor with direct water cooling: in this system, the hot processed gas is cooled down to 260 °C through direct contact with water before leaving the reactor.

- a reactor with a radiant cooler: the processed gas leaving the gasification zone passes through a radiant cooler that produces high pressure steam where it is cooled down to approximately 800 °C and then passes through a water lock, which lowers its temperature to approximately 200 °C.

The gasification pressure was assumed to be 3, 5.6 and 7 MPa for the production systems of liquid fuels, hydrogen and methanol plants, respectively (the pressure was selected to match the process condition for F-T, hydrogen and methanol production units).

Oxygen for the gasification system is supplied from an air separation system based on cryogenic separation. Oxygen purity levels of 99.5 % for liquid fuel production and 95 % for all other cases were assumed.

3.2. System layout – Fuel and chemical production plants

Fig. 4 to 6 present the process diagrams of the considered plants based on coal gasification. The data on the technological configurations are summarised in Table 2.

3.2.1. Liquid fuel production plant

Gas from the gasification system is directed to the hydrolysis reactor where, in the presence of the catalyst, carbonyl sulphide (COS) is hydrolysed to hydrogen sulphide. Gas exiting the COS reactor is cooled to approximately 38 °C in several heat exchangers fed by boiler feed water (steam production) or cooling tower water. Entrained water (condensate) is separated and used for coal-water slurry production and for slag cooling in the gasifier. Cool gas is fed to the Selexol system, where hydrogen sulphide and carbon dioxide are removed. Hydrogen sulphide is directed to the Claus system for sulfur recovery. Clean gas is heated to approximately 313 °C, deep purified from the remaining hydrogen sulphide in the reactor, filled with zinc oxide and fed to Fischer-Tropsch synthesis reactors. Fischer-Tropsch synthesis is carried out in a slurry reactor at 250 °C under a pressure of 2 MPa in a presence of cobalt catalyst. Unreacted part of syngas is fed to the carbon dioxide separation system based on chemical absorption (MDEA) and then to the dehydration and compression system. After passing through the product separation system, the gas is then recirculated to autothermic reforming and sent back to the synthesis reactor. Separated carbon dioxide from the Selexol and amine units is compressed to 12 MPa and transported to a storage location.

3.2.2. Hydrogen production plant

Partially cleaned gas from the gasification island is directed to the Water Gas Shift (WGS) reactor where approximately 97 % of the CO is converted to CO_2 and hydrogen. Gas exiting the WGS reactor is cooled to approximately 38 °C and then fed to the Selexol unit. In the two-stage Selexol system, gas is divided into three streams: sour gas (primarily H_2S), carbon dioxide and hydrogen-rich processed gas. Sour gas from the first stage of the Selexol absorber is directed to the sulfur recovery unit (Claus, Scot). CO_2 is compressed to 12 MPa in preparation for transport and storage. Cleaned processed gas with approximately 90 % hydrogen content is fed to a PSA

(Pressure Swing Adsorption) system, where hydrogen with >99 % purity is produced. The off gas from the PSA system is combusted in a steam boiler, and then steam from the boiler and from the gasification system is used for power generation in the steam turbine.

3.2.3. Methanol production plant

Partially cleaned gas from the gasifier is divided in two streams. One of them, which accounts for approximately 65 % of the total flow, is fed to the high temperature CO shift reactor, where, at temperatures between 400 °C and 410 °C, a carbon monoxide and steam reaction occurs, generating hydrogen and carbon dioxide and producing the required hydrogen concentration in syngas, which is directed to the methanol synthesis reactor. After being cooled to approximately 250 °C, the gas is then joined with the second stream and directed to the COS hydrolysis reactor. Next, hydrogen sulphide and carbon dioxide are removed in the Selexol system from the gas after it is cooled to 38 °C. The hydrogen sulphide that is removed from the gas is then transported to the Claus system for sulfur recovery. Carbon dioxide is separated with 78 % efficiency (the separation level is assumed to meet the stoichiometric ratio required for methanol synthesis ((H2–CO2)/(CO+CO2) = aprox. 2) and is then compressed to 12 MPa. The composition of the syngas leaving the synthesis system enables its direct use in methanol synthesis. Syngas that is purified in the Selexol process is then joined with circulating tail gas from the synthesis unit and, after being heated to approximately 210 °C, is conducted to the adiabatic, methanol synthesis reactor. The post-reaction mixture leaving the synthesis reactor is then cooled to 38 °C while heating the gas that is being directed to the synthesis reactor, and then it is separated into a liquid methanol fraction and off gas. The liquid fraction is decompressed and transported to a degasifying tank. The raw methanol is then directed to the rectification system, where methanol of high (>99 %) purity level is obtained. Part of the tail gas is compressed and redirected to the methanol synthesis system, and after being decompressed, the remaining gas is combusted in boiler burners where steam is overheated and directed to the steam turbine. The high pressure steam generated in the gasifying system also feeds the turbine.

3.3. Polygeneration plant

A schematic diagram of the Polygeneration Plant is presented in Fig. 6. The system enables simultaneous electricity, heat and syngas generation with sequestration of the carbon dioxide formed during the production process. Joining the combined power and heat generation with syngas production enables the high efficiency of fuel primary energy conversion, low emission indicators and high economic efficiency, also in the case of CO_2 sequestration. The presented solution was developed by Institute for Chemical Processing of Coal (IChPW) and Energoprojekt-Katowice SA (EPK) for TAURON SA (power producer, Poland) and Zakłady Azotowe Kedzierzyn SA (ZAK SA, chemical works, Poland) (Chmielniak et al., 2008 ; EPK & IChPW, 2008).

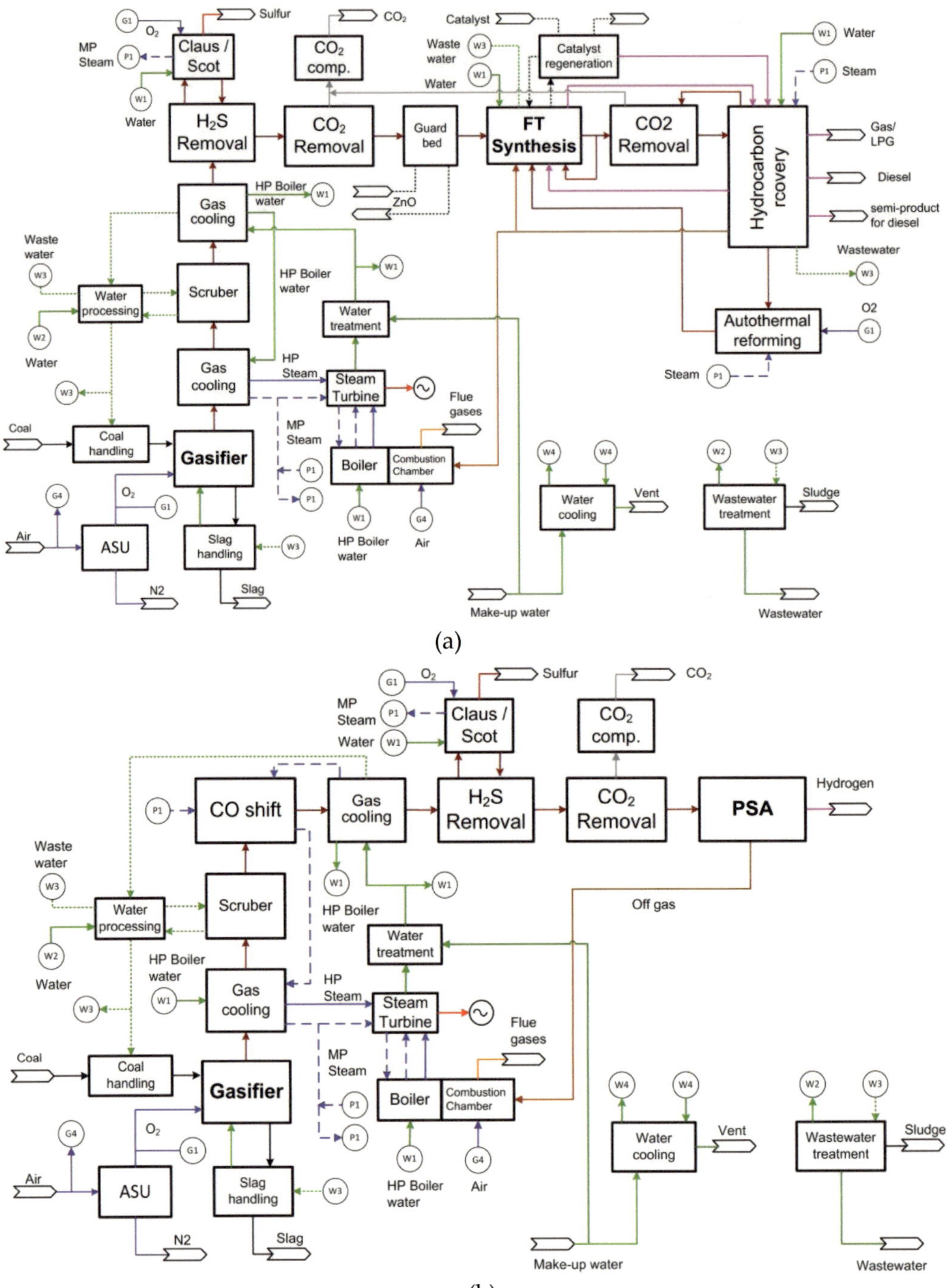

Figure 4. Process diagrams of A) a liquid fuel production plant (option I) and B) a hydrogen production plant (option II) .

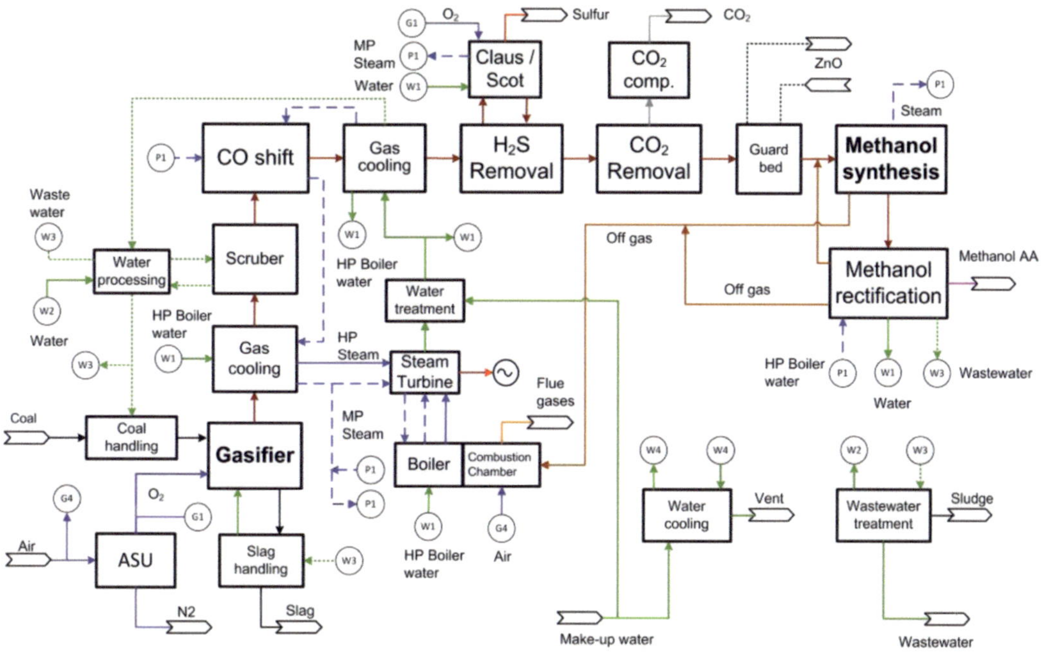

Figure 5. Process diagram of metanol production plant (option III).

To demonstrate an alternative for clean coal technology, the concept of a polygeneration facility assumes possible complete elimination of atmospheric carbon dioxide emissions. Thus, the proper configuration of the IGCC system (energy island, Fig. 6) is necessary for the efficient removal of CO_2 (CO_2 removal in a chemical island is a technological need for the production of syngas). Regarding the IGCC plants that are currently under operation (without CO_2 removal), major changes include the introduction of CO shift reactors and CO_2 separation system. The CO conversion process allows to convert gasifier product (raw gas) to hydrogen-rich syngas and to concentrate most of the carbon contained in the gas in to a CO_2 stream. This allows for the removal of carbon from the syngas before the combustion process (a CO_2 stream is removed in the subsequent stages of syngas processing). Additionally, during the conversion process, the COS hydrolysis reaction takes place without requiring additional equipment (an IGCC facility without CO_2 removal requires systems for the hydrolysis of COS). CO_2 is removed from the syngas during an absorption process. Due to the high pressures under which the gasifier is typically operated, the most energy efficient method of gas separation is by physical absorption. A double stage physical absorption system is recommended for use in a gasification system when separation of CO_2 is required.

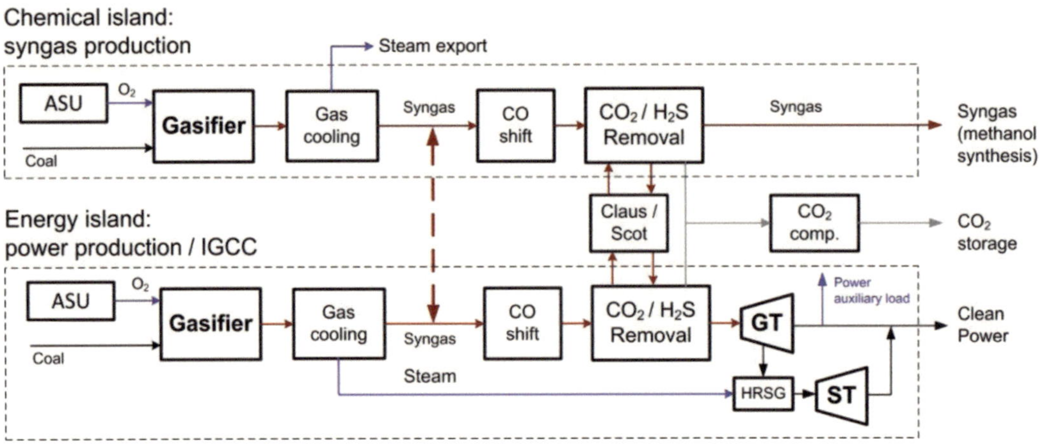

Figure 6. A scheme of Polygeneration Plant.

A Polygeneration Plant consists of three basic technological units:

- Chemical island: coal gasification system that is equipped with a gas conversion and purification system with a CO_2 separation unit and generates syngas for chemical production purposes and high pressure steam for power and heat generation. Technological configuration as in the case of the production of methanol (option III, see p. 3.2.3).
- Energy island: coal gasification system that is integrated with a combined cycle for power production (gas and steam turbines, HRSG - Heat Recovery Steam Generator) and is equipped with syngas conversion and purification systems, as well as with a pre-combustion CO_2 capture system. Configuration of gas treatment system as in the case of the production of hydrogen (option II, see p. 3.2.2) with the difference that after removal of CO_2 the gas is not enriched in hydrogen (no installation PSA) but is heated to about 240 °C and then mixed with nitrogen comes from the air separation unit[1] in order to reduce gas lower heating value (LHV) to 4.7 MJ/m_n^3 (increase of power output of gas turbine as the result of mass flow increase and lowering of gas firing temperature for, i.a., control of NOx emission).
- CO_2 transport and storage system.

The design of the Polygeneration Plant assumes that the system is coupled with a classic CHP plant (not shown in Fig. 6) consisting of a circulating fluidised bed boiler and steam turbine power generator. A CHP plant uses high temperature steam produced in the chemical island of the Polygeneration Plant for additional power and heat production. The energy production in the form of heat and power covers the needs of local consumers, the town heat distribution network and industrial users. It is assumed that the presented conceptual facility will replace two actual operating heat and power plants. Due to their

[1]If the amount of available nitrogen is not sufficient, gas is diluted through the humidification and the third option is steam injection (US DOE & NETL, 2010b)

identical production capacity, the gasifiers working in the system may complete each other, increasing the yearly availability of syngas or power production units based on the assumed production programme.

Specification	**option I**	**option II**	**option III**	**option IV**
Product	Liquid fuels	Hydrogen	Methanol	Polygeneration Plant
ASU	Cryogenic separation			
Gasification island				
Reactor	Entrained flow, slurry feed; Technology: GEE/Texaco			
Gasification Pressure	3 MPa	5.6 MPa	7 MPa	5.6 MPa
Coal conversion	98 %	98 %	98 %	98 %
Gas cooling	Radiant and convective cooler			
Oxidiser	Oxygen 99.5 % (% vol.)	Oxygen 95 % (% vol.)	Oxygen 95 % (% vol.)	Oxygen 95 % (% vol.)
Fuel	Coal-water slurry; 63% dry solids concentration in the slurry			
Gas treatment				
CO shift	No	Yes, Sour gas shift two-stage CO conversion 97 %	Yes, Sour gas shift one-stage CO conversion ~68 %[a]	Yes, IGCC: see option II Methanol: see option III
Sulfur removal	COS hydrolysis Selexol I stage (99.7 %), ZnO sulfur polishing bed (< 1 ppb)	Selexol I stage (99.7 %)	Selexol I stage (99.7 %) COS hydrolysis (raw gas bypass)	Selexol I stage (99.7 %) COS[b] hydrolysis (raw gas bypass)
Sulfur recovery	Claus, Scot; elemental sulfur			
CO_2 separation	Selexol II stage MDEA	Selexol II stage	Selexol II stage	Selexol II stage
Liquid fuel	F-T synthesis slurry reactor	-	-	-
Hydrogen	-	PSA, 85 %	-	-
Methanol	-	-	Adiabatic, fixed bed reactor	Adiabatic, fixed bed reactor[c]
Power	Steam turbine excess heat, gas (hydrocarbon recovery unit) combustion,	Steam turbine excess heat, tail gas (PSA) combustion	Steam turbine excess heat, tail gas (methanol synthesis) combustion	Combined cycle IGCC, gas turbine, HRSG, steam turbine

[a] as the result of CO Shift and by pass of the raw gas; [b] methanol line; [c] Polygeneration Plant produce syngas with composition enabling its direct use in methanol synthesis.

Table 2. Data on the process configuration of fuel production plants and a Polygeneration Plant (Dreszer & Mikulska, 2009; EPK & IChPW, 2008).

3.4. Transport and storage of CO_2

Separated carbon dioxide is compressed in a multi-stage, intercooled compressor. During the compression, the CO_2 stream is dehydrated with triethylene glycol that is introduced into the compressed stream of CO_2. Dry CO_2 is directed to an intermediate tank and then transported by a pipeline to underground storage units. The condensate from CO_2 drying is directed to a water purification system.

4. Results of process calculations

Coal: For analysis, three hard coals produced in Poland[2] were selected:

- "Ziemowit" and "Piast" coal mines: option I
- "Bogdanka" coal mines: option II
- "Janina" coal mines: options II, III and IV

For gasification, the chosen coals have acceptable water and ash contents and sufficient caloric value and ash fusion temperature. We should highlight, however, that the gasification of coals with lower quality parameters, such as high ash content, leads to gasification efficiency decrease and may cause technical problems in the slag feed system. The assessed properties of coal are presented in Table 3.

Lp.	Parameter	Coal mine			
		"Janina"	"Bogdanka"	"Piast"	"Ziemowit"
Proximate analysis					
1.	W^{ar}, %	19.1	11.3	13-16	14.8
2.	W^{ad}, %	8.6	5.5	4-6	7.3
3.	A^{ad}, %	19.8	21.0	20-25	20.1
4.	V^{ad} ,%	28.4	27.1	30.2	28.5
5.	C^{ad}_{fix}, %	43.2	46.4	43.2	44.0
6.	Q^{ar}_{i}, MJ/kg	18.16	21.28	18.0-20.0	19.83
Ultimate analysis, %					
1.	C^{ad}	54.00	59.45	55.26	56.01
2.	H^{ad}	4.04	3.47	3.56	3.50
3.	N^{ad}	0.94	1.26	0.82	0.69
	S^{ad}	2.00	1.07	0.91	0.93
4.	O^{ad}	10.62	8.20	14.32	11.40
Ash fusion temperatures, °C					
1.	Initial deformation temp. (IT)	920	900	910	910
2.	Softening temperature (ST)	1,260	1,220	1,250	1,310
3.	Hemispherical temp. (HT)	1,340	1,500	1,360	1,490
4.	Fluid temperature (FT)	1,360	1,500	1,360	1,500

^{ar} as received, ^{ad} air dried

Table 3. Properties of selected coals for analysis of coal gasification for liquid and gaseous fuel production.

[2] The dominant share - 67% of coal production in the EU27 (Lorenz, 2008).

Availability: a total yearly working time of 85 % has been assumed for all of the options, which is equal to 7,446 hours/year.

Gasifier: process calculations were made for gas generated in the gasifier using GE/Texaco technology. It was assumed that the gasification process would be carried out in a gasification reactor with 125 t/h of raw coal processing capacity. This value meets the processing capacity of operating and newly built entrained flow gasifiers, which are in the range of 100-130 t/h of coal. In typical gasification systems using GE/TEXACO technology, both radiant and convective coolers produce high pressure saturated steam. In the analysed cases, it was proposed that in the radiant cooler, the produced steam is overheated in a convective heat exchanger and then fed directly to a steam turbine for power generation.

Preparation of CO_2 for transport and storage: separated carbon dioxide is compressed to the pressure required for transport conditions, i.e., approximately 12 MPa, and then is transported to storage sites for underground storage.

Process calculation: for the considered technological options, mass balances have been determined on the basis of a calculation made in the ChemCAD v.6.0.2 process simulator for steady state conditions. For liquid fuel production by Fischer-Tropsch synthesis, process calculations were made using data from (US Department of Energy [US DOE], 1999).

4.1. Results of calculations

The summarised results of the process calculations are shown in Table 4.

Parameter	unit	option I "Ziemowit"/ "Piast" coal	option II "Janina" coal	 "Bogdanka" coal	option III "Janina" coal
Coal input	t/h	750	125	125	125
Thermal input	MWth	4,131	631	739	631
F-T liquid production	kg/h	146,200	-	-	-
Methanol	kg/h	-	-	-	62,138
Hydrogen	kg/h	-	10,941	12,197	-
Gross power output	kWe	349,920	73,470	80,040	71,965
Auxiliary load	kWe	366,957	69,204	79,864	72,778
Net power output	kWe	-17,037	4,266	176	-813
Production efficiency	%	N.A.	57.7 hydrogen LHV	54.8 hydrogen LHV	54.6 methanol LHV
CO_2 sequestration (total)	kg/h	883,660	188,448	220,039	210,462
geological	kg/h	883,660	188,448	220,039	125,022
chemical	kg/h	N.A.	-	-	85,440
CO_2 capture[a]	kg/h	62[b]	86	86	96
CO_2 emission[c]	kg/h	40,800 (56,866)	25,800 (21,777)	29,620 (29,454)	4,143 (4,910)

[a] including geological and chemical sequestration, [b] chemical sequestration not included, [c] including the necessary purchase of electricity (943 kg CO_2/MWh) (Finkenrath, 2011)

Table 4. Results of the process calculations (option I -III).

After consuming 750 t/h (5.6 million t/y) of raw coal, a plant produces 146.2 t/h of Fisher-Tropsch synthesis products, including 14.6 t/h of LPG (liquefied petroleum gas), 25.3 t/h of diesel and 106.3 t/h of components for diesel production. In addition, sulfur (6.6 t/h) and carbon dioxide (883.7 t/h) are also produced in the system. The off gas from the F-T processes and the steam generated in the system are used to produce electricity (electric power: 350 MW_e). The electricity produced covers approximately 95 % of the system needs; to balance the power consumption, an additional 127 GWh (17 MW_e) of electric energy is needed.

In the case of hydrogen production, which depends on coal, the plant produces 10.9 and 12.2 t/h of hydrogen from "Janina" and "Bogdanka" coal, respectively. The application of lower quality coal decreased the hydrogen production by approximately 11 %. The gross electricity production also decreased, but due to the growing auxiliary needs in the case of "Bogdanka" coal, which has a greater oxygen demand, a facility using lower quality fuel produces more net energy. In both cases, the electricity production covers the needs of the system. The system also produces sulfur (2.2 and 1.1 t /h) and carbon dioxide (188 and 220 t/h). The efficiency of hydrogen production is 58 % and 55 % (based on LHV) for "Janina" and "Bogdanka" coal, respectively.

A methanol production plant produces 62 t/h of methanol with a high grade purity level. The efficiency of methanol production is approximately 55 % (based on LHV). The energy generated in the system nearly covers the system needs (approximately 99 %). The sulfur production amount is 2.2 t/h. For all of the analysed options, methanol production is characterised by the lowest CO_2 emissions to the atmosphere and, consequently, the highest efficiency CO_2 removal (96 %).

This is because "chemical sequestration" takes place in the methanol production process and part of the CO_2 formed during coal gasification and the conversion of synthesis gas is "stored" in the final product, i.e., methanol.

Case IV involving the Polygeneration Plant is described and analysed in a later section of the paper.

5. Investment expenses

To calculate the investment expenses, an exponential investment assessment method was used based on the following function:

$$C_1 = C_0 \left(\frac{S_1}{S_0} \right)^f \tag{1}$$

where: C_1 is the calculated investment for the system component, C_0 is the reference investment cost, S_1 is the scale of the system component, S_0 is the base scale parameter and f is the scaling exponent.

The base scales and scaling exponents for the components of the production facilities based on coal gasification are shown in Table 5.

Capital expenditures specified for the base year were calculated for the current year using the method of indices according to equation (2):

$$C_2 = C_1 \left(\frac{I_2}{I_1} \right) \tag{2}$$

where: C_2 is the current investment. C_1 is the base investment. I_2 is the current index value and I_1 is the base index value.

The indices used in this study were from the M&S (Marshall & Swift Equipment Cost Index) and CEPCI indices (Chemical Engineering Plant Cost Index) as published in Chemical Engineering. Having assessed the main equipment investments (machines, instruments, devices), the factor analysis has been used by adding relevant coefficients to the coordinates positions and obtaining fixed assets investment estimation results. For total fixed assets investment estimation, the following equation has been used:

$$C_n = E + \sum_{i=1}^{m} f_i E \tag{3}$$

where: C_n is the fixed assets investment, E is the equipment purchasing costs, and f_i are the coefficients for instruments and devices, fittings, foundations assembly cost, etc.

Plant component	Scaling parameter	Base scale	Exponent
Coal handling	Coal feed	100 t/h	0.67
Gasifier	Coal thermal input	697 MW_{th}	0.67
Oxygen plant – ASU	O2 flow	76.6 t/h	0.50
O_2 compression	Compression power	10 MW_e	0.67
N_2 compression	Compression power	10 MW_e	0.67
Selexol –H_2S removal	Sulfur feed	3.4 t/h	0.67
Selexol –CO_2 removal	CO_2 removed	327 Mg/h	0.67
CO_2 drying and compression	Compression power	13 MW_e	0.67
CO Shift (WGS)	Thermal input	1,377 MW_{th}	0.67
Claus. SCOT	Sulfur feed	3.4 t/h	0.67
Boiler	Heat transfer surface	225, 000 m^2	0.67
Steam turbine	Turbine output	136 MW_e	0.67
Gas turbine	Turbine output	266 MW_e	–
FT synthesis reactor	Thermal output	100 MW_{th}	1.00
FT product upgrading	FT product production	286 m^3/h	0.7
MeOH synthesis reactor– w/o recirculation	Syngas flow	2.89 kmol/s	0.65
MeOH synthesis reactor – w/ recirculation	Syngas flow	10.81 kmol/s	0.65
MeOH separation and purification	Methanol production	4.66 kg/s	0.29
PSA – hydrogen separation	Hydrogen production	0.294 kmol/s	0.74
CO_2 removal	CO_2 flow	3, 280 mol/h	0.60

Table 5. Base scales and scaling exponents for coal conversion system equipment investments.

The investment costs were calculated assuming expenditures presented as "overnight costs" on the basis of the second quarter of the year 2006 and taking into account an investment cost growth of approximately 60 % by mid-2008. To determine the escalation of capital costs a 30 % increase in the cost of engineering services (60 % share in cost increase) and a 100 % increase in steel price[3] (40 % share in cost increase) were assumed.

The costs of instruments and devices include the initial equipment plus chemical substances and catalysts. Unpredictable expenses include process costs and project risk.

To calculate investment costs for CO_2 transport and storage 40 km (option I) and 100 km (options II and III) pipelines were assumed.

The investment estimation was conducted with the same accuracy as the pre-feasibility study. i.e., ± 30 %. The investment estimation results are presented in Tables 6 and 7.

Investment component	Thousands $ (1 $ =2.2531 PLN; 2008)		
	option I	option II	option III
Instruments and devices supply[a]	1,766,211	390,751	400,737
Instruments and devices assembly[b]	671,160	148,507	152,279
Instrumentation and control equipment	105,973	23,434	24,056
Electric installation	162,491	35,950	36,883
Construction works	264,931	58,630	60,095
Land development	105,973	23,434	24,056
Total direct investments	**3,076,740**	**680,707**	**698,105**
Design and supervision	370,904	82,065	84,151
Total direct and indirect investments	**3 447,644**	**762,771**	**782,256**
Unpredictable expenses	635,836	140,695	144,246
Total investment in **Fixed capital**	**4 083,480**	**903,466**	**926,501**
Start-up	68,953	15,268	15,623
Total investments	**4,152,433**	**918,734**	**942,124**
Total investments, Thousands $/TPD (Investments 10^3 $/coal input in tonne per day)	**230.7**	**306.2**	**314.0**

[a] – includes auxiliary equipment, [b] – includes foundations and piping

Table 6. The investment estimation results for the technological part of the considered plants.

[3] Steel Business Briefing Ltd, september 2008

Description	Thousands $ (1 $ =2.2531 PLN; 2008)		
	option I	option II	option III
CO_2 pipeline construction	146,082	113,419	76,932
CO_2 storing facility	47,601	17,309	11,784
Total	193,683	130,729	88,716

Table 7. Total CO_2 transport and storage related investments.

6. Financial and economic analysis

The base year for finacial and economic analysis is assumed to be 2008 (Q4). The analyses have been prepared using fixed prices, without consideration for inflation prognoses or other changes that may constitute factors influencing future prices of the elements involved in the production process. Any prognoses for the coal, gaseous and oil based fuel processing sector bears considerable risk, which convinced us to use actual prices (base year) and keep the relationships between individual assisting factor prices in our analysis. All of the prices used in the calculations are net with VAT excluded. In the calculation, the unit prices were estimated according to the prudence rule for both sales income and for enterprise working cost, which creates a safety margin in terms of possible price fluctuations and other unexpected expenses. At the time of analysis was performed 1 $ =2.2531 PLN and 1 € = 3.438 PLN. The limit value of the internal rate of return assumed at 6.4 and 8.2 % respectively for the models FCFF (Free Cash Flow to Equity) and FCFA (Free Cash Flow to Firm). The analysis was performed using the UNIDO method (COMFAR III Expert software).

Regarding the foreseen changes in compulsory CO_2 emission allowances starting in 2013, the efficiency calculation is based on three development scenarios:

- basic, assumes project functioning in the present conditions with no regulations on CO_2 (no necessity to buy rights) – hereinafter referred to as scenario 1.
- reference, where a plant owner buys 100 % of the CO_2 emission rights at a price of 39 €/t – hereinafter referred to as scenario 2.
- prospective, assumes the necessity of building CO_2 transport and storage facilities. In this scenario, we include the costs of purchasing and assembling systems for carbon dioxide sequestration, which enable the majority of emitted carbon dioxide to be stored in designated geological structures. For the remaining CO_2 emitted to the atmosphere, there is a requirement to purchase 100 % of the emission rights at a price of 39 €/t – hereinafter referred to as scenario 3.

Assumptions for the calculation are summarised in Table 8 and Table 9 show the adopted total operational costs for the chemicals, the transport and storage of CO_2 and environmental protection costs (waste disposal, emission fees: NO_x, SO_2, dust, CO_2). The results of the economic analysis are presented in Fig. 7.

The liquid fuel production does not reach the required return rate of the invested capital in the predicted scenarios. The reasons for this situation are the large initial investments for building the plant and production start-up. In case of scenarios which assume the necessity to purchase CO_2 emissions, and especially in the scenario 2 weak financial result is due to the large amounts of CO_2 formed in relation to the manufactured product which is about 6 t/t.

Hydrogen generation enables invested capital return in both analysed cases ("Janina" coal and "Bogdanka" coal); however, considering the possibility of CO_2 emission rights fee implementation, it will be necessary to build additional carbon dioxide transport and storing facilities. Whenever a project lacks these structures, there is no profitability (results – scenario 2).

The methanol production option produced the best results among all of the options analysed for scenarios 2 and 3. This is related to the lower CO_2 amount that is emitted (option II) or designed for sequestration (option III) compared to the hydrogen generation options. It is associated with the "chemical sequestration" i.e. the use of CO_2 for methanol synthesis.

A lack of economic effectiveness in scenario 2 for options I and II and, at a lower rate, for option III with respect to scenario 3 confirms the desirability of CO_2 sequestration (capture. transport and storage), particularly from the perspective of the probability of 100 % emission rights duty after 2012.

The results of the calculations of DPBT (Dynamic Pay Back Time) for the FCFF models allow us to make the following conclusions:

- liquid fuel production does not allow a return on investment expenditures in the assumed lifetime of the installation (30 years).
- for the hydrogen generation project, the discounted pay back times are the following: "Janina" coal: scenario 1 – 9 years from the operation start-up, scenario 3 – 13 years from the operation start-up; "Bogdanka" coal: scenario 1 – 8 years from the operation start-up, scenario 3 – 12 years from the operation start-up.
- methanol generation enables the achievement of financial results that guarantee invested capital return within 9 years from the operation start-up in scenario 1 and 10 years from the start-up in scenario 3.

Project profitability and liquidity assessment

In scenarios 1 and 3, the projects generate positive financial results, which constitute the basis for project stability and for getting the surplus necessary for invested capital return. Scenario 1 assures slightly higher profitability; nevertheless, we may potentially face CO_2 emission rights purchasing after 2012. For option I, the financial performance is insufficient to ensure a return on the invested capital.

Specification	Unit	Cost/ price	Comment
Unitary prices			
Liquefied petroleum gas (LPG)	$/t	1,556	The basis for the technical propane unit price calculation was its market price, less the excise tax (2008 Polish market).
Diesel	$/t	936	Fuel oil wholesale price (2008 Polish market).
Component for diesel production	$/t	749	Price was determined by the fuel oil wholesale price, decreased by 20 % for the value added for its final processing.
Sulfur	$/t	266	Sulfur prices grew considerably from 2007-2008 from 50 to 500 $, which made us choose a safe price level considering possible speculative fluctuations. Additionally, price decreases caused by an oversupply in the market are usually small in this product segment.
Hydrogen	$/t	3,106	Costs of hydrogen production from natural gas (NG) were calculated according to the equation presented in (Stiegel & Ramezan, 2006). The NG suppliers' price parameters have been used in this equation, using the prices for large buyers. As the equation structure primarily considers investment amortisation values, which drastically grew during 2007 and 2008, the results have been increased by 30 % for investment growth compensation. Chemical business specialists were consulted on the calculation methodology and estimated total production cost.
Power	$/MWh	89	The power sales price, has been accepted as competitive in comparison with prices offered by the CHP plants to the industry, (2008 Polish market).
Methanol	$/t	596	Average price on the European market for the 2007-2008 period.
Unitary costs			
Coal	$/GJ	3.99	Market price (2008 Polish market)
Power	$/MWh	111	See above
CO_2 emission cost	€/t	39	Related to data published directly by the European Commission (SEC(2008) 85/3)
Water	$/t	0.11	-
Solid gasification product	$/t	-	For the prudence rule, the solid product is given away for free, which eliminates the costs of its treatment and disposal.

Table 8. Unitary costs and prices.

Specification	Thousands $		
	option I	option IIa (IIb)	option III
Chemical substances	2,663	444 (444)	444
CO_2 pipeline operation cost (scenario 3)	4, 674	3, 629	2 ,462
CO_2 storage operation costs (scenario 3)	2 ,799	1 ,156	858
Emission fees (scenarios 1 and 2)	1,062	239 (275)	166
Emission fees (scenario 3)	421	115 (130)	74
Waste disposal	2,219	444 (444)	444

Table 9. Operational costs related to chemical consumption, CO_2 transport and storage and environmental protection.

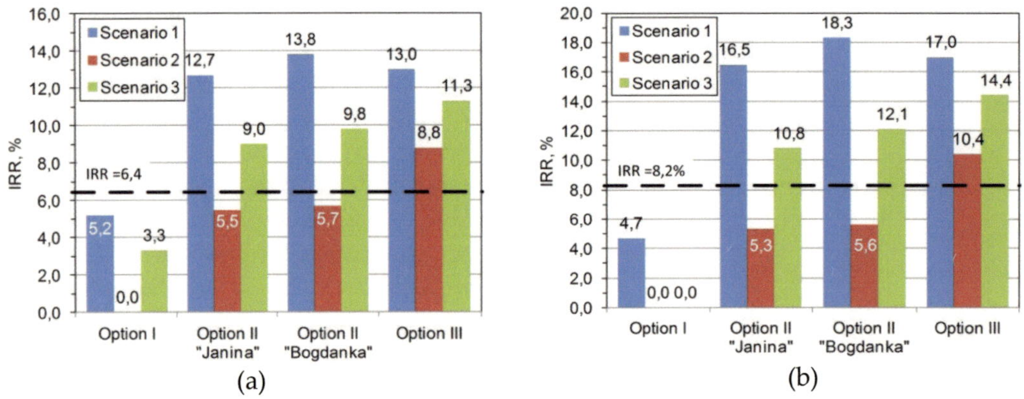

Figure 7. Internal return rate according to FCFF (A) and FCFA (B)

Risk assessment – Project sensibility

The project sensibility has been examined for all options in scenarios 1 and 3 (Fig. 8).

The following parameters have been subject to analysis:

- coal purchase prices: ± 10 % and their 20 % increase.
- investments: ± 10 % and 20 % and 30 % growth.
- basic product sales price in all of the options: ± 10 % and 20 %.
- CO_2 emission rights: 10 % and 20 % growth.

The results of the calculations enable us to formulate the following conclusions:

- coal prices changed in a given area do not implicate large deviations from the calculated efficiency indicators. A basic fuel price increase of 20 % does not cause any loss of liquidity in options II and III using both scenarios. For option I, a 15 % coal price drop in scenario 1 and a 40 % drop in scenario 3 is necessary to obtain a minimum level of profitability,
- an investment level growth of 30 % causes a loss of efficiency in option II using scenario 3. Achieving efficiency for option I is related to a necessity to reduce investments by 25 % in scenario 3 and by approximately 10 % in scenario 1,

- hydrogen sales prices drop by 20% will cause loss of efficiency in scenario 3. For the production of methanol, the lower limit for price level is 23% below the price which was assumed for the calculations. Achieving efficiency measures for option I is related to a necessity to raise sales prices by 15% and 27% respectively for scenario 1 and 3.
- thanks to a CO_2 transport and storage system, the project is not excessively price sensitive in terms of emission rights purchasing in scenario 3. Even with 20 % growth, the project efficiency is preserved. Option III is characterised by the smallest fluctuation and lowest carbon dioxide emission indicator.

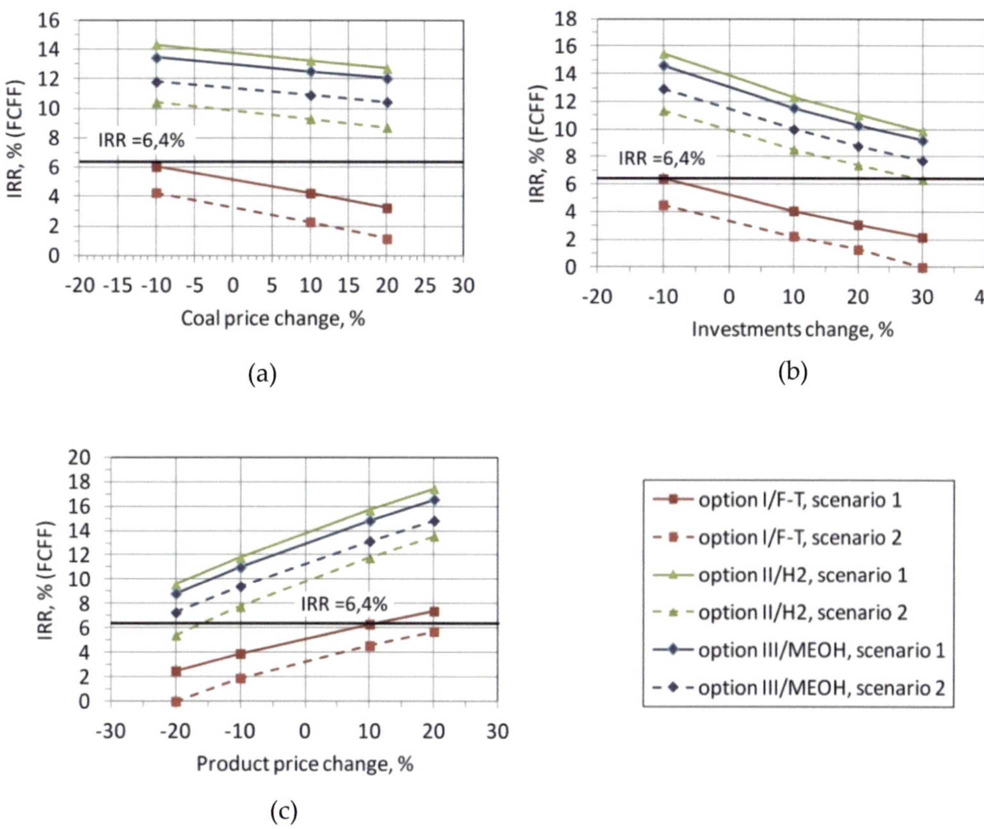

Figure 8. Sensitivity analysis A) coal price, B) investments, C) product price.

Additionally, for the defined scenarios using the considered options, the basic product minimum prices have been determined to assure profitability limit achievement. i.e., IRR equal 6.4 % (Table 10.). The prices of the analysed gasification products have been referred to the oil and natural gas prices in the following manner:

- option I – a motor oil semi-finished component is the basic product of the system. The unit price of the semi-finished motor oil component that was used in this analysis was the motor oil wholesale price decreased by 20 % (see Table 8). We have assumed that the motor oil semi-finished component will be equivalent to crude oil.

- option II – Hydrogen is the basic product of the system. The basis for the hydrogen generation cost calculation was the price of natural gas. The costs of hydrogen production from natural gas (NG) were calculated according to the equation presented in (Stiegel & Ramezan, 2006).
- option III – Methanol is the basic product of the system has been compared to the equivalent natural gas prices on the basis of available projects and consultations with Polish Chemical Industry Chamber experts. We should, however, highlight that methanol prices in the market are subject to considerable fluctuations, which are not always caused by natural gas prices changes. The calculations above may be burdened with methodological error that is difficult to define.

The results of calculations show minimum oil and natural gas prices, which assure the profitability of products included in individual options and according to assumed scenarios. The methanol production project has the best relationship in this matter as hydrogen production marketability is more dependent on natural gas prices. For coal-based liquid fuel production (motor oil semi-finished component) to be attractive with the different scenarios considered, oil prices must exceed 87 $/bbl.

Option	unit	Scenario 1	Scenario 2	Scenario 3
Liquid fuel	$/t	832	1,338	948
production	$/bbl oil equivalent	87	140	99
Hydrogen	$/t	2,173 (2,090)	3,220 (3,192)	2,699 (2,617)
production	$/1000 m^3 NG equivalent	364 (350)	553 (549)	459 (444)
Methanol	$/t	418	539	455
production	$/1000 m^3 NG equivalent	373	481	406

Table 10. Minimum selling prices of manufactured products assuming minimum profitability (IRR = 6.4 %).

7. Polygeneration plant

Polygeneration systems mediate the simultaneous production of chemicals and electricity from syngas. The purpose of these systems is to make maximum use of the chemical energy of coal by maximising the total energy efficiency of the transformation of primary fuel into useful products while minimising the capital expenditure and operating costs. Syngas may be used independently to produce chemicals and electricity, most advantageously in IGCC (integrated gasification combined cycle) systems.

Polygeneration usually include electricity production that is integrated with the generation of hydrogen, methanol or the products of Fischer-Tropsch synthesis.The principal advantages of a polygeneration system include:

- increased economic flexibility (two or more products);
- lower production costs due to more efficient use of syngas and of the technological heat produced in the course of the production process.

The integration of the processes of power and chemical production in a polygeneration system allows the achievement of high rates of fuel conversion, low emission rates and high economic efficiency, as in the case of CO_2 sequestration.

Table 11 presents the basic process data of the considered system. Consuming about 257 t/h of coal, the production of syngas amount to 85.1 t/h. This is enough to obtain 63.4 t/h of methanol. Net power and heat (in form of HP steam) output is 142 MW_e and 130 MW_{th} respectively. Geological sequestration of CO_2 will be 311 t/h. The amount of CO_2 stored in the chemical end product (methanol) will be 87 t/h.

Tables 12 and 13 show the investment costs and the minimal energy and synthesis gas prices to ensure the viability of a project (NPV> = 0 and IRR> = 7 %. with an amortisation period of 20 years). The calculation results are presented separately for the main technological units and for the whole facility of the Polygeneration Plant for the base case (CO_2 emissions within the scope of given emission rights) and in the case of CO_2 sequestration (separation; transport and storage of CO_2; fees for the remaining CO_2 emissions 39 €/t).

Fig. 9 shows the results of the calculations of produced synthesis gas prices against electricity prices (NPV> = 0 and IRR> = 7 %. with an amortisation period of 20 years) and the area of the economic attractiveness of the project.

Parameter	unit	Option IV Poygeneration Plant "Janina" coal
Coal input	t/h	257
Thermal input	MW_{th}	1,296
Syngas production	kg/h	85,079
equivalnt methanol production	kg/h	63,400
Gross power output	kW_e	282,700
Auxiliary load	kW_e	140,591
Net power output	kW_e	142,109
Thermal output[a]	kW_{th}	130,000
Production efficiency (mixed)	%	57.6[b]
Syngas production efficiency	%	73.2
Power production efficiency (IGCC)	%	31.4
CO_2 sequestration	kg/h	397,811
geological	kg/h	310,636
chemical	kg/h	87,175
CO_2 capture [c]	kg/h	88
CO_2 emission	kg/h	38,802

[a] high pressure steam from chemical module (see Fig. 6), [b] including syngas (chemical enthalpy), heat (HP steam) and power production, [c] including geological and chemical sequestration

Table 11. Results of process calculations for option IV.

The calculation results clearly indicate the attractiveness of the polygeneration process. The combination of electricity generation and synthesis gas production for the presented technological configuration (use of gasification technologies for energy production and syngas) causes a significant reduction in the minimum price of energy in comparison to the IGCC system (production of electricity) to 49 and 21 $/MWh without and with CO_2 sequestration, respectively (Tables 12 and 13).

Specification		Unit	IGCC (power island)	Syngas production unit (chemical island)	Poligeneration Plant
Investments		mln $ 10^3 $/TPD[a]	**1,105** **358.3**	**670** **217.3**	**1, 776** **287.9**
Price limits:	Power	$/MWh	**131**	111[b]	**82**
	Syngas	$/1000 m_n^3		**144**	202[c]

[a] Investments 10^3 $/coal input in tonne per day, [b] the approved purchase price of electricity reflects the price level of december 2008, [c] Adopted the maximum price of synthesis gas (Q4 2008), considered to be commercially attractive (the price of the synthesis gas produced from natural gas).

Table 12. Investments and price limits for manufactured products (power and syngas); Polygeneration Plant without CCS.

Specification		Unit	IGCC (power island)	Syngas production unit (chemical island)	Poligeneration Plant
Investments:		mln $ 10^3 $/TPD[a]	**1,256** **368**	**804** **221.5**	**2,060** **294.8**
Including:					
CO_2 Transport and Storage		mln $	121	121	242
Price limits:	Power	$/MWh	**191**	111[b]	**170**
	Syngas	$/1000 m_n^3		**167**	202[c]

[a] Investments 10^3 $/coal input in tonne per day - technological part only without CO_2 Transport and Storage, [b] and [c] see table 13.

Table 13. Investments and price limits for manufactured products (power and syngas) Case: CO_2 sequestration.

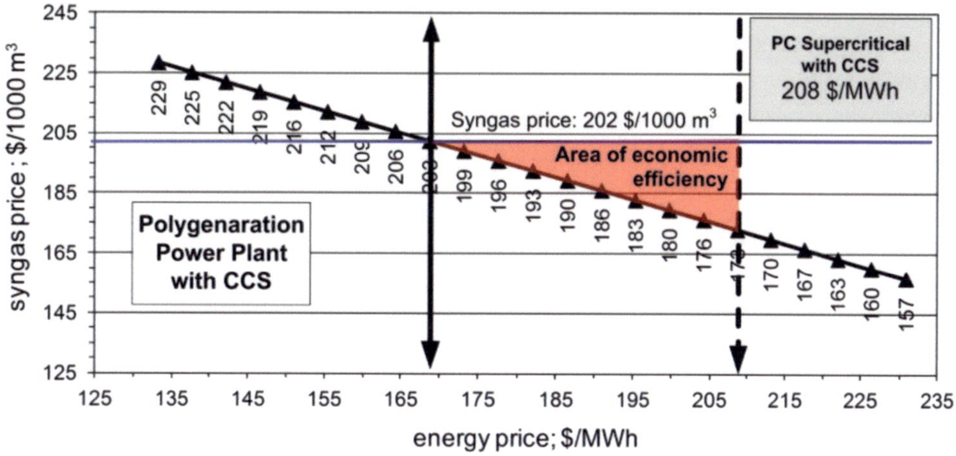

Figure 9. Estimated cost of synthesis gas in relation to the price of electricity and the area of economic efficiency of the Polygeneration Plant.

The CO_2 sequestration benefits of the proposed solution are also visible when comparing the Polygeneration Plant with a Supercritical Power Plant based on coal combustion. A comparison of the energy price limits for both cases at the same production level shows that with polygeneration we obtain lower energy prices by 38 $/MWh (energy price forecast for the supercritical coal unit with CCS amounts to 208 $/MWh). This underlines the attractiveness of the presented solution and the need to develop the proposed concept under appropriate technological conditions with the existence of a recipient for the produced synthesis gas as an alternative to traditional solutions.

8. Conclusion

The analysis concerned the installations for gaseous and liquid fuel production based on coal gasification using commercially available technologies of coal gasification, gas cleaning and conversion and chemical synthesis.

Systems for liquid fuels, hydrogen and methanol production were analysed in detail assuming three scenarios: basic (with no necessity to buy rights for CO_2 emission), reference (purchase 100 % of CO_2 emission rights at a price of 39 €/t), and prospective (assuming construction of CO_2 transport and storage facilities).

The analysis of the examined cases shows that with the adopted assumptions, the most favourable option is definitely the production of methanol, which shows economic effectiveness in all of the scenarios and, in the case of scenarios 2 and 3, gives the best results among the options analysed. The reason for this superiority among other options is related to low CO_2 emission, associated with the "chemical sequestration"i.e. the use of CO_2 for methanol synthesis.

The economic attractiveness of the production of hydrogen is significantly more dependent on natural gas prices. Hydrogen production is economically feasible only in scenarios 1 (base) and 3 (prospective). Developments in this direction and, consequently, the hydrogen economy seem to be limited due to a lack of cost-effective storage technology and transport infrastructure. At present, hydrogen from coal can effectively be used in chemical plants for the production of ammonia and fertiliser by substitution of the hydrogen produced from natural gas.

The coal to liquid fuels process based on Fischer-Tropsch synthesis is attractive only when exceed 87, 140 and 100 $/bbl for scenarios 1, 2 and 3, respectively.

Among the analysed technological options, the production of liquid fuels from coal using FT synthesis is definitely the least attractive and, on the basis of the obtained results, is not recommended as a potential direction for the application of coal gasification technology.

However, the idea of the production of liquid fuels from coal is still attractive, and the production of liquid fuels from coal using methanol seems to be a reasonable option. Methanol is used directly as motor fuel or is added to liquid motor fuels to improve their operational performance (methyl tertiary butyl ether, MTBE). Moreover, technologies for the production of motor fuels from methanol (MTG - methanol to gasoline and MTO/MOGD – methanol to olefines/Mobile olefines to gasoline and destilate) are being intensively developed and are commercially available at the industrial scale.

A lack of economic effectiveness in scenario 2 for options I and II and, at lower rates, for option III with respect to scenario 3, confirm the desirability of CO_2 sequestration (capture, transport and storage), particularly from the perspective of the necessity to purchase CO_2 emission rights after the year 2012.

The analysis of the Polygeneration Plant clearly shows the attractiveness of the solutions and the need to develop the proposed concept in appropriate technological conditions with the existence of a recipient for the synthesis gas produced as an alternative to traditional solutions. The realisation of this production process would give the possibility of significant reductions in the price of electricity generated, even in the case of CO_2 sequestration, compared to traditional technologies, including IGCC, while maintaining cost-effective production of synthesis gas for chemical applications. Also important from the economic point of view is installation flexibility in terms of the final product. i.e., the ability to design a production profile according to market demand for the manufactured products.

Author details

Marek Sciazko and Tomasz Chmielniak
Institute for Chemical Processing of Coal, Zabrze, Poland

9. References

Chmielniak, T.; Ściążko, M. & Uliniarz, M. (2008). Poligeneration power plant with CO_2 capture. *Archives of Energetics*, Vol. 38, No. 2/2008, pp. 45-54, ISSN: 0066-684X.

Dreszer, K. & Mikulska, B. (Ed.). (2009). *Feasibility study of the installation for the production of gaseous and liquid fuels from coal*, Instytut Chemicznej Przeróbki Węgla & Energoprojekt-Katowice S.A., ISBN 978-83-913434-7-0, Zabrze, Poland (in polish).

Energoprojekt-Katowice S.A (EPK) & Institute for Chemical Processing of Coal (IChPW). (2008). The concept of polygeneration plant for power, heat and syngas production to meets the needs of PKE SA Blachownia Power Plant and chemical works Kędzierzyn SA; Project no. 2007/0228/K, Contractor Południowy Koncern Energetyczny S.A. (TAURON Wytwarzanie S.A ; power producer) and Zakłady Azotowe Kędzierzyn S.A. (chemical works), (In polish).

Finkenrath, M. (2011). Cost and Performance of Carbon Dioxide Capture from Power Generation. International Energy Agency IEA 2011, 02.2012, Available from http://www.iea.org/papers/2011/costperf_ccs_powergen.pdf

Liu, H. (2010). OMB Gasification – Industrial Application Updates of Slurry Feeding & Developments of Dry Feeding, *Gasification Technologies Conference*, San Francisco 2010, 02.2012, Available from http://www.gasification.org/ library/overview.aspx

Lorenz, U. (2008). Main world steam coal exporters to the European market – some aspect of supply and prices. *Energy Policy Journal*, Vol. 11, No.1, pp. 255-272, ISSN 1429 – 6675. (In polish).

Stiegel, G.J. & M. Ramezan, M. (2006). Hydrogen from coal gasification: An economical pathway to a sustainable energy future, *International Journal of Coal Geology*, Vol.65 (2006), pp. 173– 190, ISSN: 0166-5162.

US Department of Energy (1999). Baseline Design/Economics For Advenced Fischer-Tropsch Technology (1991÷1999), Raport (vol.1-14), Project No. DE-AC22-91PC90027, Available from www.fischertropsch.org/DOE/DOE_reports /90027/90027_toc.htm.

US Department of Energy & National Energy Technology Laboratory (2002). Tampa Electric Polk Power Station Integrated Gasification Combined Cycle Project, Final Technical Report, Project no. DE-FC-21-91MC27363 US DOE/NETL, 03.2012, Available from http://www.netl.doe.gov/technologies/coalpower /cctc/cctdp/bibliography/demonstration/pdfs/tampa/TampaFinal.pdf

US Department of Energy & National Energy Technology Laboratory (2007). Gasification World Database 2007 – Current Industry Status, Robust Growth Forecast, (2007). Available from http://www.netl.doe.gov.

US Department of Energy & National Energy Technology Laboratory (2010a). Worldwide Gasification Database (Excel file); Available from http://www.netl.doe.gov/

US Department of Energy & National Energy Technology Laboratory (2010b). Cost and Performance Baseline for Fossil Energy Plants, Volume 1: Bituminous Coal and Natural Gas to Electricity, Revision 2, November 2010, DOE/NETL-2010/1397, Available from http://www.netl.doe.gov/

Section 5

Modeling and Simulation of Gasification

Chapter 11

Optimization of Waste Plastics Gasification Process Using Aspen-Plus

Pravin Kannan, Ahmed Al Shoaibi and C. Srinivasakannan

Additional information is available at the end of the chapter

http://dx.doi.org/10.5772/48754

1. Introduction

In this era of plastics dominated world, it remains a fact that there exists an ever-increasing margin between the volume of waste plastics generated and the volume recycled [1]. Of the total plastic waste, recyclable thermoplastics like polyethylene, polystyrene, polypropylene and PVC account for nearly 78% of the total and the rest is composed of the non-recyclable thermosets like epoxy resins and polyurethane [2]. Typically, plastics waste management is practiced according to the following hierarchical order: Reduction, Reuse, Recycling, and finally energy recovery. Although reuse of plastics seems to be best option to reduce plastic wastes, it becomes unsuitable beyond certain cycles due to the degradation of plastic. Mechanical recycling of plastics involves significant costs related to collection and segregation, and is not recommended for food and pharmaceutical industries. While chemical recycling focuses on converting waste plastics into other gaseous or liquid chemicals that act as a feedstock for many petrochemical processes, energy recovery utilizes the stored calorific value of the plastics to generate heat energy to be used in various plant operations. Moreover, since plastic wastes always consist of a mixture of various polymeric substances, chemical recycling and energy recovery seems to be best possible solution, both in terms of economic and technological considerations.

One of the major processes of chemical recycling involves thermal treatment of the waste plastics. The inevitable shift in world's energy paradigm from a carbon based to hydrogen based economy has revolutionized the capabilities of thermal treatment processes, viz. combustion, gasification and pyrolysis, in particular on the latter two techniques. In fact, recent technical investigations on the novel municipal solid waste (MSW) management methods reveal that a combined gasification and pyrolysis technique is more energy efficient and environmentally friendly than other processes [3].

In general the process of gasification for energy extraction from solid carbon source involves three simultaneous or competing reactions namely combustion, pyrolysis and gasification. The partial combustion of solid fuel creates an oxygen devoid, high temperature condition within the reactor which promotes the pyrolysis reaction, breaking the fuel into products that are a mixture of char and volatiles containing small and long chain hydrocarbons. The presence of gasifying agent (steam) drives the water shift reaction converting the carbon sources in to a mixture of valuable chemicals, tar, fuel gases and some residual particulate matter. The products undergo various downstream operations in order to separate and purify the valuable gaseous products that are later utilized for energy generation. This auto thermal feature makes the gasification process an economically viable and efficient technique for recovery of energy from waste plastics.

Gasification in commercial scale is practiced based on batch, semi batch and continuous modes of operation depending upon the processing capacity of the plant. Typically a plant processing large throughput utilizes fluidized beds due to the advantages such as enhanced gas-solid contact, excellent mixing characteristics [4], operating flexibility [5], and ease of solids handling [6] that lead to a better overall gasification efficiency. Fluid beds are preferred as it offers high heat and mass transfer rate and a constant reaction temperature which results in a uniform spectrum of product in a short residence time. It is important to keep the good fluidization characteristics of the bed, since introduction of material with different properties than the original components of the bed affect the quality of fluidization. Introduction of plastic material in fluidized beds demand additional attention due to its softening nature and possibility of blocking the feeding line. As soon as the plastic enters the hot reaction zone, it thermally gets cracked and undergoes a continuous structural change until it is eliminated from the bed. The sequence of interaction between the inert particle in the fluidized bed and the plastic material has been narrated by Mastellone et al., [7].

Gas-solid fluidization is the operation by which a bed of solid particles is led into a fluid-like state through suspension in a gas. Large scale gasifiers employ one of the two types of fluidized bed configurations: bubbling fluidized bed and circulating fluidized bed. A bubbling fluidized bed (BFB) consists of fine, inert particles of sand or alumina, which are selected based on their suitability of physical properties such as size, density and thermal characteristics. The fluidizing medium, typically a combination of air/nitrogen and steam, is introduced from the bottom of the reactor at a specified flow rate so as to maintain the bed in a fluidization condition. The dimension of the reactor section between the bed and the freeboard is designed to progressively expand so as to reduce the superficial gas velocity which prevents solid entrainment, and to act as a disengaging zone. A cyclone is provided at the end of the fluidized bed either to return fines to the bed or to remove fines from the system. The plastic waste is introduced into the fluidized bed at a specified location, either *over*-bed or *in*-bed using an appropriately designed feeding system. Pyrolysis experiments by Mastellone et al. [7] has shown that when the feed is introduced over the bed (from the freeboard region), it results in uniform surface contact with the bed material, thus enhancing transfer properties. The bed is generally pre-heated to the startup temperature either by direct or indirect heating. After the bed reaches the ignition temperature, plastic wastes are slowly introduced into the bed to raise the bed temperature to the desired operating temperature which is normally in the range of 700-900 °C. The plastic wastes are simultaneously pyrolyzed as well as partially combusted. The exothermic combustion reaction provides the energy to sustain the bed temperature to promote the pyrolysis reactions.

One of the main disadvantages of fluidized bed is the formation of large bubbles at higher gas velocities that bypass the bed reducing transfer rates significantly. If the gas flow of a bubbling fluidized bed is increased, the gas bubbles become larger forming large voids in the bed entraining substantial amounts of solids. The bubbles basically disappear in a circulating fluidized bed (CFB) and CFB the solids are separated from the gas using a cyclone and returned back to the bed forming a solids circulation loop. A CFB can be differentiated from a BFB in that there is no distinct separation between the dense solids zone and the dilute solids zone. The residence time of the solids in the circulating fluid bed is determined by the solids circulation rate, attrition of the solids and the collection efficiency of the solids in the cyclones. The advantages of the circulating fluidized bed gasifiers are that they are suitable for rapid reactions resulting in high conversion The disadvantage being, i) temperature gradients in the direction of the solid flow, ii) limitation on the size of fuel particles iii) high velocities resulting in equipment erosion. Although there are many different types of fluidized beds available for gasification and combustion, bubbling fluidized type is the most preferred type whenever steam is used as a gasifying medium [8]. The advantages of steam gasification have been well addressed in the literature [9].

A wide variety of plastics are in use depending upon the type of application, of which the most widely utilized are polyethylene (PE), Polypropylene (PP), Polyvinyl chloride (PVC), Polystyrene (PS) and Polyethylene terephthalate (PET). Each type differs in physical and chemical properties, and so do their applications. In general, the combustion of most of the plastics is considered safe with the exception of PVC that generates dioxins due to the presence of chlorine compound in its structure. In contrast with combustion, pyrolysis and gasification are endothermic process which require substantial amount of energy to promote the reactions. The pyrolysis process generally produces gas, liquid and solid products, the proportions of which depends on the operating conditions, while the gasification is predominantly reactions involving carbon or carbon-based species and steam, producing syngas (CO and H_2) and minor higher molecular weight hydrocarbons [6].

Cracking of PE either into its constituent monomer or other low molecular weight hydrocarbons has become a vital process due to the increased amounts of polyethylene wastes in the present world. Pyrolysis and/or gasification of PE serve as an appropriate tool for the recovery of energy and for waste plastic disposal simultaneously. Compared with other alternative feedstock like biomass and coal, PE possesses relatively higher heating value, and is much cleaner in terms of fuel quality attributing to lesser fuel pre-processing costs. Pyrolysis or gasification of PE results in a product stream rich in hydrogen and minimal CO or CO_2 content as compared to cellulose based wastes that yields relatively higher carbon monoxide and lower hydrogen product composition mainly due to the presence of oxygen in cellulose based feedstock.

Irrespective of the type of reactor and type of waste being handled, the key operating parameters that play a vital role in the gasification process are the equivalence ratio, reactor temperature, steam to fuel ratio, gasifying medium and residence time. In order to exert better reliability of the system, the operating variables have to be optimized and controlled with significant accuracy. The cheapest and most effective technique to qualitatively understand the effect of each operating variable and to identify possible optimal conditions is through process simulation. Such attempts to develop simulation models for process optimization has been reported in open literature of fuel sources such as, tyre [6], coal [10-13], and biomass [8, 14-16] using various computer simulation packages. However, the utility of any process simulation tool has not been well explored or recorded in the literature for modeling plastics gasification.

This chapter discusses recent work by the authors on Aspen Plus based process model to analyze the performance of a plastics gasification process under equilibrium conditions. The primary goal of this work is to successfully test and demonstrate the applicability of Aspen Plus to simulate the gasification process for one of the most abundantly used plastic, polyethylene (PE). This study will serve some preliminary qualitative and quantitative information on the overall behavior of the gasification process including the sensitivity of process parameters.

2. Model development

2.1. Modeling the gasification process

The gasification process models available in literature can be generally classified under steady state or quasi-steady state or transient state models. The steady state models do not consider the time derivatives and are further classified as kinetics *free* equilibrium models or kinetic rate models [17]. The following is a list of few researchers who have used the above-mentioned models for modeling the gasification process of various fuels; transient model for coals by Robinson [18], steady state kinetic model for biomass by Nikoo [14], steady state kinetic model for plastic wastes by Mastellone [7], kinetics *free* equilibrium model for biomass by Doherty [15], Paviet [17], and Shen [8], kinetics *free* equilibrium model for tyre by Mitta [6]. Of these, the kinetics *free* equilibrium steady state model is the most preferred for predicting the product gas composition and temperature, and more importantly for studying sensitivity analysis of the process parameters. Table 1 shows a summary of a few gasification simulation models developed in Aspen Plus for various materials.

The model used in this work to investigate the simulation of PE gasification in fluidized bed reactor is based on the model previously developed by Mitta et al. [6] for simulating tyre gasification. The *simplified* tyre gasification equilibrium model was simulated using Aspen Plus and it was successfully validated using the experimental data. Such an equilibrium type of approach considers only the equilibrium products, namely methane, hydrogen, carbon monoxide, carbon dioxide, water, sulphurous and nitrogen compounds formed within the reactor. Any other high molecular weight hydrocarbons, such as tars and oils, are less likely to form under equilibrium conditions and hence are not included in the simulation. More importantly, the equilibrium condition facilitates an exhaustive optimization study focusing on key process parameters, including the gasification temperature, equivalence ratio, steam to fuel ratio, and gasifying medium, thereby neglecting the complexities of the gasifier hydrodynamics and reaction kinetics.

Material	**Model**	**Process Variable**	**Range**	**Findings / Remarks**	**Ref.**
Biomass	Equilibrium (*volatile rxns.*) & kinetic (*char gasification*)	Temperature (°C)	700-900	Higher temperature, lower ER and higher steam-to-fuel ratio favors hydrogen and CO production Boudouard and methanation reactions were not considered	[14]
		ER	0.19-0.27		
		Steam to fuel ratio	0-4		
Tyre	Equilibrium	Temperature (°C)	750 -1100	Higher temperature, higher fuel/air ratio and lower steam/fuel ratio favors hydrogen and CO production All components listed in gasification reactions, along with H_2S, are considered as possible products	[6]
		Fuel to air ratio	0.2-0.8		
		Steam to fuel ratio	1.25-5		
Biomass	Restricted Equilibrium	Temperature (°C) (achieved by changing ER between 0.29 - 0.45)	674-1195	Air preheating effective at ER's less than 0.35 Without air preheating, optimum conditions for ER is 0.34 and gasification temperature between 837 to 874 °C Only Reactions (1-8) along with reactions for the formation of H_2S and NH_3 were considered for Gibbs free energy minimization	[15]
		Air Preheating Temperature (°C)	25-825		

Table 1. Summary of gasification simulation of various materials using Aspen Plus from literature.

The following assumptions are made in the current study for developing the process model.

1. All the chemical reactions were assumed to have reached equilibrium within the gasifier.
2. Only methane, hydrogen, carbon monoxide, carbon dioxide, oxygen, nitrogen, H_2S, and water were considered to be present in the product stream.
3. The primary components of char are only carbon and ash.

The entire gasification process was modeled using Aspen's *built-in* unit operation library in two stages; pre-processing and gasification. The two stages are discussed separately in the following sections.

2.1.1. Fuel pre-processing

Figure 1 illustrates the process flow sheet of the *simplified* PE gasification model. The first stage corresponds to fuel preprocessing where the polyethylene sample was processed or conditioned to remove any moisture present before the start of the gasification process. Drying and separation are the unit operations grouped in this stage and are represented by the respective modules in Aspen Plus. The fuel polyethylene stream labeled as "PE" was defined as a non-conventional stream and the ultimate and proximate analysis are provided as input to the model, refer Table II for parameter values. Polymer NRTL/Redlich-Kwong equation of state with Henry's law "POLYNRTL" and "POLYSRK" was chosen as parameter models to calculate the thermo physical properties of the components.

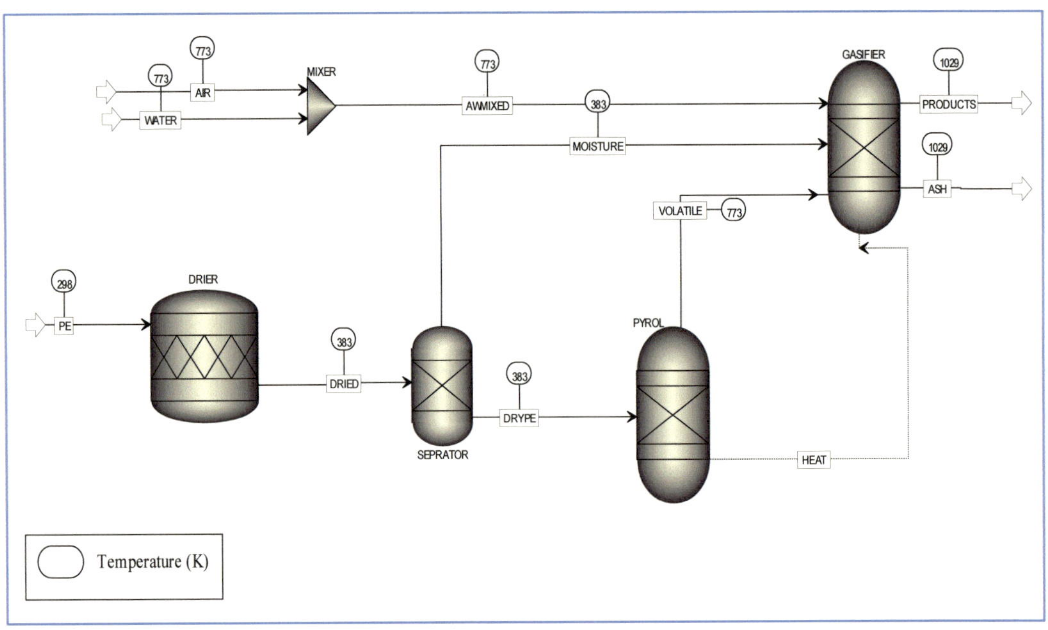

Figure 1. Process flow diagram of a PE gasification process in Aspen Plus

Sample	Proximate Analysis				Ultimate Analysis						
	Moisture	FC	VM	Ash	Ash	C	H	N_2	Cl_2	S	O_2
PE	0.02	0	99.85	0.15	0.15	85.81	13.86	0.12	0	0.06	0
Tyre	0.94	31.14	65.03	3.83	3.83	85.65	8.26	0.43	0	1.43	0.4

Table 2. Proximate and Ultimate analysis of the fuels used in this study.

At first, the fuel stream was first introduced into a drying unit "DRIER", which was modeled in Aspen Plus using an RSTOIC module. A temperature of 110 ºC and a pressure of 1 atm were selected as drier operating conditions. The stream leaving the drier, labeled "DRIED" contains the dried PE in solid phase and the removed moisture in vapor phase. This stream was fed to a separation unit "SEPARATOR" that splits the feed stream into product streams, labeled as "DRYPE" and "MOISTURE".

2.1.2. Volatiles and char gasification

In a typical gasification process, the fuel is first pyrolyzed by applying external heat where it breaks into simpler constituent components. These volatile components, along with char are then combusted, and the heat liberated from the combustion reactions would be used up by the subsequent endothermic gasification reactions. In the Aspen plus model, the dried portion of the fuel "DRYPE" exiting from the "DRIER" enters a pyrolyzer "PYROL" modeled as a RYIELD block in Aspen Plus. Based on the ultimate analysis of PE shown in Table II, the product yield distribution was calculated in the RYIELD module using Aspen Plus built-in calculator. An operating temperature of 500 ºC and a pressure of 1 atm were chosen in order to set the exiting stream "VOLATILE" to a pre-heated temperature of 500 ºC.

Parameter	Type	Value / Range
Fuel feed rate	constant	6 kg/h
Air flow rate	variable	5 – 30 kg/h
Steam flow rate	variable	0.3 – 30 kg/h
Air temperature	constant	773 K
Steam temperature	constant	773 K
Pyrolyzer temperature	constant	773 K
Drier temperature	constant	383 K

Table 3. List of process parameters provided as input to the model.

No.	Gasification Reactions	Heat of Reaction (kJ/mol) T =1000 K, P = P_o	Type
1	$C + ½ O_2 \leftrightarrow CO$	-112	Reactions with oxygen
2	$CO + ½ O_2 \leftrightarrow CO_2$	-283	
3	$H_2 + ½ O_2 \leftrightarrow H_2O$	-248	
4	$C + H_2O \leftrightarrow CO + H_2$	136	Reactions with water
5	$CO + H_2O \leftrightarrow CO_2 + H_2$	-35	
6	$CH_4 + H_2O \leftrightarrow CO + 3H_2$	206	
7	$C + CO_2 \leftrightarrow 2CO$	171	Boudouard reaction
8	$C + 2H_2 \leftrightarrow CH_4$	- 74.8	Methanation reactions
9	$CO + 3H_2 \leftrightarrow CH_4 + H_2O$	-225	
10	$CO_2 + 4H_2 \leftrightarrow CH_4 + 2H_2O$	-190	

Table 4. Summary of Gasification Reactions.

The volatiles stream, along with char was then passed to a gasifying unit "GASIFIER" that was modeled as a RGIBBS module. As it can be noticed in the model, the combustion and gasification reactions are allowed to take place within the "RGIBBS" module itself. The RGIBBS module calculates the equilibrium composition of the system using Gibbs free energy minimization technique. It provides an option to either consider all the components present in the system as equilibrium products or restrict the components based on some specific reactions or restrict it based on a temperature approach. In this study, all components from the gasification reactions, listed in Table IV, along with H_2S were included as possible fluid phase or solids products in the RGIBBS module. The gasifying mediums, air and steam, are preheated and mixed before it is sent to the gasifier. The outlet stream labeled as "PRODUCTS" contains product gases resulting from the gasification process while the "ASH" stream contains any residual solids.

The flow rate of fuel stream was held constant at 6 kg/h for all simulations. The two key parameters that influence the reactor temperature and the product distribution are equivalence ratio and the steam-fuel ratio, and hence were the only variables considered in the simulation. Equivalence ratio can be defined as the ratio of mass of oxygen/air supplied to the mass of oxygen/air necessary for complete combustion of all the carbon and hydrogen present in the feed to carbon dioxide and water respectively.

2.2. Model validation

The base case model for the gasification process was developed using Aspen plus built in modules based on the simulations popularly adopted in literature. In order to validate the appropriateness of the present model, simulations have been performed for gasification of tyre and the results were compared with the work due to Mitta et al. [6]. The ultimate and proximate analysis data used for tyre simulation in this study has been listed in Table II. However since the simulation parameters were not fully detailed by the authors, the parameters utilized in the present simulation is not the same as reported by Mittal et al. Therefore, only a qualitative comparison of the effect of parameters on the product distribution was considered for comparison purposes. Results showed good agreement in terms of the trends of the composition *versus* temperature plots and that serves as a basis for model validation.

In this work, a similar kind of study was performed to investigate the performance characteristics of the PE gasification process. In the case of isothermal gasification studies, it is challenging to include the temperature variation effects resulting from the entering steam flow, and exclusion of which results in significant deviation in the simulation results [14]. Hence, in this work, an adiabatic type of gasification reactor was modeled to investigate the effects of two key parameters, namely the equivalence ratio and steam-to-fuel ratio. The response variables include the gas composition, Carbon monoxide efficiency, hydrogen efficiency, and combined CO and hydrogen efficiency.

The carbon monoxide efficiency measures the extent of conversion of carbon present in the fuel to carbon monoxide. The definition of hydrogen efficiency and the combined efficiency follows the same. Van den Bergh [18] has reported expressions to calculate the CO, H_2, and combined CO and H_2 efficiencies. A similar definition was introduced in this work to estimate carbon dioxide efficiency as shown below.

$$\text{CO efficiency} = \left[\frac{f_{CO} \times F}{V_m}\right] \times \frac{1}{r_C} \times 100\% \tag{1}$$

$$\text{CO2 efficiency} = \left[\frac{f_{CO2} \times F}{V_m}\right] \times \frac{1}{r_C} \times 100\% \tag{2}$$

$$\text{Hydrogen efficiency} = \left[\frac{n_{H2}^{H} \times f_{H2} \times F}{V_m}\right] \times \frac{1}{r_H} \times 100\% \tag{3}$$

$$\text{Combined CO and H2 efficiency} = \frac{\left\{\frac{(f_{CO}+f_H)\times F}{V_m}\right\}}{\left\{\frac{r_H}{n_{H2}^{H}}+r_C\right\}} \times 100\% \tag{4}$$

where, f_{CO} and f_{CO2} represents the volume fraction of CO and CO_2 in the product gas respectively, r_C is the rate of carbon feeding [moles of carbon/min], F is the total gas flow rate [L/min], V_m is the standard molar volume [24.1 L/mol at 293 K and 1 atm], r_H is the rate of elemental hydrogen feeding [moles of elemental H/min], n_{H2}^{H} is the number of H atoms in PE monomer, *and* f_{H2} is the volume fraction of hydrogen in the gas. The combined efficiency represents the fraction of the maximum possible conversion or production achievable by the system. This maximum limit is considered when all the available carbon and hydrogen present in the fuel is converted to CO and H_2 [18]. The performance of the gasifier is also analyzed in terms of cold gas efficiency (CGE) that is defined as:

$$\text{CGE} = \frac{V_g\, Q_g}{M_b\, C_b} \tag{5}$$

Where
V_g = Gas generation rate (m^3/sec)
Q_g = heating value of the gas (kJ/m^3)
M_b = fuel consumption rate (kg/sec)
C_b = heating value of the fuel (kJ/m^3)

3. Results and discussion

3.1. Effect of steam-to-PE ratio

The effect of steam-to-PE mass ratio on PE gasification process was investigated in the range of 0.05 to 5 (corresponding to a mole ratio of 0.04 to 3.9) with a constant PE feed rate of 6 kg/h and an equivalence ratio of 0.15 (air flow rate of 15 kg/h). It can be expected that at low concentrations of water, oxidation reactions *via* Reactions (1-3) would dominate resulting in a higher temperature. The resulting temperature rise in turn would propel Reactions (4 and 6), which according to chemical equilibrium principle would shift forward, resulting in formation of CO and hydrogen. When the partial pressure of the reactant steam was increased, Reactions (4-6) would exhibit a tendency to shift forward, thus leading to a higher CO_2 and hydrogen content with simultaneous drop in CO molar composition. Due to the participation of the endothermic reactions at higher steam composition, the overall equilibrium temperature would show a decreasing trend. At some point, when there is enough hydrogen available to react with the carbon, the formation of methane would be favored as per Reactions (8–10). Subsequently, the methane formed would react with the excess steam to form back CO and hydrogen, as depicted by reaction (6). Overall, at any steam-to-PE ratio, the equilibrium system temperature and product composition would be a result of the competing simultaneous endothermic and exothermic reactions.

Figure 2 illustrates the variation of product molar composition and the equilibrium reactor temperature as a function of steam-to-PE mass ratio. The simulation *predicted* equilibrium temperature resulting from the gasification process helps to deduce certain qualitative conclusions on the overall gasification reaction and thus validate the theoretical explanations. From the simulation results, it can be noticed that when steam content is much less than the stoichiometric amount required for Reaction (4), which is equivalent to a steam-to-PE mass ratio of 1.33, the composition of hydrogen displays a sharp increasing trend while that of methane decreases. The high temperature and high methane content at lower steam-to-PE ratios are a result of the methanation and oxidation reactions. Above the stoichiometric point, hydrogen along with carbon monoxide shows a gradual decreasing tendency with a simultaneous increase in CO_2 content. This is in agreement with the theoretical explanation, wherein it was predicted that an increase in the amount of steam would strongly favor the forward endothermic reaction forming carbon monoxide and hydrogen. With higher steam content, the oxidation of CO is favored resulting in a steady increase of carbon dioxide during the gasification process. The steam composition in the product stream is a result of the excess and unreacted steam entering and exiting the reactor. As expected, above the stoichiometric point, the temperature of the reactor remains constant at around 850 K, possibly balanced by the complicated endothermic and exothermic gasification reactions.

Figure 3 shows the effect of the steam-to-PE ratio on the fractional efficiency of CO, CO_2 and H_2. It is evident that at around a steam-to-PE ratio of 0.4, the production of CO and hydrogen peaks while that of carbon dioxide is at a minimum. This is a favorable condition for any waste gasification process where it is desired to minimize as much as carbon dioxide as possible. Hence, it can be concurred that the favorable steam-to-PE mass ratio for the gasification process should be between 0.4 and 0.6, where the combined as well as the individual compositions of CO and H_2 are at a maximum. Furthermore, the cold gas efficiency (CGE) of the process seems to be affected only at lower steam-to-PE ratio. The predicted CGE values are much higher than those obtained in typical waste gasification process which is about 60%. It can be expected that under equilibrium conditions, as considered in this study, the gas yield is significantly higher than real process which directly contributes to increased efficiency.

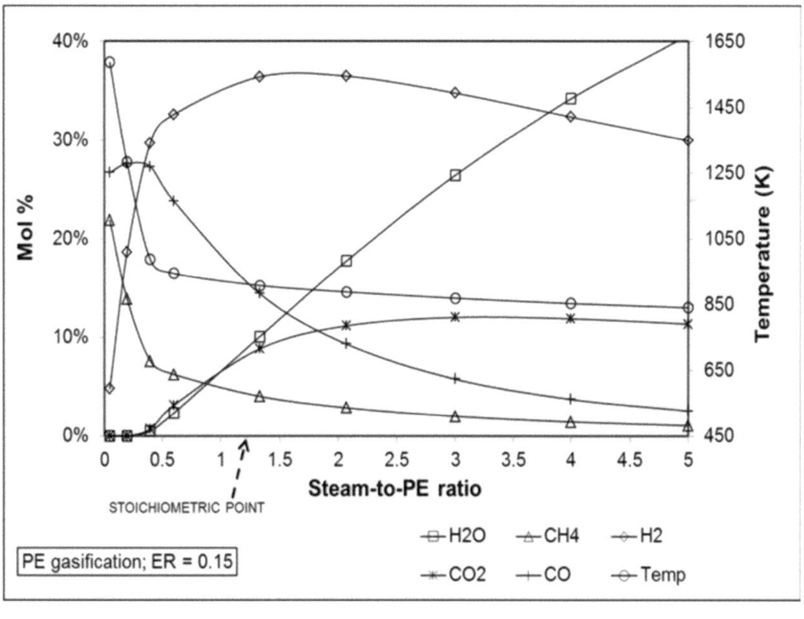

Figure 2. Product molar composition and temperature at various steam-to-PE ratios.

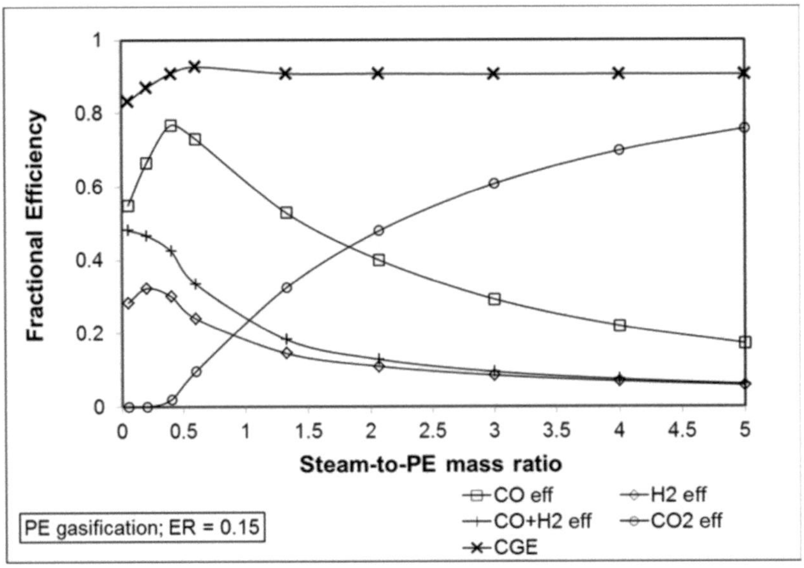

Figure 3. Fractional efficiencies at various steam-to-PE ratios.

3.2. Effect of equivalence ratio

The effect of equivalence ratio on the overall gasification efficiency was studied at two different steam-to-PE ratios. Typically, a commercial biomass gasifier is operated at an ER value of 0.25 in order to maintain auto thermal conditions (van den Bergh, 2005). Hence, a range of 0.05 to 0.3 was selected for this study in order to determine the optimum ER for PE gasification process. The cases for the two different steam-to-PE ratios have been presented and discussed separately below.

The oxidation reactions of carbon, CO and hydrogen, depicted by Reactions (1-3) are spontaneous and exothermic, resulting in release of significant amount of heat energy. It can be expected through Reaction (1) that at low values of ER (low values of stoichiometric air), only incomplete combustion of carbon would take place leading to the formation of CO with release of heat. Therefore, for the range of ER considered in this study, only Reactions (1) and (3) are the possible oxidation reactions, and thus any heat released during the combustion process will be directly attributed to these two reactions.

In general, at any fixed steam-to-PE ratio, the other parameters that drive the gasification process would be the ER and consequently the heat released from the combustion reactions. The intensity of the heat released controls the temperature, which in turn affects the directional shift in equilibrium of the gasification reactions. For example, the endothermic reactions (4, 6, and 7) would tend to shift in the forward direction with an increase in temperature and *vice versa*. Hence with increasing ER, it can be expected that the conversion of carbon to CO and hydrogen would be highly favored to other products such as carbon dioxide and methane.

Case 1: Steam-to-PE ratio 0.6

At low ER and low steam content, Reactions (4, 5 and 7) would be possibly controlled by the temperature and the partial pressure of steam. At such conditions, it could be expected that Reaction (5) would not be driven forward resulting in lower carbon dioxide formation. Furthermore, at low ER values, reactions with water would significantly compete with the oxidation reactions, thus limiting the resulting equilibrium temperature. At high ER and low steam content, this effect would be compounded such that temperature would be the primary variable that would determine the direction of the gasification reactions. In addition, at higher ER the composition trend of CO could be expected to fall down due to the subsequent combustion and methanation reactions of CO.

Figure 4 illustrates the variation of product gas composition and temperature as a function of various equivalence ratios. Between ER values of 0.05 and 0.2, reactor temperature, CO content, and hydrogen content increases steadily while the composition of methane decreases very sharply. In addition, the composition of carbon dioxide shows a steady decrease whereas the molar composition of water remains a constant. At ER values higher than 0.2, it can be observed that the temperature increases very sharply along with a steady decrease of hydrogen and carbon monoxide. It can also be noticed that beyond this point, only hydrogen, CO, and water are the major components of the product stream. The low values of carbon dioxide predicted throughout the range can be explained by the fact that at such low ER and steam-to-PE ratios considered in this study, neither complete oxidation nor steam gasification of carbonaceous components, depicted by reactions (2) and (5) respectively, proceeds at any significant rate. The sharp increase in the temperature beyond ER = 0.2 is due to the domination of the exothermic combustion reactions over others. The simulation results are very much in agreement with the theoretical expectations discussed earlier in this section.

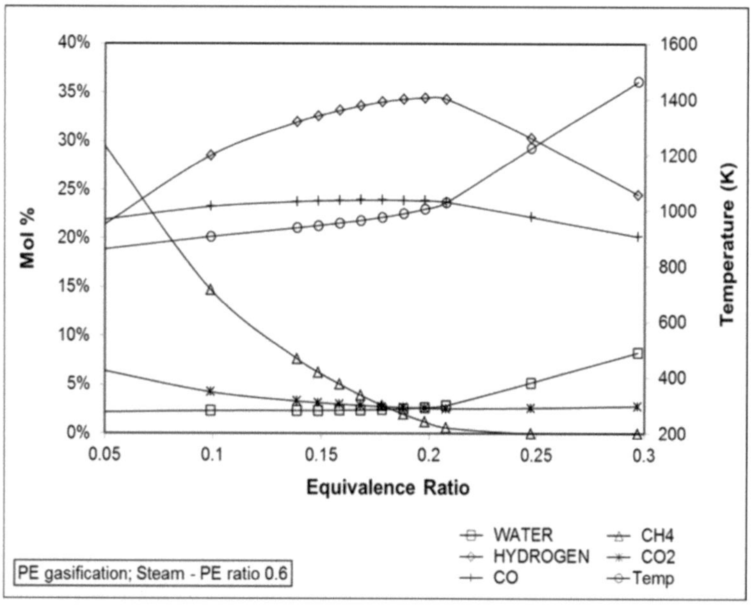

Figure 4. Illustration of the effect of equivalence ratio on product composition and temperature.

Figure 5 illustrates the variation of the fractional efficiencies with the equivalence ratios. It is clear that the efficiency of the conversion proceeds rapidly at lower ER's and reaches a maximum at ER of 0.2 and at a fixed steam-to-PE ratio of 0.6. The effect of ER on CGE is not significant at lower values since the composition of CO, hydrogen and methane that directly contribute to the heating value of the product gas increases until ER = 0.2. Beyond this point, since the yield of the above products decreases, CGE follows a decreasing trend and records a value of about 75% at an ER value of 0.3.

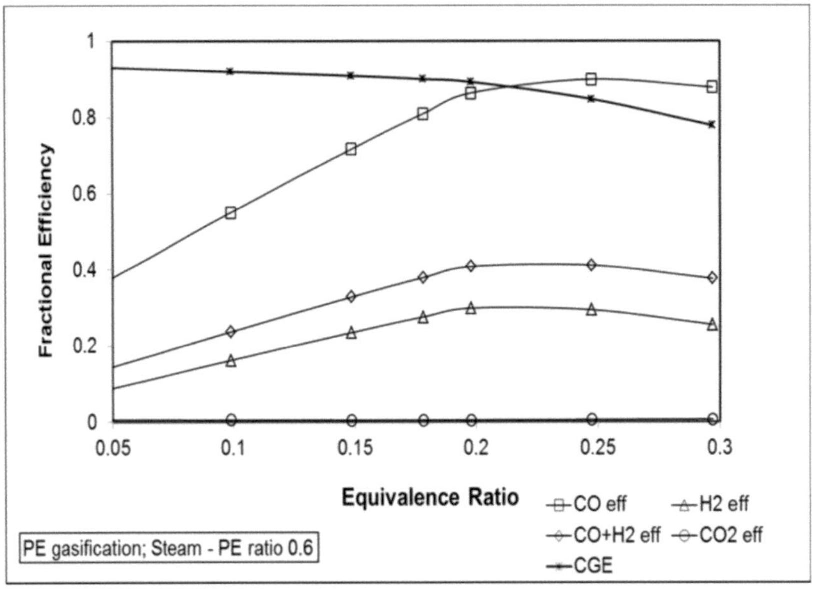

Figure 5. Illustration of the effect of equivalence ratio on gasification efficiency at a fixed steam-to-PE ratio of 0.6.

Case 2: Steam-to-PE ratio 4

An additional study of the effect of ER on the gasification process at a higher steam-to-PE ratio was included to provide better and comprehensive understanding of the sensitivity of equivalence ratio. In this case, the gasification reactions would not only be driven by the heat released by the preceding combustion reactions, but also by the partial pressure of steam. At a higher steam-to-PE ratio, it could be expected that Reaction (4) would significantly compete with Reaction (1) to consume the carbon present in the feed. Hence, the absolute value of the equilibrium temperature would be lower when compared to the previous case, steam-to-PE ratio of 0.6. Although high ER values would restrict the forward shift of the exothermic Reaction (5), the presence of higher steam content would favor the equilibrium to shift in the forward direction resulting in higher net carbon dioxide content.

Referring to Figures 4 and 6, it is evident that the trends of composition and temperature follow the same as case 1, but with different absolute values. It should be noted that the simulations predicted a temperature of about 800 K at an ER of 0.1 for case 2 compared to a value of ca. 850 K for case 1. It can also be observed that the composition of carbon dioxide was slightly higher and that of carbon monoxide was significantly lower than the results reported earlier in Case 1.

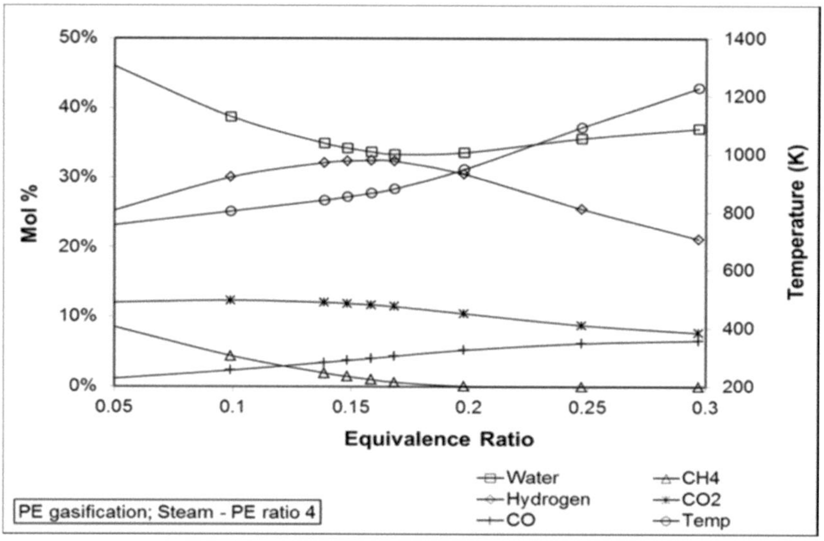

Figure 6. Illustration of the effect of equivalence ratio on product composition and temperature at a fixed steam-to-PE ratio of 4.

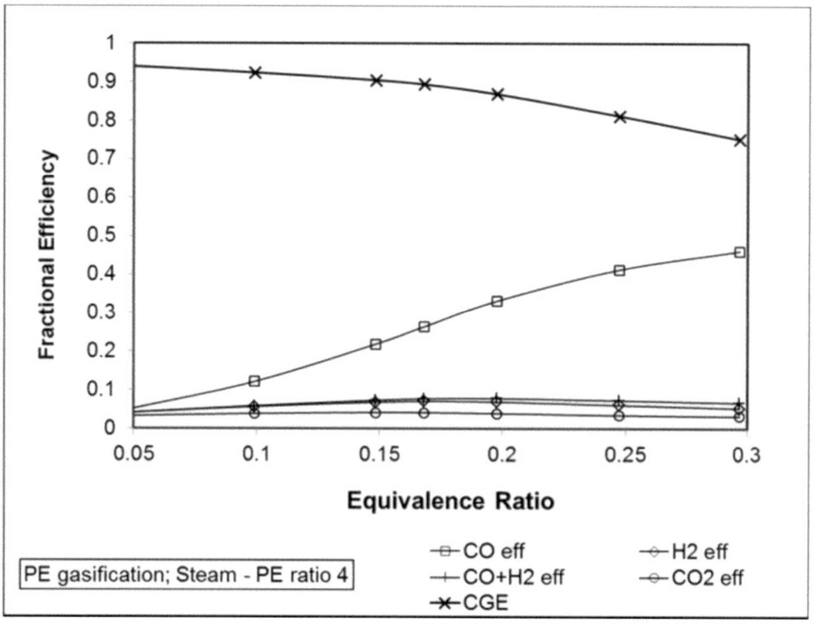

Figure 7. Illustration of the effect of equivalence ratio on gasification efficiency at a fixed steam-to-PE ratio of 4.

It can also be noticed from Figures 5 and 7 that the absolute maximum value of the combined CO and H_2 efficiency is significantly different among the two cases, which are predicted as 40% for case 1 and 7% for case 2. The composition of carbon dioxide in the product gases is very negligible at lower steam content, while it reaches about 4% for the case of higher steam content. Nevertheless, in both the cases, the maximum fractional efficiency of all the components occurs at an ER value of ca. 0.2. Furthermore, as discussed earlier in section 3.1, the effect of steam-to-PE ratio ion CGE is remarkable only until 0.6. Thus, the trend of CGE in Figure 7 for the case of higher steam-to-PE ratio resembles the same as that of Figure 5.

Hence, it can be concluded that an ER value of 0.2 and steam-to-PE ratio of 0.4 to 0.6 would yield a product stream containing 35% hydrogen, 25% CO, and negligible CO_2 at a temperature of 1000 K. These values seem acceptable for all practical purposes and are very much in agreement with the literature data, where a value steam-to-fuel value of 0.42 and an ER value of 0.15 were reported as the optimum parameters for co-gasification of wood and polyethylene [18].

4. Conclusions

The gasification process of waste polyethylene was successfully modeled using a combination of various unit operation modules available in Aspen Plus simulation package. The model used in this work to investigate the simulation of PE gasification in fluidized bed reactor is based on the model previously reported in literature for simulating waste tyre gasification. The equilibrium model developed in this study enables one to predict the behavior of PE gasification process under various operating conditions. Moreover, the results obtained are easy to interpret and thus could be directly corroborated with actual plant data.

Although temperature plays a vital role in controlling the conversion and product composition, it has been treated as a *free* variable in this study. Other process conditions were optimized in order to attain the appropriate temperature suitable for different applications that ideally lies between high temperature low calorific value and low temperature high calorific value product gas. The product distribution was the result of many competing simultaneous reactions mainly dictated by the temperature and the steam flow. The effect of the equivalence ratio and steam-to-PE ratio on the gasification efficiency was investigated in the range of 0.05 to 0.3 and 0.05 to 5 respectively. Based on the simulation results, the behavior of the conversion process was characterized and the values of the combined and individual fractional efficiencies have been presented. The following results summarize the findings from this study:

- Optimum steam-to-PE ratio was determined to be between 0.4 and 0.6 for low temperature applications. Under this condition, the yield of syngas and cold gas efficiency reaches a maximum.
- Product gas temperatures as high as 1273 K could be attained at higher steam-to-PE ratio at the expense of decrease in calorific value
- Sensitivity analysis on ER proposes an optimum value of about 0.2. Both CGE and syngas efficiency reaches a maximum at this point.

Due to the lack of detailed experimental data on waste PE gasification for various process conditions, the predicted data could not be validated. Although the results from this work heavily depend on the assumption made, i.e. thermodynamic equilibrium, significant qualitative results were deduced that would help to establish a sound reference for any detailed process optimization studies. Furthermore, this model can be used to estimate the final gas composition and other parameters, including gas yield and temperature for other solid waste fuels and mixtures. Upon including the hydrodynamics and gasification kinetics, this model could be used to evaluate the performance and behavior of many types of gasifiers under different process conditions.

Author details

Pravin Kannan, Ahmed Al Shoaibi and C. Srinivasakannan
Department of Chemical Engineering, The Petroleum Institute, Abu Dhabi, U.A.E

5. References

[1] http://www.gasification.org/uploads/downloads/GTC_Waste_to_Energy.pdf (Last accessed 20/3/2012)

[2] Panda A K, Singh K, Mishra D K (2010) Thermolysis of waste plastics to liquid fuel: A suitable method for plastic waste management and manufacture of value added products—A world prospective. Renewable Sustainable Energy Rev. 14(1): 233-248.

[3] Malkow T (2004) Novel and innovative pyrolysis and gasification technologies for energy efficient and environmentally friendly MSW disposal. Waste Manage. 24: 53-79.

[4] Sadaka S S, Ghaly A E, Sabbah M A (2002) Two phase biomass air-steam gasification model for fluidized bed reactors: Part I--model development. Biomass Bioenergy 22(6): 439-462.

[5] Arena U, Zaccariello L, Mastellone, M L (2009) Tar removal during the fluidized bed gasification of plastic waste. Waste Manage. 29: 783-791.

[6] Mitta N R, Ferrer-Nadal S, Lazovic A M, Parales J F, Velo E, Puigjaner L (2006) Modelling and simulation of a tyre gasification plant for synthesis gas production. Comput. Aided Chem. Eng. 21: 1771-1776.

[7] Mastellone M L, Arena U, Barbato G, Carrillo C, Romeo E, Granata S (2006). A Preliminary Modeling Study of a Fluidized Bed Pyrolyzer for Plastic Wastes. Paper presented at the 29th Meeting on Combustion, Italian Section of the Combustion Institute, Pisa, Italy.

[8] Shen L, Gao Y, Xiao J (2008) Simulation of hydrogen production from biomass gasification in interconnected fluidized beds. Biomass Bioenergy 32(2): 120-127.

[9] Franco C, Pinto F, Gulyurtlu I, Cabrita I (2003) The study of reactions influencing the biomass steam gasification process. Fuel 82(7): 835-842.

[10] Robinson P J, Luyben W L (2008) Simple Dynamic Gasifier Model That Runs in Aspen Dynamics. Ind. Eng. Chem. Res. 47(20): 7784-7792.

[11] Yan H M, Rudolph V (2000) Modeling a compartmented fluidized bed coal gasifier process using ASPEN PLUS. Chem. Eng. Commun.183: 1-38.

[12] Lee H G, Kim C, Han S H, Kim H T (1992) Coal gasification simulation using Aspen Plus. Paper presented at the US - Korea Joint Workshop on Coal Gasification Technology.

[13] Douglas P L, Young B E (1991) Modelling and simulation of an AFBC steam heating plant using ASPEN/SP. Fuel 70(2): 145-154.

[14] Nikoo M B, Mahinpey N (2008) Simulation of biomass gasification in fluidized bed reactor using ASPEN PLUS. Biomass Bioenergy 32(12): 1245-1254.

[15] Doherty W, Reynolds A, Kennedy D (2009) The effect of air preheating in a biomass CFB gasifier using ASPEN Plus simulation. Biomass Bioenergy 33(9): 1158-1167.

[16] Mansaray K G, Al-Taweel A M, Ghaly A E, Hamdullahpur F, Ugursal V I (2000) Mathematical modeling of a fluidized bed rice husk gasifier: part I - model development. Energy Sources 22(1): 83-98.

[17] Paviet F, Chazarenc F, Tazerout M (2009) Thermo Chemical Equilibrium Modelling of a Biomass Gasifying Process Using ASPEN PLUS. Int. J. Chem. Reactor Eng. 7: 18-.

[18] van den Bergh A (2005) The co-gasification of *wood and polyethylene; The influence of temperature, equivalence ratio, steam and the feedstock composition on the gas yield and composition.* Eindhoven University of Technology.

Chapter 12

Neural Network Based Modeling and Operational Optimization of Biomass Gasification Processes

Maurício Bezerra de Souza Jr., Leonardo Couceiro Nemer, Amaro Gomes Barreto Jr. and Cristina Pontes B. Quitete

Additional information is available at the end of the chapter

http://dx.doi.org/10.5772/48516

1. Introduction

Gasification processes are rather complex and difficult to model as they include gas-solid two-phase flow, mass and heat transfer, pyrolysis, homogeneous gas phase reactions and heterogeneous gas-solid reactions. Modeling of these phenomena based on basic principles of conservation is still at an incipient stage of development [1]. Additionally, most of the works have focused on coal gasification (for example, see [2-3]). Consequently, the development of a mechanistic model demands that many idealizations and suppositions are made [4], resulting in a very simplified model with little predictive capability.

Artificial neural networks (ANNs) are universal approximators [5] and have received numerous applications [6]. The literature, as indicated by [7], points out their ability to recognize highly nonlinear relations and to organize disperse data in a nonlinear mode in the context of empirical or hybrid modeling. These characteristics of the ANNs are very interesting and useful, motivating their use in the modeling of biomass gasification processes.

Hence, the present work aims to investigate – through the use of artificial neural networks (ANNs) and literature data – the correlation between the composition of the produced gas and the characteristics of different biomass for several operating conditions employed in fluidized bed gasifiers. Additionally, the neural network based developed model is employed to find conditions that maximize the yield of a given component of the produced gas.

This work is structured as follows. In section 2, fundamental aspects concerning biomass gasification and the modeling of the process are briefly reviewed. Section 3 focus on the modeling based on ANNs, while section 4 presents the optimization investigations using the developed models. Finally, the main conclusions are presented in section 5.

2. Biomass gasification

2.1. Fundamental aspects

Gasification is a process in which a solid or liquid fuel is converted into a gaseous fuel with contact with a gasifying agent. Coal, biomass, petroleum coke and other materials can be used in the process. The produced gas is mostly composed of hydrogen (H_2), carbon monoxide (CO), carbon dioxide (CO_2) methane (CH_4), traces of heavier hydrocarbons (as ethane and ethylene), water, nitrogen (when air is used as gasifying agent) and some contaminants. Besides the gaseous products, there are also subproducts as tar and solid non-converted residual carbon (char) [8].

The gas composition and the production of subproducts depend on several factors as: energy delivered to the process, type of gasifier, operating conditions and type of biomass employed. The gas produced can be used in different applications such as: gas turbines or internal combustion enginees, production of syngas and, after an adequate cleaning up and reforming, production of hydrogen or direct use on fuel cells [9-10]. The reactions inside the gasifier can be divided in four stages according to the temperature [8]: drying (> 150°C); pyrolysis (150 -700°C); combustion (700 - 1500°C) and reduction (800 -1100°C).

Some characteristics of the biomass have a significative effect on the performance of the gasifier. For this reason, proximate and ultimate analyses are used in order to characterize the biomass [7].

Because of their flexibility for use with different types of biomass [8], only fluidized bed (bubbling and circulating) gasifiers were studied in the present work. Different gasification agents can be considered, such as: air, oxygen, steam or a combination of them. In the case of use of air or oxygen, the heat released by the exothermic reactions between the oxygen and the fuel is used to keep the gasifier in the operating temperature and as heat source for the endothermic reactions. When steam is employed, it is necessary to use an external heat source [8].

2.2. Modeling of the process

The availability of accurate biomass gasification process models would help the operation and optimization of these processes. However, as noted by reference [11], the majority of the works have been developed for coal. In comparison with coal, biomass is made up not only of lignin, but also of cellullose and hemicellulose, each one having its own thermal behavior, what makes the biomass gasification even more difficult [11].

Reference [12] commented the difficulties of developing a model based on the kinetic equations of the different reactions, together with the mass and energy balances and hydrodynamic considerations for a circulating fluidized bed biomass gasifier. In this work, the authors cite the objective of developing a model "as good as possible".

Previous literature papers employed neural networks to predict characteristics of combustion, pyrolisys or gasification processes. In reference [3], a hybrid gasification model using ANNs was developed to estimate reactivity parameters of different types of coal with relative success. Reference [13] used ANNs to predict the emission of pollutant gases in the combustion process of a mixture of coal and urban solid residuals with a good agreement between experimental and predicted data.

In 2001, reference [1] developed a hybrid model for the gasification of biomass in a fluidized reactor that employed steam as gasification agent. The authors used multilayer ANNs to estimate parameters of a phenomenological model. The hybrid model was used to determine the production rate of the gas and its composition in terms of H_2, CO, CO_2 e CH_4. However, the neural networks were trained for each biomass separately.

In 2009, reference [7] used ANNs to predict LHV (lower heating values) of the gas and of the gas with tar and char and gas yields using the following input variables: type of residual (paper, wood, kitchen garbage, plastic and textile materials), gasification temperature and equivalence ratio. The results indicate that ANNs are a viable alternative for the modeling of the studied process.

3. Modeling using ANNs

3.1. Fundamental aspects

ANNs (Artificial Neural Networks) area a computational paradigm in which a dense distribution of simple processing elements is used to provide a representation of complex processes (and/or ill-defined and/or nonlinear).

ANNs are nowadays a standard modeling tool, being the feedforward paradigm named MLP (Multilayer Perceptron) the most popular one. Their fundamentals will not be discussed here as they can be found in several references [5, 14-16], only main aspects concerning topology and training of MLPs will be briefly commented in the following, as these ANNs were the ones chosen here.

The MLP paradigm is usually composed of an input, a hidden and an output layer of neurons. The neurons in the input layer are typically linear, while the ones in the hidden layer have nonlinear (often sigmoidal) activation functions. The neurons in the output layer may be linear or nonlinear. Each interconnection between two layers of neurons has a parameter associated with it that weights the feedforwardly passing signal. Additionally, each neuron in the hidden and output layers has a threshold parameter, also known as bias.

Typically, the neurons in the input layer simply forward the signals to the hidden neurons. The behavior of the neurons in the other layers will be explained using Figure 1.

Figure 1 exhibits the j-th neuron of the (k+1) layer of a multilayered neural network. This j-th neuron of the (k+1) layer receives a set of information $s_{pi,k}$ (i = 1, ..., n_k) – corresponding to the outputs (also called activations) of the n_k neurons of the previous layer – weighted, each one, by the weight $w_{j\ i\ k}$ corresponding to its connection. The neuron sums up these weighted inputs and the resulting value is added to a internal limit, a bias that can be represented by $\theta_{j,k+1}$. The neuron 'j' produces a response for the set of signals, according to an activation function f( ) [5, 14-16]:

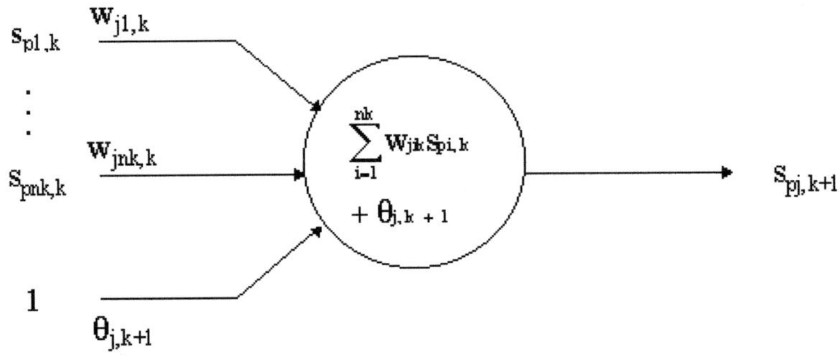

Figure 1. j-th neuron of layer (k+1).

The behavior of a neuron can be mathematically expressed by:

$$\lambda_{pj,k+1} = \left[\sum_{i=1}^{n_k} w_{jik} s_{pi,k} \right] + \theta_{j,k+1} \tag{1}$$

Some examples of activation functions are given below:

Linear function:

$$f\left(\lambda_{pj,k+1}\right) = \lambda_{pj,k+1} \tag{2}$$

Sigmoidal function:

$$f\left(\lambda_{pj,k+1}\right) = \left[1 + \exp\left(-\lambda_{pj,k+1}\right)\right]^{-1} \tag{3}$$

Hyperbolic tangent function:

$$f\left(\lambda_{pj,k+1}\right) = \tanh\left(\lambda_{pj,k+1}\right) \tag{4}$$

The training of an ANN is the determination of its parameters (weights and biases) using input-output data patterns. Typically, a function that gives the error of the network for the training patterns is minimized using multidimensional indirect optimization techniques. For MLP networks, an efficient approach is to start the optmization iterations using the backpropagation technique (which employs gradient descent search) and then proceed to a conjugate gradient search until a sufficiently small error function is obtained [17].

In order to guarantee the ability of the neural network to generalize when presented to new data, the available input-output patterns are randomly divided into two sets: one for training (usually 2/3 of the available set) and the other for validation (the remaining 1/3).

3.2. Methodology

First, a literature search was carried out for experimental biomass gasification data. Data from several references for gasification of different biomass were collected [4, 7, 9-10, 18-34].

The data search had three main information focuses: the gasification system (technical and operating aspects), type of biomass and the characteristics of the produced gas, as described in the following:

- Gasification system: type of gasifier as well as the dimensions of the reactor, the operating conditions and the gasification agent employed.
- Biomass: proximate and ultimate analysis data were collected. Additionally, when the heating value was not provided, its value was estimated based on the ultimate analysis data.
- Produced gas: the main information collected was the composition in terms of H_2, CO, CH_4 and CO_2.

Some characteristics of the gasification system (as the type of bed and the operation pressure) were restricted in the search for building the database that would be further used to train the neural networks. So, the ANNs were trained for fluidized bed gasifiers, using sand as bed and operated at atmospheric pressure. Only laboratory and pilot dimensions were used in this study. Initially, data for all the gasification agents were included in the training of the ANNs.

The complete database built had 181 input-output experimental patterns taken from references [4, 7, 9-10, 18-34] and can be obtained from the corresponding author under request. In the following, some observations are presented regarding the collected data:

- The contents of ashes, volatile components and fixed carbon were determined on a dry matter basis.
- The variable S/B indicates the ratio between the values of steam and biomass feed mass flows.
- The variables C, H, N, O and S indicate the mass percentage of carbon, hydrogen, nitrogen, oxygen and sulfur, respectively in the biomass fuel.
- The composition of the produced gas is given in volumetric percentage.
- The variable C_2H_n indicates the sum of the hydrocarbons with two atom of carbon that are formed in the process (mostly, ethylene and ethane).

At first, the choice of the input variables for the ANNs model was made heuristically, considering the analysis of the studied problem and the influence of the input variables in the prediction of the composition of the produced gas. Later, a sensitivity analysis was also implemented in order to help in that task.

In this work, the Statistica Neural Networks – SNN (Statsoft®) software was used in order to train and validate the neural networks. The ANNs that presented the best performance were of the kind MLP (Multilayer Perceptron). MLPs are feedforward, multilayered neural networks that typically present one input, at least one hidden layer and one output layer of computational nodes (the neurons). Details can be found in several references, including [5]. The MLPs employed in this work had one hidden layer of hyperbolic tangent neurons.

4. Results

4.1. Modeling results using ANNs

ANNs were trained using partial information about the gasification system and the biomasses in order to make predictions about the produced gas (composition in terms of H_2, CO, CH_4, CO_2, C_2H_n). The following information concerning the operating conditions of the gasifier was used as input variables to the neural network: equivalence ratio, steam/biomass ratio (S/B), temperature of the gasifier (T) and gasification agent used (the categories: air, steam, air/steam, steam/oxygen). Additionally the following information about the biomass was also used as input variables to the neural network: proximate (specifically, moisture, ash and volatile contents) and ultimate (specifically, C, H and O percentages) analysis data.

It must be emphasized that only partial information was used in order to avoid a large number of input variables to the neural network and, consequently, a large number of neurons and of parameters, that could lead to an overdimensioned neural network, with little predictive capability [5], considering the limited amount of literature data used for training.

A very detailed study was carried out, concerning the design and comparison of multilayered (MLP) neural network models models. ANNs with multiple and individual outputs were trained. The inclusion of an input categorical variable that classified the gasifier as 'bubbling' or 'circulating fluidized bed' was also investigated. Comparison with multi-regression linear models was performed and the MLP outperformed the linear models in all the cases studied here, due to the nonlinear nature of the data. This study is fully described in reference [35]. Here, due to space reasons, selected results are shown. The selection aimed to provide illustrative results of the application of ANNs in the modeling and optimization of gasification of different type of biomass in fluidized gasifiers.

In the following, preliminary results considering both bubbling and circulating fluidized bed gasifiers are presented. For the choice of this 'universal' neural network, three hundred ANNs were compared employing 2/3 of the patterns in the built database for training and 1/3 for validation. A total of 131 patterns were available for the prediction of the output variables (H2, CO, CH4, CO2 and C2Hn). The data for the continuous input variables were between the following limits: $7.5 < Moisture < 9.4$; $71.02 < Volatiles < 82$; $0.32 < Ash < 26.4$; $36.57 < C < 48$; $4.91 < H < 6.04$; $39 < O < 45.43$; $0 < Equivalence\ ratio < 0.9$; $0.113 < S/B < 4.7$ and $650 < T < 900$. The gasification agent was considered as a categorical input variable for this 'universal' neural network.

A total of 300 MLPs, with different topologies, was trained using the Statistica Neural Networks. The ANN selected was the one that presented the smallest error for validation. It presented a topology consisting of 13 neurons in the input layer (4 for the categorical variable and 9 for the continuous input variable), 13 in the hidden layer and 5 in the output layer. This configuration can be seen in Figure 2.

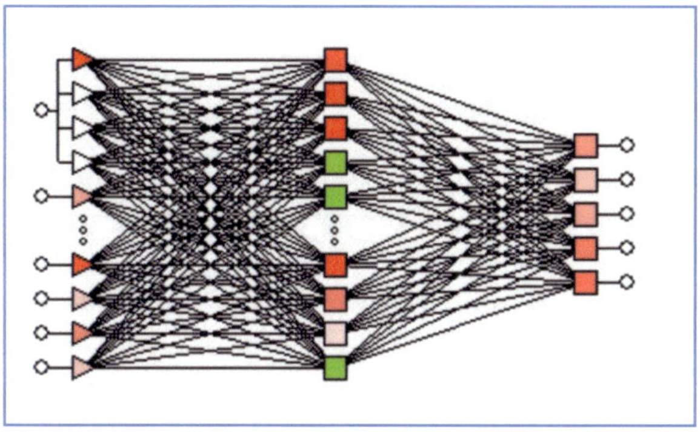

Figure 2. Topology of the multiple output neural network 13:13:5 MLP

Table 1 presents an analysis of the performance of the neural network. One statistical parameter used in the analysis was the SD-ratio parameter, which calculates the ratio between the standard-deviation of the ANN model and the standard-deviation of the training and validation (or selection) data. If the SD-Ratio is 1.0, then the network does no better than a simple average. A low (lower than 0.25) SD-ratio for the validation (or selection data) is indicative of a very good generalization capability of the ANN. This criterium (named here Select Performance) was used to select the best neural network during the training phase. It can be seen that the values of the SD-Ratio are between 0.13 and 0.18, with the exception of the result for C_2H_n, which was a bit higher (0.34).

Table 1 also show that high standard Pearson-R correlation coefficient between the actual and predicted outputs for the five output variables. In order to have accurate predictions, this parameter should be as close to one as possible. The high correlation between the predicted concentrations and the observed ones can also be observed in Figures 3 to 7. The lowest value for the correlation coefficient was observed for the prediction of C_2H_n, which shows the highest dispersion. This was similar to what had been obtained for the SD-Ratio and can be explained by the fact that the concentration of these components is very small in the produced gas; for that reason, many authors do not take their presence into account.

	H_2	CO	CH_4	CO_2	C_2H_n
Data mean	38.62	27.71	4.84	27.20	1.46
Data S.D.	17.99	14.32	4.58	12.47	1.87
Error mean	-0.23	-0.16	-0.04	0.11	0.03
Error S.D.	2.29	2.13	0.82	2.15	0.63
Absolute Error mean	1.68	1.47	0.50	1.56	0.28
SD-Ratio	0.13	0.15	0.18	0.17	0.34
Correlation	0.99	0.99	0.98	0.99	0.92

Table 1. Analysis of the Performance of the Multiple Output 13:13:5 MLP

A correlation of 1 indicates only that a prediction is perfectly linearly correlated with the observed outputs. So, here, in order to judge the quality of the predictions, a high Pearson-R correlation coefficient will be required together with small SD-Ratio parameter.

A sensitivity analysis was also carried out in order to evaluate the importance of each input variable to the predictive performance of the neural network. In this analysis, if one specific variable is considered 'unavailable' (that is, only its means value is used) the performance of the network should deteriorate and its Error should increase. Based on this fact, the sensitivity analysis, calculates the ratio between the Error and the Baseline Error (i.e. the error of the network if all variables are 'available'). If the Ratio is one or lower, then making the variable 'unavailable' either has no effect on the performance of the network or enhances it. This way, the higher the Ratio, the most important is that particular input variable to the performance of the ANN.

Table 2 presents the results of the sensitivity test, where the Rank lists the variables in order of importance. The results in Table 2 indicate that all listed 10 input variables are important; thus, it can be concluded these variables are needed in order to perform accurate predictions, being the gasification agent the most important one.

Even though these first results were quite satisfactory, improved results were sought. In order to obtain more parsimonious (in terms of number of neurons) models, without harming the statistical parameters (SD-Ratios and correlation parameters), individual – one for the prediction of each component gas – MLPs, with only one output variable, were trained.

Again, for the sake of conciseness, the results will be summarized, being their complete description found in reference [35].

A total of 300 MLPs with different topologies was trained for each individual output MLP. The best one for each case was considered as the one that presented the smallest SD-Ratio for the validation patterns. Table 3 presents the results of the individual output MLP against the multiple output MLP.

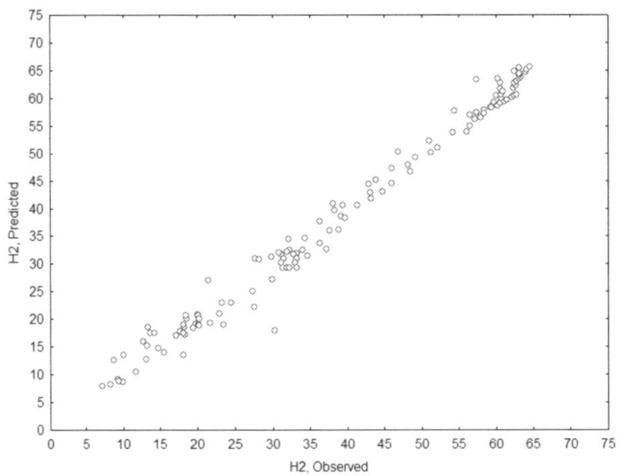

Figure 3. Prediction of H2 concentration in the output gases for the Multiple Output 13:13:5 MLP

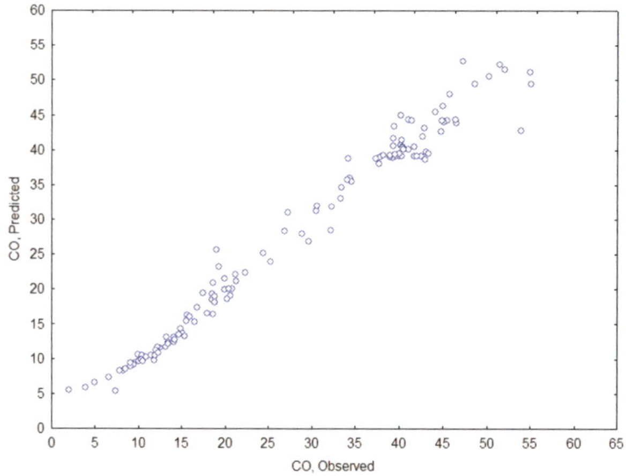

Figure 4. Prediction of CO concentration in the output gases for the Multiple Output 13:13:5 MLP

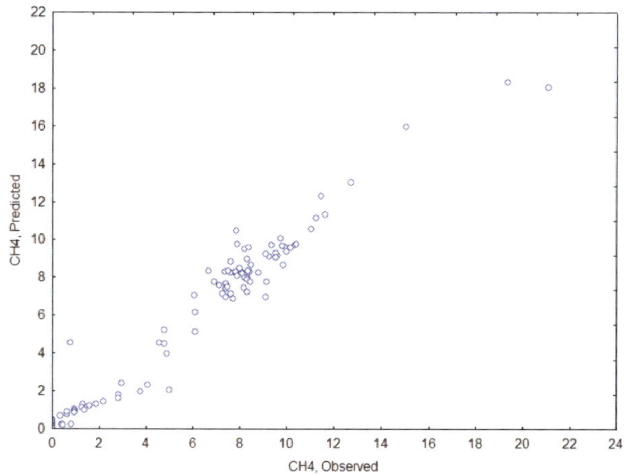

Figure 5. Prediction of CH4 concentration in the output gases for the Multiple Output 13:13:5 MLP

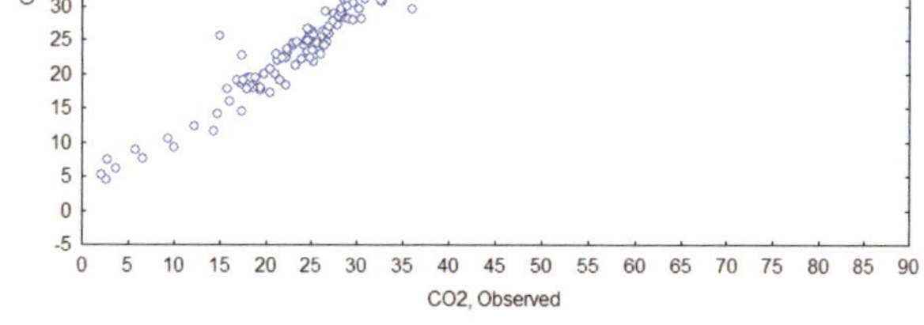

Figure 6. Prediction of CO2 concentration in the output gases for the Multiple Output 13:13:5 MLP

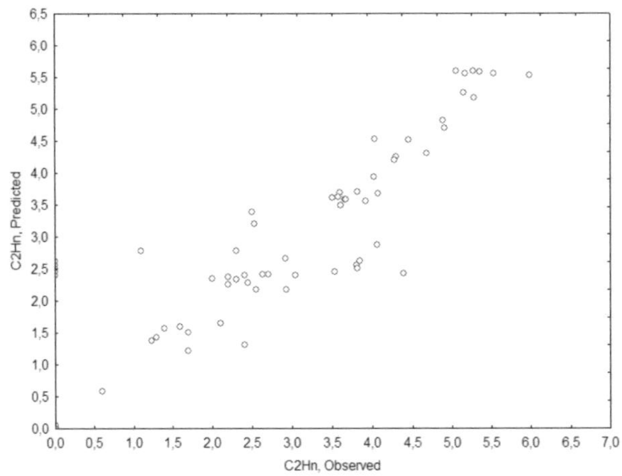

Figure 7. Prediction of C2Hn concentration in the output gases for the Multiple Output 13:13:5 MLP

	Gasification agent	Moisture	Ash	Volatiles	C	H	O	RE	S/B	T
Ratio	3.74	1.12	1.88	2.24	2.76	1.97	3.27	2.33	2.49	1.80
Rank	1	10	8	6	3	7	2	5	4	9

Table 2. Multiple output 13:13:5 MLP: sensitivity analysis

The analysis of the results presented in Table 3 show that it was possible to keep the predictive performance of the ANNs using individual output instead of multiple output models. The SD-Ratio and correlation parameters were of the same magnitude but the individual models had less neurons in the layers. It should be clarified here that for part of the models (as the ones for CO and CO_2) less variables were used in the input layer as sensitivity analysis showed that some variables were not necessary for an accurate prediction and were discarded as inputs [35].

	Multiple MLP			Individual MLP		
	SD-Ratio	Correlation	Topology	SD-Ratio	Correlation	Topology
H_2	0.13	0.99	13:13:5 MLP	0.13	0.99	11:9:1 MLP
CO	0.15	0.99	13:13:5 MLP	0.14	0.99	10:6:1 MLP
CH_4	0.18	0.98	13:13:5 MLP	0.13	0.99	13:9:1 MLP
CO_2	0.17	0.99	13:13:5 MLP	0.20	0.98	10:11:1 MLP

Table 3. Performance of the Individual Output MLP

Additionally, sensitivity tests – as the ones shown in Table 2 – revealed that the gasification agent was again the most important variable for the individual output MLPs as was the case for the multiple output ones. This motivated the development of 'specialized' MLPs for the prediction of the percentage composition of the four most important components (H_2, CO, CH_4, CO_2) in the output gas of bubbling, fluidized gasifiers.

High correlations values (ranging from 0.94 to 0.99) were obtained for these 'specialized' ANNs. In the following, results are shown for the neural network that predicts the hydrogen percentage in the produced gas of a bubbling fluidized gasifier, using steam as the gasification agent.

The obtained MLP presents 7 linear neurons in the input layer (for the input variables: 1. moisture (%wt); 2. volatile content (%wt); 3. C (%wt); 4. H (%wt); 5. O (%wt); 6. S/B; 7. T (°C)); 10 hyperbolic neurons in the hidden layer and 1 linear neuron in the outpout layer. It was trained using the backpropagation method during the 100 initial epochs and the conjugate gradient during the 127 last ones [5, 35]. This ANN will be further cited here as 7:10:1 MLP.

Figure 8 illustrates the topology of the 7:10:1 MLP (a) and its results (b). A very high correlation (0.99) between predicted and observed value was obtained. Table 4 presents the results of the sensitivity test for the input variables. It can be seen that the mass percentage of hydrogen in the biomass fuel is the most important input variable for the prediction of hydrogen in the produced gas, as expected.

4.2. Preliminary results of operational optimization

A preliminary investigation was also conducted of the optimization of the operation of a particular gasifier using the gasification model provided by the neural network. The neural model described in the previous section for a bubbling gasifier using steam as the gasificant agent was employed. For this study, the biomass was fixed and the operating conditions (in terms of T and S/B) were varied, according to the data present in the built database [35] in order to maximize the yield of a given component in the produced gas.

The results for wood and straw biomasses and maximization of the production of hydrogen are described in the following to illustrate the procedure. Initially, the response surfaces using the neural model and the data for each biomass were plotted as shown in Figure 9 (a) and (b) for wood and straw, respectively. Analyzing these surfaces, the most adequate directions for changes in the operational variables can be chosen if the objective is to increase the production of hydrogen.

In Figure 9, the operating variables were varied considering the availability of data in that operating range. So, temperature was varied between 800 and 850 °C, for wood, and 650 and 900 °C, for straw. For the ratio S/B, the considered ranges were $1.1 < S/B < 4.7$, for wood, and $0.4 < S/B < 0.9$, for straw. It can be seen in Figure 9 that, if the operating values are restricted to those ranges, the maximization of H_2 in the produced gas demands higher S/B ratios and opposite directions for T (lower T for wood and higher T for straw). So, the model provided by the ANN provides information that could give the operator the right trends to maximize the production of a given product of interest.

Advancing a further step in the optimization, just for the sake of a preliminary investigation, the ability of the neural network to generalize (interpolating the training data) was also evaluated.

For the bubbling gasifier, with steam as the gasificant agent, the database training data included the operating variables in the range 650 < T < 900 °C and 0.113 < S/B < 4.7 and three different biomasses (wood, straw and pine sawdust). A stochastic optimization method, the

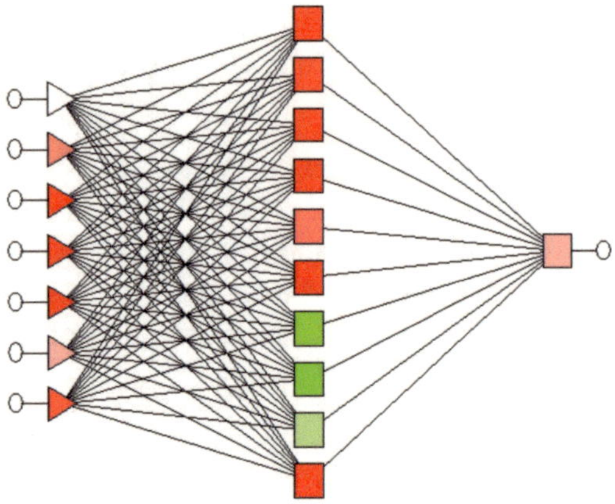

(a)

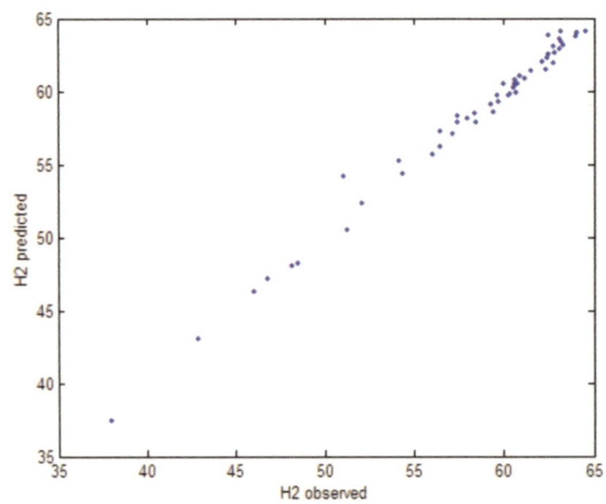

(b)

Figure 8. MLP 7:10:1 for prediction of hydrogen in the produced gas (bubbling gasifier, gasificant agent: steam): (a) illustration; (b) predicted vs. experimental values

	Moisture	**Volatiles**	**C**	**H**	**O**	**S/B**	**T**
Ratio	8.00	4.45	4.85	8.877	8.160	8.27	6.49
Rank	4	7	6	1	3	2	5

Table 4. MLP 7:10:1 for prediction of hydrogen in the produced gas (bubbling gasifier, gasificant agent: steam): sensitivity analysis

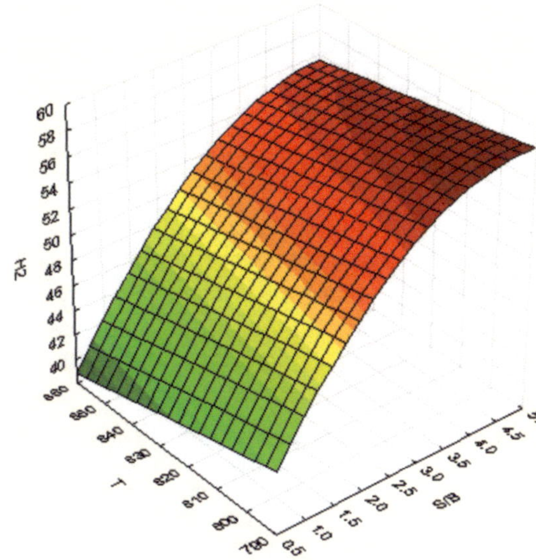

(a)

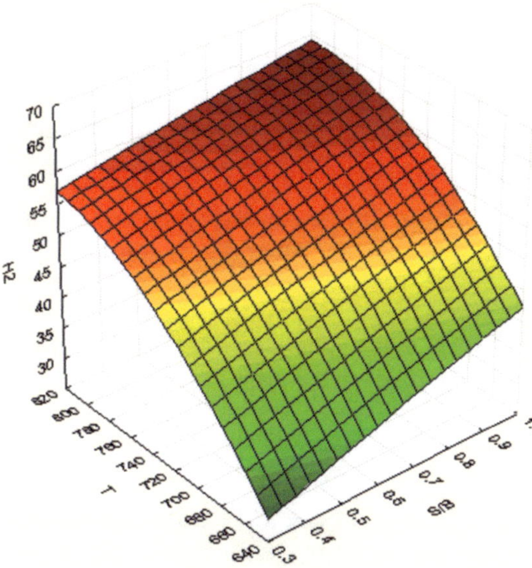

(b)

Figure 9. MLP 7:10:1 for prediction of hydrogen in the produced gas (bubbling gasifier, gasificant agent: steam): response surface for wood (a) and straw (b).

Particle Swarm Optimization (PSO) [36], was applied, in order to find the optimum (maximum yield of H_2 in the produced gas), for a given biomass, considering the whole operating range in the training database. When that approach was applied, as an example, for straw biomass, the PSO algorithm found an optimum for hydrogen production of 81.07 for S/B = 4.7 and T = 700.95 °C. This result should be analysed very cautiously; the percentage of hydrogen seems very high – as literature report percentage of 72 % [37] – but it indicates for the operator a region that should be further examined experimentally in order to reach higher percentages of the component of interest in the produced gas. Additional results and details may be found in [35].

5. Conclusions

ANNs are able to capture the latent characteristics present in the experimental data used for training, including nonlinearities [5]. Hence, multilayer perceptron neuron networks were proposed here as an alternative tool for the empirical modeling of biomass gasification. Specifically, ANN models were developed to correlate operating conditions of the gasifier and biomass data with characteristics of the produced gas, using experimental data given in the literature. It was verified that the developed model showed a good performance with a parsimonious number of units, when it was specifically built for a particular gasifier and a particular gasification agent [35]. Very high correlation rates between predictive and observed data were obtained.

The resulting trained ANN model is an algebraic mapping between input-output data, demanding little computational time. That fact makes the use of neural network very attractive in real time control and/or optimization of the process. The preliminary optimization investigation carried out here showed that the ANNs may supply the operator with information of tendencies that should be further experimentally checked in order to reach the target of maximizing the amount of a given component in the produced gas.

Calibration of the ANNs is easily performed, that is, whenever additional data are available, they may be added to the database and the ANN may be retrained, improving its predictive ability.

Presently, works based on hybrid neural-phenomenological [38] are being developed by the group as done before with a biotechnological process [39].

Author details

Maurício Bezerra de Souza Jr., Leonardo Couceiro Nemer and Amaro Gomes Barreto Jr.
Rio de Janeiro Federal University, Chemical Engineering Department, Centro de Tecnologia, Ilha do Fundão, Rio de Janeiro, Brazil

Cristina Pontes B. Quitete
Petrobras, R&D Center, R&D in Gas, Energy and Sustainable Development, Av. Horário de Macedo, Cidade Universitária, Rio de Janeiro, Brazil

Acknowledgement

This work was sponsored by Brazilian State Oil Company, PETROBRAS. Professor Maurício Bezerra de Souza Júnior acknowledges CNPq for a research fellowship.

6. References

[1] Guo B., Li D., Cheng C., Lü Z.-a., & Shen Y. (2001). Simulation of biomass gasification with a hybrid neural network model . *Bioresource Technology*, 76, 77-83.

[2] Nougués J.M., Pan Y. G., Velo E. *et al.* (2000). Identification of a pilot scale fluidised-bed coal gasification unit by using neural networks. *Applied Thermal Engineering, 20,* 1561-1575.

[3] Guo B., Shen Y., Li D., & Zhao F. (1997). Modelling coal gasification with a hybrid neural network. *Fuel, Vol. 76, No. 12* , 1159-1164.

[4] Li X., Grace J., Lim C. *et al.* (2004). Biomass gasification in a circulating fluidized bed. *Biomass & Bioenergy*, 26, 171-193.

[5] Haykin S. (1999). *Neural Networks, A Compreensive Foundation*, 2nd ed., New Jersey, Prentice-Hall.

[6] Himmelblau D. H. (2000) Applications of Artificial Neural Networks in Chemical Engineering, *Korean Journal of Chemical Engineering*, 17, 4, 373-392.

[7] Xiao G., Ni M.-j., Chi Y. *et al.* (2009). Gasification characteristics of MSW and an ANN prediction model. *Waste Management*, 29, 1, 240-244.

[8] Basu P. (2006). *Combustion and gasification in fluidized beds.* Taylor & Francis Group, USA.

[9] Chen G., Andries J., Spliethoff H. *et al.* (2004). Biomass gasification integrated with pyrolysis in a circulating fluidised bed. *Solar Energy*, 76, 1-3, 345-349.

[10] Baratieri M., Baggio P., Fiori L., & Grigiante M. (2008). Biomass as an energy source: Thermodynamic constraints on the performance of the conversion process. *Bioresource technology* , 99, 7063-7073.

[11] Nemtsov D., Zabaniotou A. (2008). Mathematical modelling and simulation approaches of agricultural residues air gasification in a bubbling fluidized bed reactor. *Chemical Engineering Journal*. 143, 1-3, 10-31.

[12] Corella J., & Sanz A. (2005). Modeling circulating fluidized bed biomass gasifiers. A pseudo-rigorous model for stationary state. *Fuel Processing Technology 86* , 1021-1053.

[13] Dong C., Jin B., Zhong Z., & Lan J. (2002). Tests on co-firing of municipal solid waste and coal in a circulating fluidized bed. *Energy Conversion and Management*, 43, 2189-2199.

[14] Rumelhart D. E., McClelland J. L. (1986) Parallel distributed processing: explorations in the microstructure of cognition., 1 ed., MIT Press, Cambrigde.

[15] Hecht-Nielsen R. (1989). Theory of the backpropagation neural network. IEEE International Conference on Neural Networks. 1, 593-605.

[16] Baughman D. R., Liu Y. A. (1995) Neural networks in bioprocessing and chemical engineering. Virginia Polytechnic Institute and State University, Blacksburg, VA.

[17] Leonard J., Kramer M. A. (1990). Improvement of the Backpropagation Algorithm for Training Neural Networks. Computers and Chemical Engineering. 14, 337-341.

[18] Doherty W., Reynolds A., Kennedy D. (2009). The effect of air preheating in a biomass CFB gasifier using ASPEN Plus simulation, *Bioresource Technology*, 33, 9, 1158-1167.

[19] He M., Xiao B., Liu S. *et al.* (2009). Hydrogen-rich gas from catalytic steam gasification of municipal solid waste (MSW): Influence of steam to MSW ratios and weight hourly space velocity on gas production and composition. *International Journal of Hydrogen Energy*, 2174-2183.

[20] Nikooa M. B., Mahinpeya N. (2008). Simulation of biomass gasification in fluidized bed reactor using ASPEN PLUS. *Biomass and Bioenergy*, 32, 12, 1245-1254.

[21] Shen L., Gao Y., Xiao J. (2008), Simulation of hydrogen production from biomass gasification in interconnected fluidized beds. *Biomass and Bioenergy*, 32, 120-127.

[22] Souza-Santos M. L. (2008a). Comprehensive Simulator (CSFMB) Applied to Circulating Fluidized Bed Boilers and Gasifiers. *The Open Chemical Engineering Journal*, 2, 106-118.

[23] Souza-Santos M. L. (2008b). CSFB applied to fluidized-bed gasification of special fuels. *Fuel*, 88, 5, 826-833.

[24] Siedlecki M., Simeone E., De Jong W. *et al.* (2007). Characterization of gaseous and condensable components in the product gas obtained during steam-oxygen gasification of biomass in a 100 KWth CFB gasifier. *15th European Biomass Conference and Exhibition - From Research to Market*, 1120-1127.

[25] Cao Y., Wang Y., Riley J. T. *et al.* (2006). A novel biomass air gasification process for producing tar-free higher heating value fuel gas. *Fuel Processing Technology*, 87, 343-353.

[26] García-Ibañez P., Cabanillas A., Sánchez J. M. S. (2004). Gasification of leached orujillo (olive oil waste) in a pilot plant circulating fluidised bed reactor. Preliminary results. *Biomass and Bioenergy*, 27, 183-194.

[27] Dellepiane D., Bosio B., Arato E. (2003). Clean energy from sugarcane waste: feasibility study of an innovative application of bagasse and barbojo. *Journal of Power Sources*, 122, 1, 47-56.

[28] Lv P., Chang J., Xiong Z et al. (2004), An experimental study on biomass air–steam gasification in a fluidized bed, Bioresource Technology, 95, 95–101.

[29] Schuster G., Loffler G., Weigl K. *et al.* (2001), Biomass steam gasification – an extensive parametric modeling study. *Bioesource Technology*, 77, 71-79.

[30] van der Drift A., van Doorn J. and Vermeulen J. W. (2001), Ten residual biomass fuels for circulating fluidized-bed gasification. *Biomass and Bioenergy*, 20, 45-56.

[31] Turn S., Kinoshita C., Zhang Z. *et al.* (1998). An experimental investigation of hydrogen production from biomass gasification. *Int. J. Hydrogen Energy*, 23, 8 , 641-648.

[32] Faaij A., van Ree R., Waldheim L. *et al.* (1997), Gasification of Biomass Wastes and Residues for Electricity Production. *Biomass and Bioenergy*, 12, 6, 387-407.

[33] Boateng A. A., Walawender W. P., Fan L. T., Chee L. T. (1992). Fluidized-bed steam gasification of rice hull. *Bioresource Technology*, 40, 235-239.

[34] Ergudenler A., & Ghaly A. E. (1992). Quality of gas produced from wheat straw in a dual-distributor type fluidized bed gasifier. *Biomass and Bioenergy, Vol.3, No. 6* , pp. 419-430.

[35] Nemer L. C. (2010), Modelagem do Processo de Gaseificação de Biomassa Empregando Redes Neuronais, Monograph, Escola de Química, UFRJ (*in Portuguese*).

[36] Kennedy J. and Eberhart R. (1995). Particle Swarm Optimization. *Proceedings of the IEEE International Conference on Neural Networks*, Perth, Australia, 1942-1948.

[37] Corella J., Aznar M. P., Caballero M. A. et al. (2008). 140 g H_2/kg biomass d.a.f. by a CO-shift reactor downstream from a FB biomass gasifier and a catalytic steam reformer. *International Journal of Hydrogen Energy*, 3, 7, 1820-1826.

[38] Psichogios D.C. and Ungar L.H. (1992). A hybrid neural network - first principles approach to process modeling. AIChE Journal, 38, 10, 1499-1512.

[39] Boareto A. J. M., De Souza Jr. M. B., Valero, F., Valdman, B. (2007) A Hybrid Neural Model (HNM) for the On-Line Monitoring of Lipase Production by Candida rugosa. *Journal of chemical Technology and Biotechnology*.

Chapter 13

Lower Order Modeling and Control of Alstom Fluidized Bed Gasifier

L. Sivakumar and X. Anithamary

Additional information is available at the end of the chapter

http://dx.doi.org/10.5772/48674

1. Introduction

Integrated coal gasification combined cycle (IGCC) is a technology wherein coal is converted to fuel gas also referred as syngas or synthesis gas. Powdered coal is made to be in contact with a mixture of oxygen(or air) and steam to produce fuel gas. This fuel gas is burnt in a gas turbine coupled with generator to produce power. The waste heat from the gas turbine is used to produce steam and the steam is sent to a steam turbine for additional power generation (Ramezan and Stiegel, 2006).

Though, IGCC has a number of technical advantages, but until recently, its application has been limited due to its higher capital costs plus the availability of cheap natural gas. However, with pollution limits becoming more stringent and natural gas prices increasing, the performance of IGCC will become more attractive and its technical advancement will further reduce its cost.

Gasification is a technology that had its beginnings in the late 1700s. In the 19th century, gasification was widely used for the production of "town gas" especially for urban areas (Ramezan and Stiegel, 2006). But due to the widespread availability of natural gas, it got vanished in the 20th century. Today, the IGCC technology is being widely used throughout the world. 250MW IGCC demonstration plants are being constructed at Tianjin in china. In India, Andhra Pradesh Power Generation Corporation Ltd in association with Bharat heavy Electricals Limited proposed 125 MW IGCC plant at Vijayawada. In USA, 262 MW Wabash River IGCC power plants in Indiana (later acquired by Conoco Philips) and 250MW Tampa Electric Co. Polk Power Station IGCC in Florida (later acquired by GE Energy) are the two main commercial IGCC coal based power plants. Even though a number of IGCC projects exist, the UK's Clean Coal Power Generation Group, ALSTOM has undertaken a detailed study on the development of a small-scale prototype integrated plant (PIP), based on the air blown gasification cycle with 150 MW output (Pike *et al.*, 1998). This type of prototype plant is useful in understanding the physics of the process, designing control systems for integrated operation.

2. Mathematical modelling

In general, mathematical modeling has been a useful tool for performance analysis, control system design, optimization and diagnosis of plants [Sivakumar and Ganapathiraman 2006]. The approach towards mathematical modeling depends upon the purpose for which the modeling is done. A detailed nonlinear mathematical model for a power boiler had been developed [Sivakumar and Bhattacharya 1979] using first principles approach – conservation of mass, energy and momentum to study the boiler transients for different types of disturbances. A furnace model with detailed calculations on the heat flux falling on different zones of furnace had been developed to study on the water wall tube failures [Sivakumar et.al 1980]. Low order transfer function models for power plant had been developed to study the performance of the proposed controllers and to design training simulators [Sivakumar et.al 1983]. This chapter deals with the development of low order mathematical models for ALSTOM gasifier which will be available to research community to study the efficiency of different control algorithms for specified disturbances. Further the suitability of conventional PID controllers for ALSTOM gasifier is investigated by the authors.

3. Air blown gasification cycle

ABGC is a hybrid combined cycle power generation technology. It was first conceived by British Coal Corporation (BCC) and developed in 1990s by Clean Coal Power Generation Group (CCPGG). Later the ABGC technology is purchased by Mitsui Babcock Energy Limited (Mitsui Babcock). Advanced design for this gasification is later done by the combined industrial collaborators - GEC Alsthom, Scottish Power plc and Mitsui Babcock with support from the European Commission's (EC's) THERMIE Programme and Department of Trade and Industry (DTI) (Pike *et al.*, 1998). Figure 1 shows the block diagram of ABGC.

Coal, steam and air react within the gasifier operating at 22bar pressure and 1150k temperature conditions in order to produce fuel gas with low calorific value. Limestone is also added in order to remove sulphur. This fuel gas is burnt in a gas turbine coupled with generator to produce electricity.

Approximately 20% of carbon in the coal does not react in gasifier which is extracted through ash removal system. This unburned carbon is fed to circulating fluidized bed combustor (CFBC) operating under atmospheric pressure and 1150k temperature conditions. Here the remaining unburned carbon is combusted completely. The water/steam (two phase mixture) absorbs heat from CFBC water walls. The steam separated by drum internals goes through different stages of super heaters receiving heat from exhaust gas coming from gas turbine (Pike *et al.*, 1998). The resulting high pressure steam is given to steam turbine coupled with generator to produce additional power generation. The total capacity of commercial ABGC is 525 MW approximately.

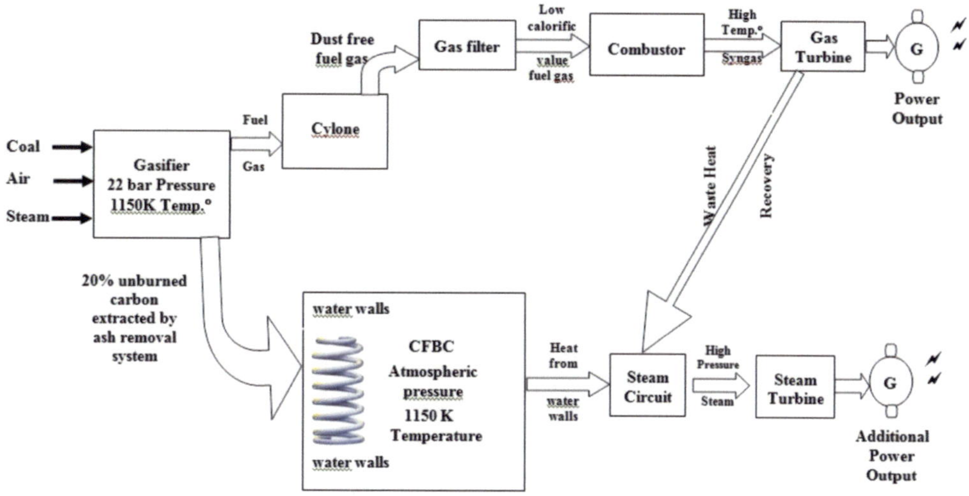

Figure 1. The Air Blown Gasification Cycle

4. Types of gasifier

There are three types of gasifier namely fixed bed, fluidized bed and entrained flow (Phillips, 2006).

4.1. Fixed bed gasifier

Here coal enters at the top of the reactor and air or oxygen enters at the bottom. As the coal moves slowly down the reactor, it is gasified and the remaining ash drops are collected at the bottom of the reactor. Example: British Gas Lurgi(BGL), Lurgi (Dry Ash) The figure 2 shows moving bed gasifier.

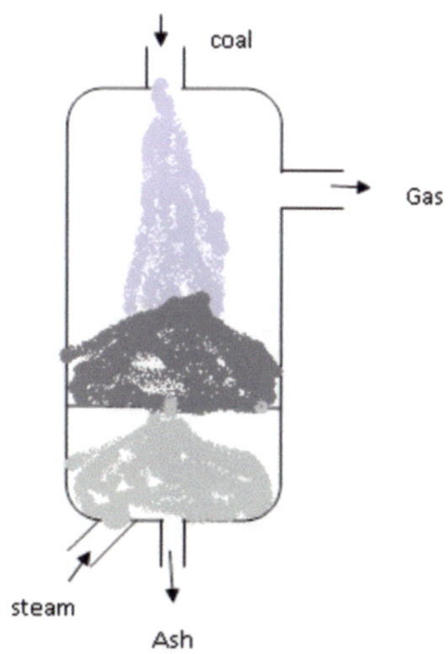

Figure 2. Moving bed gasifier

4.2. Entrained flow

Finely-ground coal is injected in co-current flow with the oxidant. The coal rapidly heats up and reacts with the oxidant. Gas is collected at the bottom. Most entrained flow gasifiers use oxygen rather than air. Example: GE entrained flow gasifier(Polk Station), E-Gas, Mitsubish Figure 3 shows entrained flow gasifier.

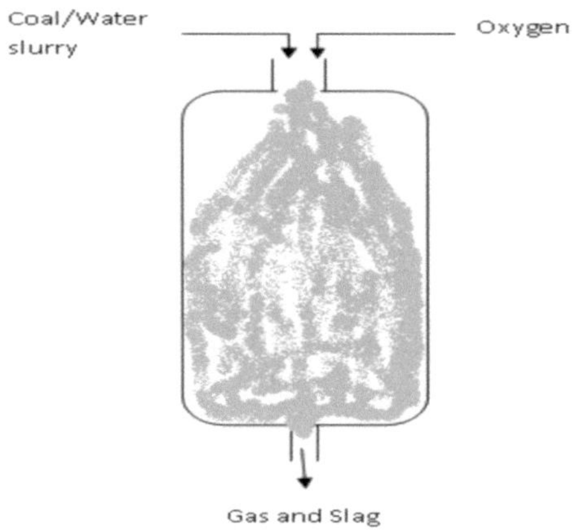

Figure 3. Entrained Flow Gasifier

4.3. Fluidized bed gasifier

A fluidized bed gasifier is a well-stirred reactor in which new coal particles is mixed with older, partially gasified and fully gasified particles. The mixing gives uniform temperatures throughout the bed. The flow of gas into the reactor (oxidant, steam, recycled syngas) must be sufficient to float the coal particles within the bed. However, as the particles are gasified, they will become smaller and lighter and will be entrained out of the reactor. Example: HT Winkler, KRW (Kellogg –Rust-Westinghouse) and ALSTOM gasifier.

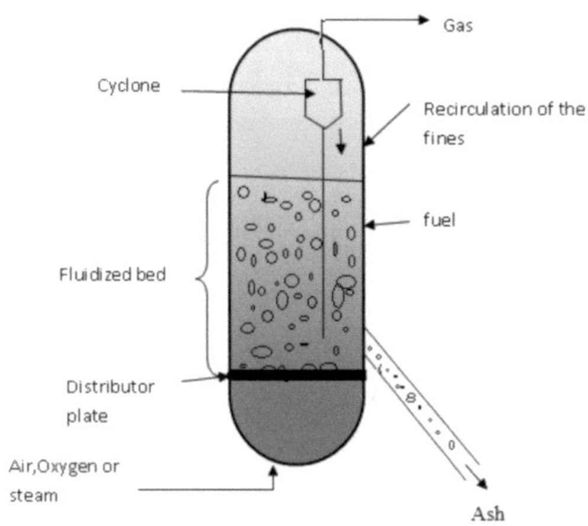

Figure 4. Fluidized bed gasifier

5. ALSTOM gasifier model

Gasifier model is the most complex one in coal gasification. It was first started by CRE Group Ltd in 1992. Later it was continued at GEC ALSTHOM mechanical Engineering Centre. The incoming coal is dried and de-volatilized to yield char, ash and volatile gases. The oxygen in fluidized air reacts with carbon in the char to form carbon monoxide and carbon dioxide. Both exothermic and endothermic reactions occur simultaneously in the gasifier. The main equations in gasifier are

$$C+O_2 \rightarrow CO_2 \quad (1)$$

$$C+{}^1/_2\,O2 \rightarrow CO \quad (2)$$

Equation 1 and 2 are exothermic gasification.

The carbon-dioxide reacts more with carbon to form carbon-monoxide. Also steam reacts with carbon to form carbon-monoxide and hydrogen.

$$C+CO_2 \rightarrow 2CO \quad (3)$$

$$C+H_2O \rightarrow CO+H_2 \quad (4)$$

Equation 3 and 4 are endothermic reactions.

The un-reacted char is added to the bed which is maintained at a constant height by char extraction system.

5.1. Alstom gasifier: Input and output variables

Alstom gasifier represents a difficult process for control because of its multivariable and non-linearity in nature with significant cross coupling between the input and output variables (Dixon 2004).

The controllable input variables to the gasifier are

- Char off-take (u1) WCHR(kg/s)
- Air flow rate(u2) WAIR(kg/s)
- Coal flow rate(u3) WCOL (kg/s)
- Steam flow rate(u4) WSTM(kg/s)
- limestone flow rate (u5) WLS(kg/s)

The Controlled output variables are:

- Gas calorific value (y1) CVGAS(J/kg)
- Bed mass (y2) MASS(kg)
- Fuel gas pressure (y3) PGAS(N/m^2)
- Fuel gas temperature (y4) TGAS(K)

One of the inputs, limestone mass (WLS) is used to absorb sulphur in the coal and its flow rate is set to a fixed ratio of 1:10 against another input coal flow rate.(WCOL).This leaves effectively 4 degrees of freedom for the control design. Fig 5 shows gasifier with input and output variables.

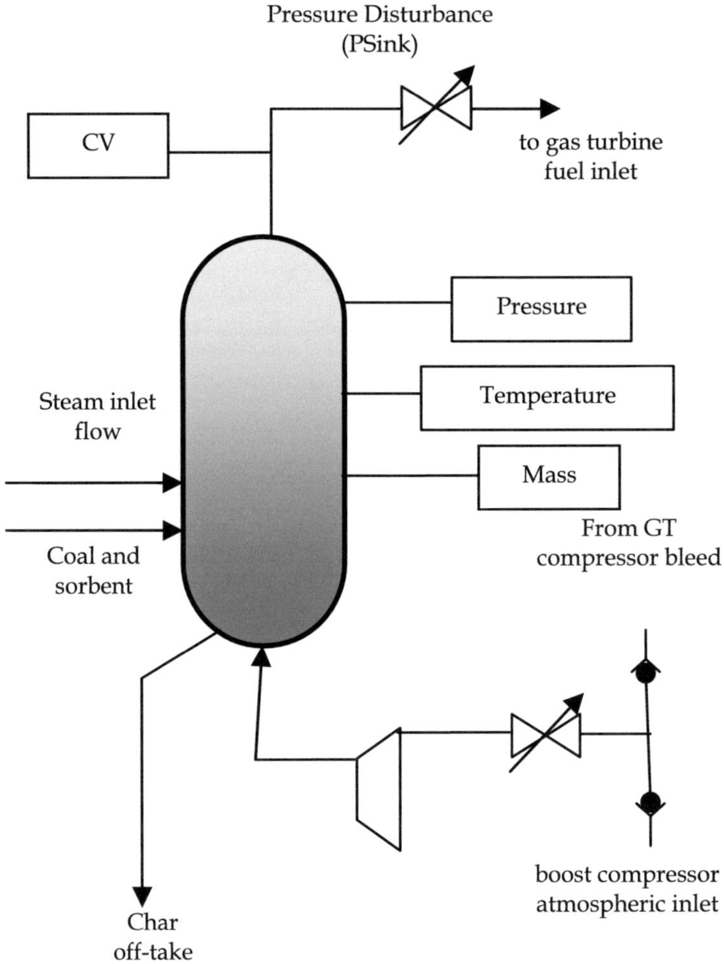

Figure 5. Gasifier with input and output variables

5.2. Load demand on gasifier

The flow rate of syngas to gas turbine is controlled through a valve at the inlet of turbine (also referred as controlled input disturbance to the gasifier). The pressure at the inlet of turbine called as PSink is the controlled variable. The control problem is to study the transient behavior of gasifier process variables such as pressure, temperature of the syngas for typical variations in gas flow drawing rate to gas turbine through appropriate changes in the throttle valve. Any proposed control system should control the pressure and temperature of the syngas at the inlet of gas turbine for any variation in gas turbine load – which in turn will affect throttle valve moment-without undue overshoots and undershoots. In fact this particular aspect has been posed as a control challenge problem for gasifier by ALSTOM.

6. ALSTOM benchmark challenges

The demand for clean air and stringent environmental regulations are forcing us to look for an alternate technology with reduced pollution emission and higher power generation. As a result of this, IGCC power plants are being developed all over the world. ALSTOM small-scale prototype (PIP) based on air-blown gasification cycle is one such IGCC. One of the component in ABGC called gasifier, is difficult to be controlled. For this reason, ALSTOM Power technology center issued a bench mark challenge to research community

- To come out /propose a suitable control strategy/algorithms so as to have an efficient control of pressure and temperature of syngas without having an undue overshoot and undershoot values equal or less than those specified in the constraints by ALSTOM for specified load disturbance through the throttle value for different operating loads such as 100%, 50%and no-load.

The ALSTOM gasifier is modeled in state space form given by

$$\dot{X} = Ax+Bu$$

$$Y=Cx+Du$$

Where

x = Internal states of gasifier, a column vector with dimension 25x1
u = Input variables, a column vector with dimension 6x1
A = system matrix governing the process dynamics, a square matrix with dimension 25x25
B = Input matrix with dimension 25x6
Y = Output variables, a column vector with dimension 4x1
C = Observable matrix with dimension 25x4
D = disturbance matrix with dimension 4x6

Towards this purpose, ALSTOM has made it available the following :

- A, B, C, D, x(0), Y for three different loads- 100%, 50% and no-load.

A virtual gasifier mathematical model is made available with the above quantities **(http://www.ieee.org/OnComms/PN/controlauto/benchmark.cfm.)** and researches can attempt different control philosophies to meet the challenge posed by ALSTOM.

The input and output variables, allowable limits on output variables during load transients for three different loads (100%, 50% and no-load) as given by ALSTOM are reproduced in Tables 1 and 2 for ready reference.

6.1. Input and output constraints

The plant inputs and outputs with their limits are given in Tables 1 and 2 respectively (Seyab *et al.*, 2006)

Inputs	Description	Maximum Value	Rate	Steady state values		
				100%	50%	0%
WCHR(kg/s)	Char extraction flow rate	3.5	0.2 kg/s^2	0.9	0.89	0.5
WAIR (kg/s)	Air flow rate	20	1.0 kg/s^2	17.42	10.89	4.34
WCOL(kg/s)	Coal flow rate	10	0.2kg/s^2	8.55	5.34	2.136
WSTM(kg/s)	Steam flow rate	6.0	1.0kg/s^2	2.70	1.69	0.676
WLS(kg/s)	Limestone flow rate	1.0	0.02kg/s^2	0.85	0.53	0.21

Table 1. Input Variables and Limits

Outputs	Description	Allowed fluctuations	Steady state values		
			100%	50%	0%
CVGAS(MJ/kg)	Fuel gas calorific value	± 0.01	4.36	4.49	4.71
MASS(kg)	Bedmass	± 500	10000	10000	10000
PGAS(N/m^2)	Fuel gas pressure	$\pm 1 \times 10^4$	2×10^6	1.55×10^6	1.12×10^6
TGAS(K)	Fuel gas temperature	± 1.0	1223.2	1181.1	1115.1

Table 2. Output variables and limits

6.2. Researchers attempt in the first phase (1997-2001)

The first round challenge was issued in the year 1997. It included three linear models operating under 0%, 50%and 100% load conditions respectively. The model includes state space equation with A,B,C and D values. The challenge requires a controller which controls the gasifier at three load conditions with input and output constraints in the presence of step and sinusoidal disturbances. Many controllers have been suggested for the first challenge (Dixon, 1999).

1. Dixon (1999) used multivariable P and I controllers using multi-objective optimal tuning technique and model based predictive control design to meet the constraints.
2. Rice et al. (2000) proposed predictive control that uses linear quadratic optimal inner loop and it is supervised by an outer predictive controller loop.
3. Proportional integral plus (PIP) by Taylor et al. (2000) from Lancaster University was based on discrete time model of the plant.
4. Prempain et al. (2000) demonstrated the use of loop shaping H-infinity control design method.
5. The multi-objective Genetic algorithm (MOGA) was proposed by Griffin et al. (2000) which performed a loop-shaping H-infinity design.
6. A sliding mode, nonlinear design approach was suggested by Sarah Spurgeon. Here switching surface is designed to move the plant from one operating point to the other.

7. Neil Munrom decomposed the original problem into a series of much simpler schemes in an effort to divide and conquer rule.
8. Munro (2000) combined sequential loop closing with a high –frequency decoupling approach along with divide and conquer method

But none of the controller met all the objectives specified in the challenge – more so with particular reference to the transient limits imposed on output variables during load variations.

6.3. Second challenge

The second round challenge was issued in the year 2002. In the second round challenge, ALSTOM specified nonlinear simulation model in MATLAB/SIMULINK [10] and desired the controller capability during load changes and coal quality disturbance. Recently, a group of control solutions for the benchmark problem were presented at Control-2004 Conference at Bath University, UK in September 2004. Most of controllers were reported as capable of controlling the system at disturbance tests.

The author, Dixon (2002) used multi-loop PI controller to the gasifier control. He used system identification technique to obtain the linear model from the non – linear plant data. The base line controller was used by the other researchers for comparison purposes. The following controllers were suggested to meet the performance criteria (Dixon, 2004).

1. Multi objective optimization approach suggested by Anthony Simms from Nottingham University needs further improvement by the addition of proportional control loops.
2. H-infinity design approach given by Sarah Gatley from Leicester University used loop shaping combined with anti-windup compensator. It produced a robust design because of its simple design process and without the need for detailed knowledge of the plant.
3. Multiple PID controller design using penalty based multi objective genetic algorithms by Adel Farag from Technical University of Hamburg gave excellent results that satisfied reasonable input output constraints.
4. A novel controller by Tony Wilson from Nottingham University used state estimators to improve on the base line performance. Kalman filters are used to estimate the pressure disturbance and coal quality change.
5. Proportional integral plus controller by James Taylor of Lancaster University used discrete time linear model of the gasifier.
6. Model Predictive controller using a linear state space model of the plant was a collaborative effort from Cranfield and Loughborough.

All the papers had achieved reasonable success in terms controlling the gasifier model. But none of the controller met the overall performance criteria and still this benchmark challenge is left for the academicians for further research.

The difficulty in meeting the performance criteria appears to necessarily work with the higher order model for control system design. This motivates the authors to derive low order transfer function models for control system study.

7. Low order transfer function models

On analyzing the ALSTOM gasifier model, the model is found to be more complex and it contains very high cross-coupling between input and output (Dixon 2004). It necessitates low order model for further control research. The state space equation is converted to transfer function models using MATLAB command sys = ss(a,b,c,d) and [num,den]=ss2tf(a,b,c,d,1). After conversion by Matlab command, the system is described in s- domain as follows:

$$\begin{bmatrix} y1(s) \\ y2(s) \\ y3(s) \\ y4(s) \end{bmatrix} = \begin{bmatrix} G11(s) & G12(s) & G13(s) & G14(s) \\ G21(s) & G22(s) & G23(s) & G24(s) \\ G31(s) & G32(s) & G33(s) & G34(s) \\ G41(s) & G42(s) & G43(s) & G44(s) \end{bmatrix} \begin{bmatrix} u1(s) \\ u2(s) \\ u3(s) \\ u4(s) \end{bmatrix} + \begin{bmatrix} Gd1(s) \\ Gd2(s) \\ Gd3(s) \\ Gd4(s) \end{bmatrix} \text{P}_{\text{SINK}}$$

where

yi(s) = output variables ; i={1,4 }
Gij(s) = transfer characteristic between jth output due to ith input ; i= {1,4} j={1,4}
ui(s) = input variable ; i={1,4}
Gdi(s) = describing the impact of variation in Psink on output variable; i= {1,4}
Psink = sink gas pressure at gas turbine inlet.

It is to be noted that the denominator polynomial of each element G_{ij} is of 24th order while the numerator is of order less than or equal to 23rd. A typical transfer characteristic between an output (pressure) due to all inputs shown diagrammatically as follows:

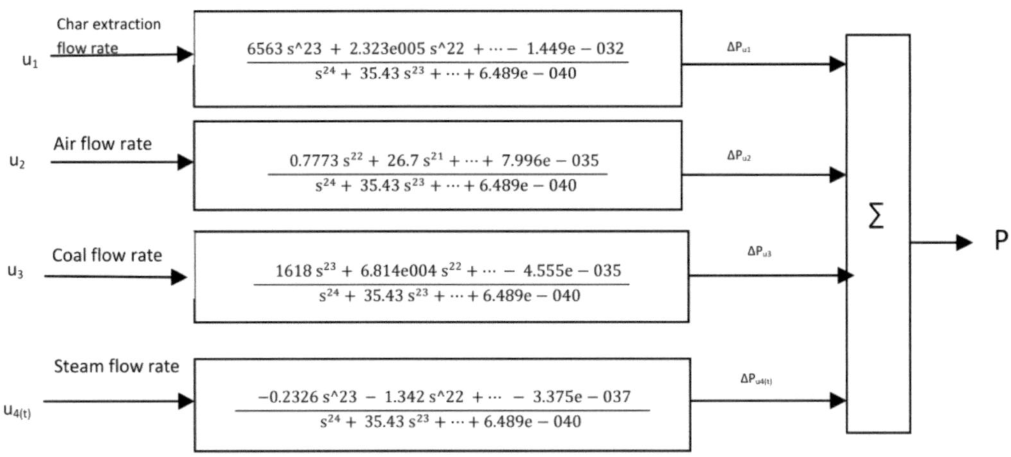

Figure 6. Transfer characteristic between pressure due to all inputs

Here ΔP_{ui} is the incremental change due to different inputs u_i. Thus

ΔP_{u1} is the incremental change in pressure due to steady state change in char extraction flow rate,

ΔP_{u2} is the incremental change in pressure due to steady state change in Air flow rate,

ΔP_{u3} is the incremental change in pressure due to steady state change in Coal flow rate and

ΔP_{u4} is the incremental change in pressure due to steady state change in steam extraction flow rate. The output is given below

$$P(t) = P_{steady\ state} + \Delta P_{u1} + \Delta P_{u2} + \Delta P_{u3} + \Delta P_{u4}$$

Now the problem boils down to the reduction of higher order transfer function models obtained by MATLAB command to lower order transfer function models by the application of different methods.

It is observed that author Haryanto et al. (2009) developed an equivalent lower order transfer function models towards the development of integrated plant simulator. In this chapter, the authors have developed lower order transfer function models using algebraic and reduced order approximation methods (Sivakumar and Anithamary, 2011).

7.1. Reduced order approximation (RSYS)

The matlab command RSYS = BALRED(SYS,ORDERS) computes a reduced order approximation(RSYS) of LTI system. The desired order (number of states) is specified by ORDERS. BALRED uses implicit balancing techniques to compute the reduced-order approximation RSYS. The second order transfer function is obtained using Henkel Singularity approximation method. The transfer function for typical block G11 corresponding to 100% load is given below:

$$G11 = \frac{-1.197e004\ s^2 + 330.4\ s + 0.001125}{s^2 + 0.0008608\ s + 2.075e{-}007}$$

All the transfer function blocks G_{ij} : (i = {1,4},j={1,4}) evaluated using reduced order approximation by the authors corresponding to 100%, 50% and no-load are given in Appendix A, Appendix B and Appendix C.

7.2. Algebraic method

The higher order transfer function is equated with the lower order model:

$$\frac{a_{n-1}s^{n-1} + a_{n-2}s^{n-2} + \cdots + a_0}{b_n s^m + b_{n-1}s^{m-1} + \cdots + b_0} = \frac{A_2 s^2 + A_1 s + A_0}{B_2 s^2 + B_1 s + BA_0}$$

On cross multiplying, the equation becomes

$$(a_{n-1}s^{n-1} + a_{n-2}s^{n-2} + \cdots + a_0)(B_2s^2 + B_1s + B_0) = (b_ns^m + b_{n-1}s^{m-1} + \cdots + b_0)(A_2s^2 + A_1s + A_0)$$

The ALSTOM transfer function for G_{11} is given below

$$G11=\frac{\begin{gathered}0.03215s^{22}+14.45s^{21}+1289s^{20}+1467s^{19}+721.8s^{18}+208.5s^{17}+\\ 40.02s^{16}+5.452s^{15}+0.547s^{14}+0.04129s^{13}+0.002367s^{12}\\ +0.0001031s^{11}+3.388e-006s^{10}+8.235e-008s^{9}+1.435e-009s^{8}\\ 1.695e-011s^{7}+1.246e-013s^{6}+4.869e-016s^{5}+8.221e-019s^{4}\\ +6.532e-022s^{3}+2.442e-025s^{2}+3.437e-029\,s+1.126e-034\end{gathered}}{\begin{gathered}s^{24}+35.38s^{23}+78.31s^{22}+68.51s^{21}+32.81s^{20}+9.998s^{19}+2.106s^{18}+\\ 0.3225s^{17}+0.03703s^{16}+0.00325s^{15}+0.0002203s^{14}+1.156e-005s^{13}\\ 4.687e-007s^{12}+1.45e-008s^{11}+3.36e-010s^{10}+5.64e-012s^{9}\\ 6.52e-014s^{8}+4.785e-016s^{7}+1.95e-18s^{6}+3.99e-021s^{5}+\\ 4.505e-024s^{4}+2.982e-024s^{3}+1.148e-030s^{2}+2.389e-034s+\\ 2.078e-038\end{gathered}}$$

The a_0 can be obtained by the formula (Poongodi *et al.*, 2009)

$$a_0 = \frac{b_{m-1}/b_m \pm a_{n-2}/a_{n-1}}{m \pm n}$$

$$a_0 = \frac{35.38/1 \pm 14.45/0.03215}{24 \pm 22}$$

a_0 = 10.5403, 242.4178, -9.0014, -207.0325

Taking the appropriate value of a_0, equating the powers of s, and solving the equation, the unknown values of B0,B1,B2,A1,A2 can be obtained. Thus,

$$G11= \frac{-43.210273s^2-32.8849432314s+10.5403}{-0.0083690166s^2+0.067824414s+0.0019433}$$

Similarly lower order models G12 to G44 corresponding to higher order models specified by ALSTOM can be obtained.

All the transfer function blocks Gij : (i = {1,4},j={1,4}) evaluated using algebraic method by the authors corresponding to 100%, 50% and no-load are given in Appendix A, Appendix B and Appendix C.

In order to evaluate the reduced order transfer function models obtained through different methods, the unit step response of ALSTOM model has been taken as reference response and the responses obtained through different methods as in figure 7 are compared and shown in figures 8-11 for typical transfer function blocks namely

G11 – the transfer characteristic between change in calorific value due to change in char extraction flow rate.

G24 – the transfer characteristic between change in temperature due to change in air flow rate.

G33 – the transfer characteristic between change in pressure due to change in coal flow rate.

G42 – the transfer characteristic between change in bedmass due to change in steam flow rate.

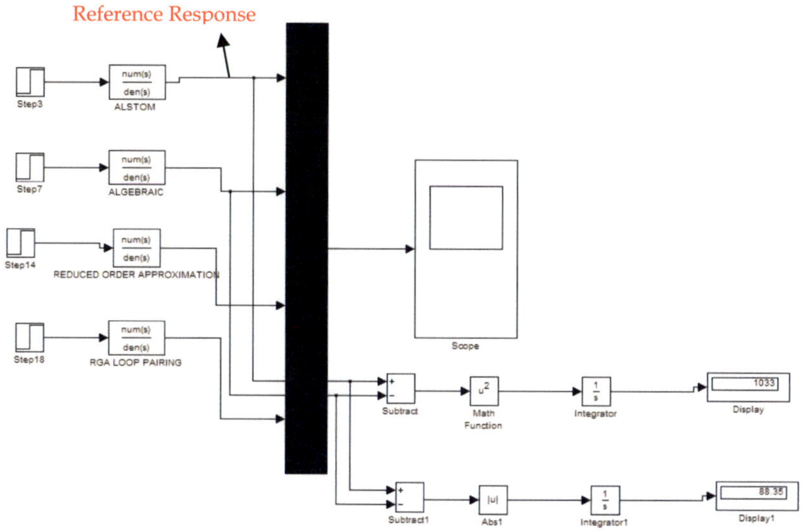

Figure 7. Matlab SIMULINK model to evaluate the IAE and ISE error

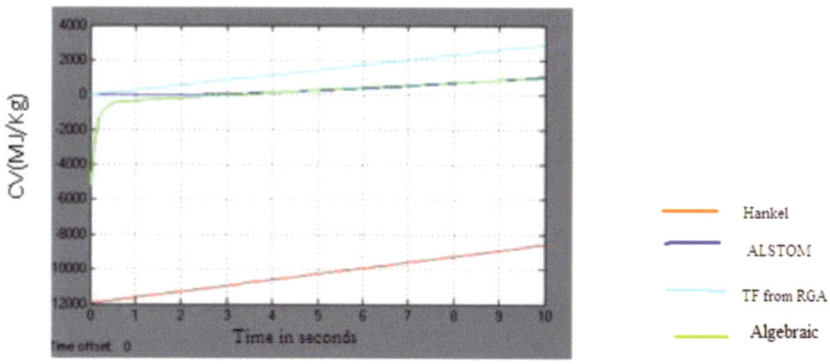

Figure 8. Variation of calorific value(y1) with char extraction flow rate (u1) keeping u2,u3,u4 constant

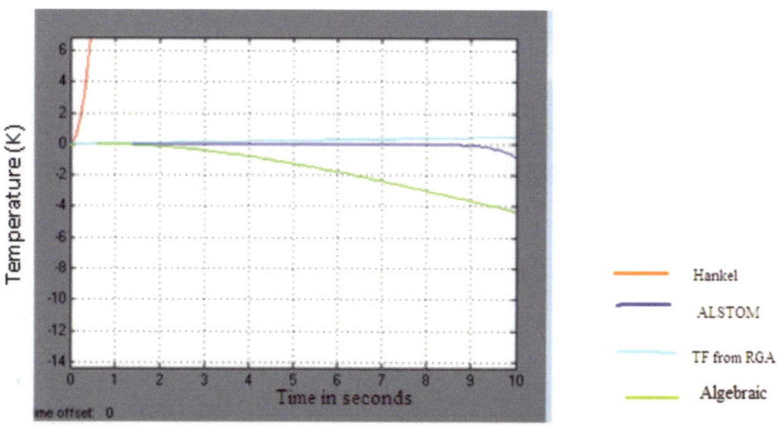

Figure 9. Variation of fuel gas temperature(y4) with air flow rate (u2) keeping u1,u3,u4 constant

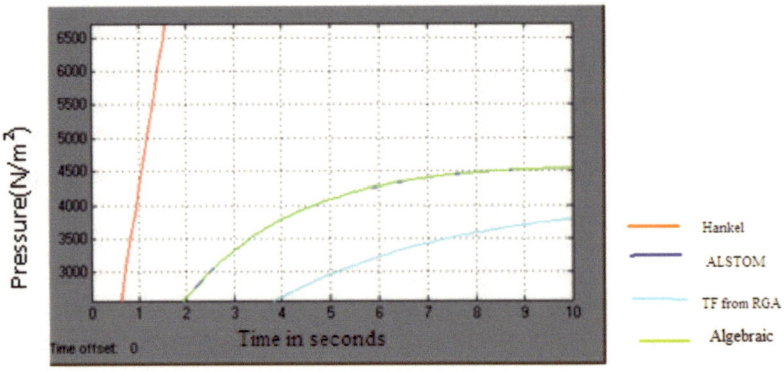

Figure 10. Variation of fuel gas pressure(y3) with coal flow rate (u3) keeping u1,u2,u4 constant

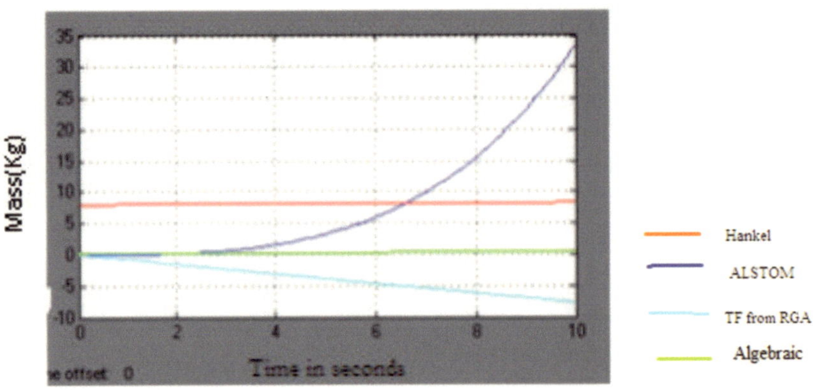

Figure 11. Variation of Bed mass(y2) with change in steam flow rate (u4) keeping u1,u2,u3 constant

The errors on the basis of IAE (Integral Absolute Error) and ISE (Integral Squared Error) are computed for each transfer function block obtained by algebraic method, reduced order approximation and RGA loop pairing over a period of time (little above the rise time) are shown in Table 3 for 100% load.

Transfer function	INTEGRAL ABSOLUTE ERROR			INTEGRAL SQUARED ERROR		
	Algebraic method	Reduced order approximation	TF using RGA loop pairing	Algebraic method	Reduced order approximation	TF using RGA loop pairing
G11	1644	1.062e+005	1.087e+004	2.16e+006	1.133e+009	1.455e+007
G12	7.09	751.5	2.954e+005	7.606	1.013e+005	1.12e+010
G13	4.828e+004	4.48e+004	8.039e+004	7.98e+008	7.98e+008	7.784e+008
G14	5.096	88.35	2.308e+005	5.85	1033	6.955e+009
G21	2.868e+005	5.23e+006	8.71e+004	1.157e+10	3.598e+12	8.637e+009
G22	11.5	1.19e+004	20.97	20.74	2.549e+007	57.78
G23	50.56	4.638e+004	6.8e+004	1018	2.668e+008	5.145e+008
G24	73.09	76.29	114.2	1412	830.3	2555
G31	9.128e+006	6.606e+006	8.799e+006	2.166e+13	1.009e+13	2.519e+013
G32	0.4021	1747	6.277e+004	0.0362	3.051e+005	4.58e+008
G33	35.04	1.78e+005	9250	283.1	4.443e+009	9.1e+006
G34	2.549	141.8	1.086e+005	0.8598	3622	1.344e+009
G41	1.437e+007	2.407e+007	1.434e+007	8.005e+13	1.411e+014	7.98e+013
G42	15.18	1.103e+004	2.695	39.14	0.1632	1.213
G43	462.3	5.714e+004	1.133e+005	3.812e+004	5.035e+008	1.46e+009
G44	1.683	508.8	0.4994	0.3358	4.662 e+004	0.1532

Table 3. Integral Absolute and Squared error criteria for 3 models

It is observed that the low order models derived using algebraic methods is much superior to one proposed by Haryanto et.al., using RGA loop pairing and reduced order approximation proposed by authors.

7.3. Lower order modeling using genetic algorithm

Out of 16 transfer functions using algebraic method, four transfer functions G21, G31, G41 and G13 (shown in bold) are found to have higher ISE and IAE error criterion than the lower order models obtained using RGA loop pairing. This observation has motivated the authors to obtain further reduced order transfer function models with minimum ISE and IAE error criterion using genetic algorithm. Appendix D gives the auxiliary scheme for low order model (Sivanandam and Deepa, 2009).

The ALSTOM higher order transfer function for G13 is given below:

$$G13=\frac{\begin{array}{c}-1.1s^{23}-12.44s^{22}+2893s^{21}+2764s^{20}+1212s^{19}+324.2s^{18}+59.29s^{17}+\\ 7.864s^{16}+0.7827s^{15}+0.05961s^{14}+0.003508s^{13}+0.00016s^{12}\\ +5.623e-006s^{11}+1.505e-007s^{10}+3e-009s^{9}+4.293e-011s^{8}\\ 4.155e-013s^{7}+2.4446e-015s^{6}+6.972e-018s^{5}+4.046e-021s^{4}\\ +8.036e-024s^{3}-.1.1772e-026s^{2}-5.48e-030\,s+8.511e-034\end{array}}{\begin{array}{c}s^{24}+35.38s^{23}+78.31s^{22}+68.51s^{21}+32.81s^{20}+9.998s^{19}+2.106s^{18}+\\ 0.3225s^{17}+0.03703s^{16}+0.00325s^{15}+0.0002203s^{14}+1.156e-005s^{13}\\ 4.687e-007s^{12}+1.45e-008s^{11}+3.36e-010s^{10}+5.64e-012s^{9}\\ 6.52e-014s^{8}+4.785e-016s^{7}+1.95e-18s^{6}+3.99e-021s^{5}+\\ 4.505e-024s^{4}+2.982e-024s^{3}+1.148e-030s^{2}+2.389e-034s+\\ 2.078e-038\end{array}}$$

The second approximation is given as

$$G13=\frac{5.48e-30s\ -8.511e-034}{1.148e-30s^{2}+2.389e-034s\ +2.076e-038}$$

The transient and steady state gain for G13 is

$$TG/_{G13(s)}=\frac{-1.1}{1}\ =\ -1.1$$

$$SSG/_{G13(s)}=\frac{8.511e-34}{2.076e-38}\ =\ 4.0997e{+}04$$

The auxiliary scheme given in appendix E is used to find R(s) from G(s)

$$R(s)=\frac{-5.48e-030\ s+8.511e-034}{1.148e-030s^{2}+2.389e-034s+2.078e-038}$$

The above equation should be tuned to satisfy the transient and steady state gain so that R(s) reflects the characteristics of G(s)

$$R(s)=\frac{-1.1\ s-7.4137631e-04}{s^{2}+2.081e-04s+1.8083624e-08}$$

$$=\frac{B_1 s+B_0}{b_2 s^{2}+b_1\, s+b_0}$$

The parameters B0 = -7.4137631e-04, b1= 2.081e-04 and b0= 1.8083624e-08 are used as seed value for genetic algorithm with ISE error as the objective function. The ISE error (E) can be obtained by taking the sum of the square of the difference between the step response of higher and lower order transfer function. The ISE error is given by

$$E=\sum_{t=0}^{\tau}(Y_t\ -y_t\ )2$$

where, Yt is the unit step time response of the higher order system at the tth instant in the time interval $0\leq t\leq\tau$, where τ is to be chosen and y_t is the unit step time response of the lower order system at the tth time instant. The matlab commands

```
options =gaoptimset('InitialPop', [B1 B2 B3])
[x fval output reasons] = ga(@objectivefun, nvars,options)
```

are used with ISE error as objective function. Here the population is set at 20 individuals and the maximum generation is 51. The crossover fraction is 0.8. Similarly the lower order models G31, G21 and G41 corresponding to higher order models specified by ALSTOM can be obtained. Table 4 shows the IAE and ISE error using genetic algorithm is further reduced than using algebraic method. Figure 12 shows the flowchart for lower order modeling using Genetic Algorithm.

Reduced transfer function using genetic algorithm	B0	b1	b2	IAE using genetic algorithm	IAE using Algebraic method	ISE using genetic algorithm	ISE using Algebraic method
G13= $\frac{-1.1\,s+2.8705}{s^2 - 0.0922s+0.0339}$	2.8705	-0.0922	0.0339	**3.001e+004**	4.828e+004	**2.575e+008**	7.98e+008
G21= $\frac{-9207\,s+4.7874}{s^2+\ 350.5581-0.3551}$	4.7874	350.5581	-0.3551	**8.718e+004**	2.868e+005	**8.634e+009**	1.157e+10
G31= $\frac{6563\,s+6.3912}{s^2+0.0581s+1.8084e-08}$	6.3912	0.0581	1.8084e-08	**8.57e+006**	9.128e+006	**2.442e+013**	2.166e+13
G41= $\frac{-8868\,s+2.3705}{s^2- 0.2803s+0.0939}$	2.3705	-0.2803	0.0939	**1.399e+007**	1.437e+007	**7.782e+013**	8.005e+13

Table 4. Reduced errors due to genetic algorithm in the evaluation of G13,G21,G31,G41

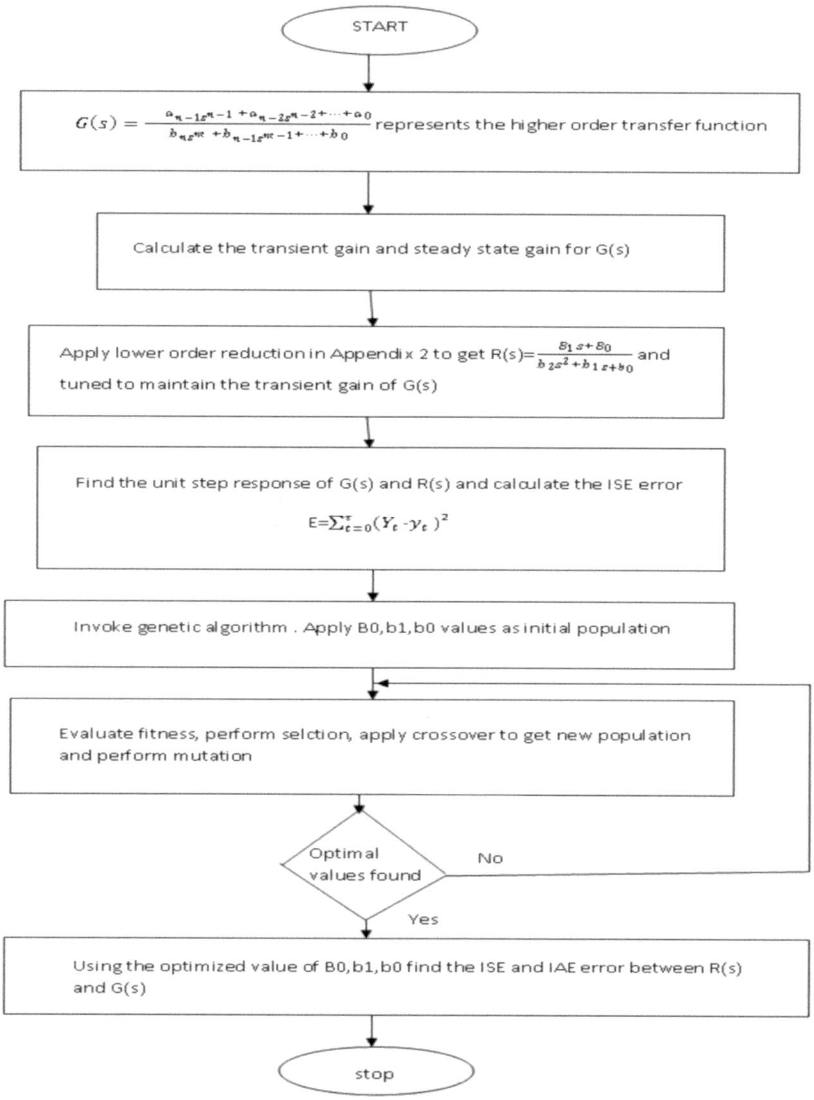

Figure 12. Flowchart for lower order modeling using Genetic Algorithm

Transfer function obtained using Genetic Algorithm seems to be the most effective method for obtaining lower order models. Though the transfer functions for G11, G31,G41, G22 have been obtained through genetic algorithm to illustrate the superiority over other methods, all the transfer function blocks can be obtained in the same way as explained earlier.

8. Gasifier control and simulation

Even though many advanced control algorithms are proposed for complex process and systems, the authors are strongly of the opinion that PID control will also meet the control requirements using appropriate controller constants and feed forwards if necessary. Hence the PID controller is considered as a tool for gasifier control and simulation studies are done.

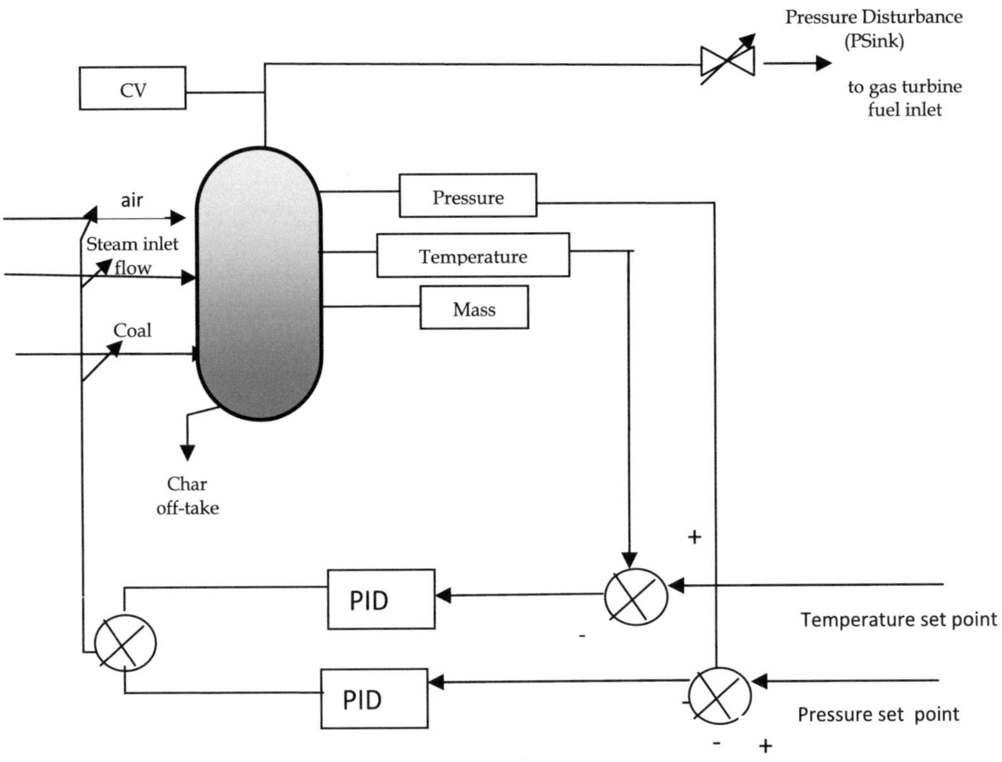

Figure 13. PID controller for pressure and temperature output variables

Here PID controller is used to vary the steam and coal inputs for syngas pressure and coal and air is varied for syngas temperature. Table 5 gives the PID parameters for pressure and temperature of the syngas

P-Psink error	Kp	Ki	Kd
PID(temperature)	0.5	0.25	0.001
PID (pressure)	7.5	4	3

Table 5. PID constants for syngas temperature and pressure

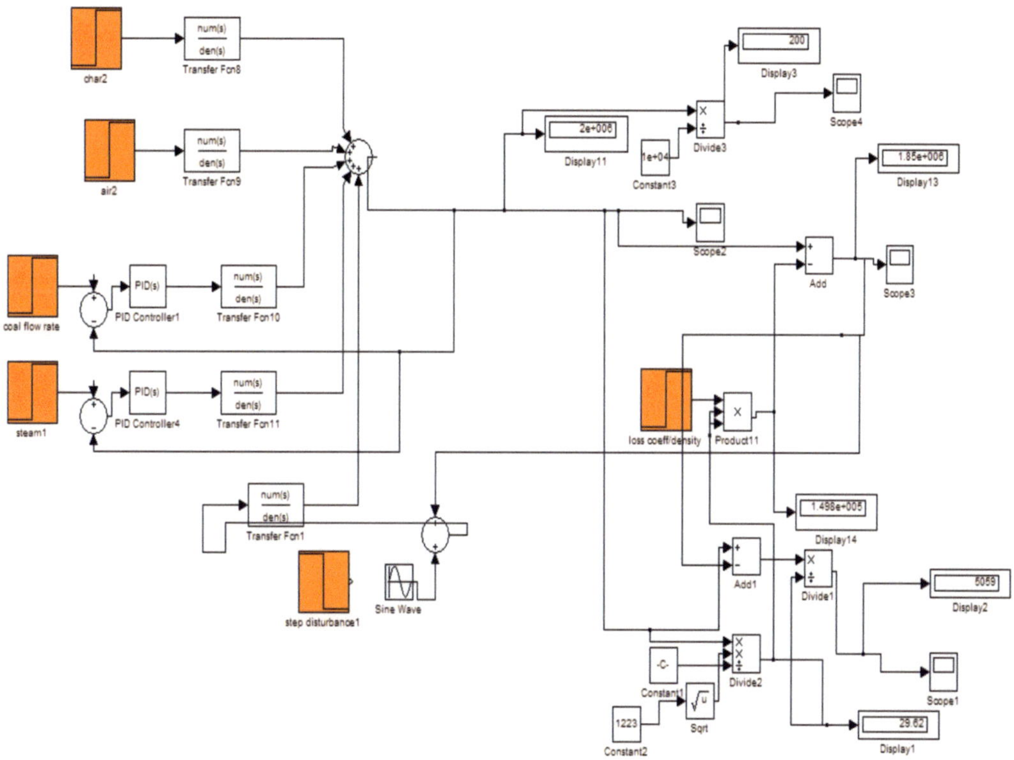

Figure 14. SIMULINK model for syngas pressure in the presence of step and sinusoidal disturbances

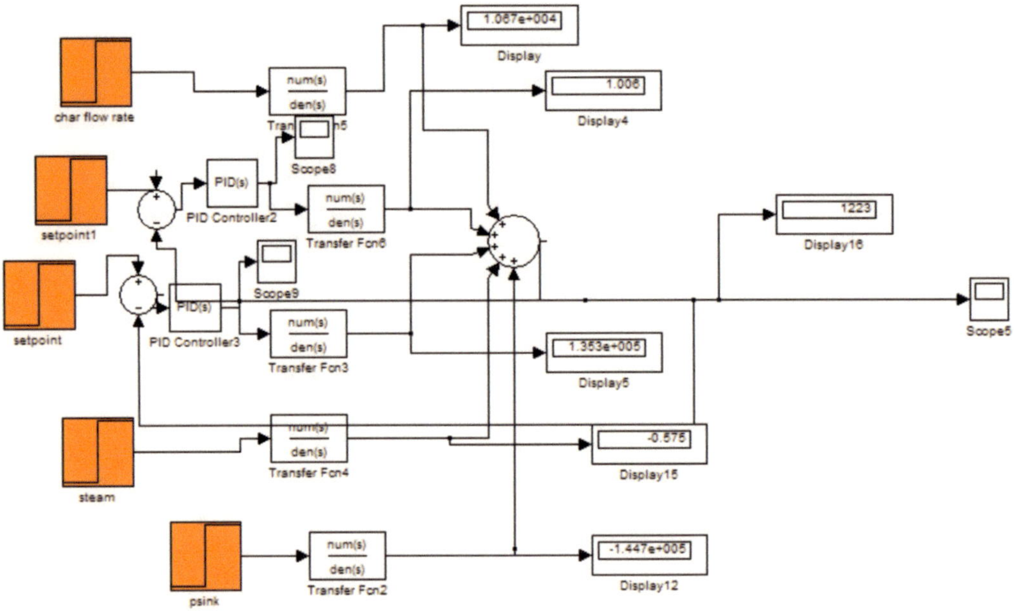

Figure 15. SIMULINK model for syngas temperature in the presence of step and sinusoidal disturbances

Figure 16. Syngas pressure maintaining at $2*10^6 N/m^2$ in the presence of disturbance

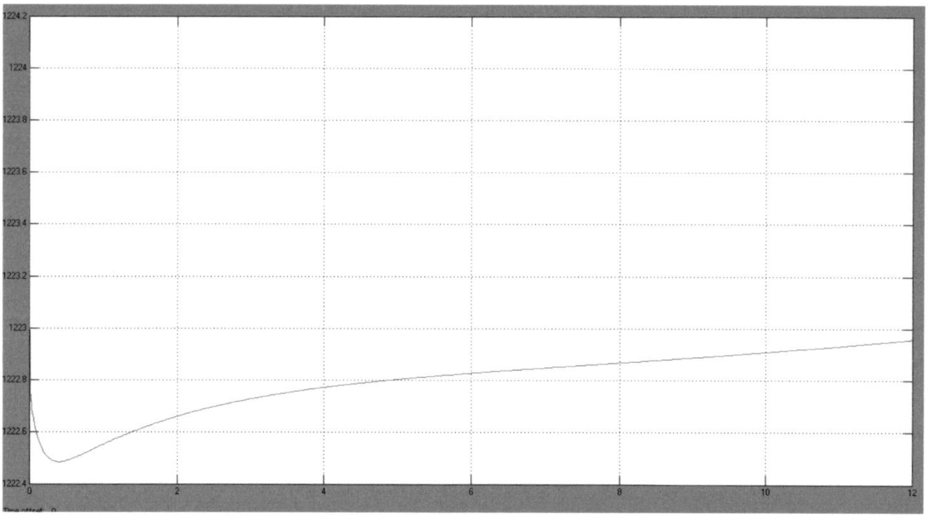

Figure 17. Syngas temperature maintaining at 1223K in the presence of disturbance

9. Conclusion

The development of low order transfer function model are required due to the difficulties encountered in the development of control strategies on ALSTOM benchmark challenge. In this direction, the authors have developed low order transfer function models using Algebraic method and reduced order approximation. The performance of these models has been evaluated on the basis of ISE and IAE error criteria. It is observed that the low order models derived using algebraic methods is much superior to one proposed by Haryanto et.al., and reduced order approximation. Some lower order transfer functions obtained using algebraic method are found to have higher error criterion than RGA loop pairing. Using Genetic Algorithm these errors are minimized and it is believed that the models proposed by algebraic method with Genetic Algorithm will become basis for further research on Gasifier control.

The authors have applied PID control algorithms for gasifier control around 100% load. As desired in the challenge problem, step and sinusoidal disturbances have been given in Psink. Preliminary simulation results show that the pressure and temperature of the syngas are controlled within the permissible constraint limits. However the authors intend to do extensive simulations for 100%, 50% and no-load with error due to pressure and temperature setpoints modulating different input variables.

Author details

L. Sivakumar

Department of Mechatronics, Sri Krishna College of Engineering & Technology, Coimbatore,
Formerly with Corporate R&D, BHEL, Hyderabad, India

X. Anithamary

Department of Electronics and Instrumentation Engineering, Karunya University, Coimbatore, India

Acknowledgement

The authors would like to express their thanks to Dr. Yongseung Yun, the book editor on Gasification for his valuable suggestions. Further the authors are grateful to Karunya University and Sri Krishna College of Engineering and Technology, Coimbatore for the support and encouragement.

Appendix A: Transfer function matrix of Alstom plant for 100% load

Transfer function blocks	algebraic method	reduced order approximation
G11	$\frac{-43.210273s^2 - 32.8849432314s + 10.5403}{-0.0083690166s^2 + 0.067824414s + 0.0019433}$	$\frac{-1.197e004\ s^2 + 330.4\ s + 0.001125}{s^2 + 0.0008608\ s + 2.075e - 007}$
G12	$\frac{0.67268851s^2 + 0.22784337s + 1.36739}{8.7409426609s^2 - 6.32996277s - 0.0002336}$	$\frac{-3.468\ s^2 - 1.063s - 0.001214}{s^2 + 0.0008608s + 2.075e - 007}$
G13	$\frac{-29.294957767s^2 + 58.590928399s + 0.99338}{0.9053252009s^2 + 0.217203375s - 0.00002424}$	$\frac{108.4\ s^2 + 6.901s - 0.008504}{s^2 + 0.0008608s + 2.075e - 007}$
G14	$\frac{11.42165811s^2 - 18.197458774s + 2.892}{-298.17003810s^2 + 56.7756422s - 2.2915}$	$\frac{-1.851s^2 + 0.06763\ s - 2.618e - 007}{s^2 + 0.0008608\ s + 2.075e - 007}$
G21	$\frac{0.7699194835s^2 - 0.4621252416s + 1.4975}{-0.00005375534s^2 - 0.0000999029s - 3.530 * 10^{\wedge} - 5}$	$\frac{-1.068e005\ s^2 + 43.57s - 0.008799}{s^2 + 0.0008608\ s + 2.075e - 007}$
G22	$\frac{0.1119962125834s^2 - 0.335052778707s + 1.5892}{-10.666078439387s^2 - 3.7028097164s - 1.016069 * 10^{\wedge} - 3}$	$\frac{-71.72\ s^2 - 0.1481s - 0.0003245}{s^2 + 0.0008608\ s + 2.075e - 007}$
G23	$\frac{-38.1754867787s^2 - 606.4765403s + 1.03212}{-0.176233518s^2 - 0.067401899s + 0.0004235}$	$\frac{1.142e004\ s^2 - 14.05s - 0.0005055}{s^2 + 0.0008608\ s + 2.075e - 007}$
G24	$\frac{7.589742045s^2 + 3.20126491848s + 0.80506}{66.2192853528s^2 + 13.8974102235s + 0.39192}$	$\frac{8.026s^2 + 0.03455s + 4.229e - 006}{s^2 + 0.0008608\ s + 2.075e - 007}$
G31	$\frac{7.31943261016s^2 - 83.3609061793s + 0.76028}{-0.011722633s^2 - 0.0005261888s + 2.49106 * 10^{\wedge} - 5}$	$\frac{1.507e005s^2 - 171.7s + 0.01169}{s^2 + 0.0008608\ s + 2.075e - 007}$
G32	$\frac{-8.34856920133s^2 + 15.2823278158s + 1.4825}{26.0070991592s^2 + 2.5943768447s + 0.000366}$	$\frac{-175.2s^2 + 0.4962s + 0.0008401}{s^2 + 0.0008608\ s + 2.075e - 007}$
G33	$\frac{-2.8969959854S^2 - 269.839875362S + 1.645}{-0.1417932867S^2 - 0.05582682S + 0.538723 * 10^{\wedge} - 4}$	$\frac{4288s^2 - 4.413s - 0.006334}{s^2 + 0.0008608\ s + 2.075e - 007}$
G34	$\frac{-0.149422569149s^2 + 0.605489884s + 0.8755}{-13.277800585s^2 - 15.075170803s - 0.0538}$	$\frac{0.9117s^2 + 0.0606s - 3.372e - 006}{s^2 + 0.0008608\ s + 2.075e - 007}$
G41	$\frac{0.989892606658S^2 - 6.37153721233S + 1.5006}{0.0005803934141S^2 + 0.000131142815S - 3.06317 * 10^{\wedge} - 5}$	$\frac{1.941e005s^2 + 48.24s - 0.01016}{s^2 + 0.0008608\ s + 2.075e - 007}$
G42	$\frac{0.863152338637S^2 - 1.69330243903S + 2.3304}{-12.329066005S^2 - 3.02684037S - 0.00315996}$	$\frac{1.941e005s^2 + 48.24s - 0.01016}{s^2 + 0.0008608\ s + 2.075e - 007}$
G43	$\frac{-15.4009119304S^2 - 2940.056236928S + 2.31138}{-0.566750514100S^2 - 0.1874561152S + 0.0021771}$	$\frac{1.709e004s^2 + 6.082s + 0.0002203}{s^2 + 0.0008608\ s + 2.075e - 007}$
G44	$\frac{201.4423617140688S^2 + 275.7771961791S + 0.81865}{-4192.317426968S^2 - 1162.912156389S - 0.01737}$	$\frac{3.079s^2 - 0.02195s - 9.775e - 006}{s^2 + 0.0008608\ s + 2.075e - 007}$

Appendix B: Transfer function obtained of 50% load

Transfer function blocks	algebraic method	reduced order approximation
G11	$\frac{82973.885826s + 1.311852}{1978.394989s^2 + 1.308653s + 1.568416e - 05}$	$\frac{-1330s^2 + 395.4s + 0.006024}{s^2 + 0.0005765s + 7.203e - 008}$
G12	$\frac{-4.151117e - 06s^2 + 0.182109e - 03s + 2.547485}{-2.550036s^2 - 2.54957s - 0.91584e - 04}$	$\frac{2.476s^2 - 1.01s - 0.0007864}{s^2 + 0.0005765s + 7.203e - 008}$
G13	$\frac{727.803446s^2 - 31191.578014s + 1.266519}{9.782304s^2 - 7.184683s + 0.001618}$	$\frac{199.6s^2 + 5.647s + 5.637e - 005}{s^2 + 0.0005765s + 7.203e - 008}$
G14	$\frac{633.894001s^2 - 26444.64477s + 1.280635}{11207.461122s^2 - 350.714185s + 0.216796}$	$\frac{0.3227s^2 + 0.06097s + 4.256e - 007}{s^2 + 0.0005765s + 7.203e - 008}$
G21	$\frac{-7.638891e + 18s + 2.495387}{6.3235e + 14s^2 + 4.938082e + 13s - 7.626576e - 05}$	$\frac{-1.839e005s^2 + 56.37s - 0.002357}{s^2 + 0.0005765s + 7.203e - 008}$
G22	$\frac{1.016135e - 04s^2 + 0.00445778s + 2.676878}{-29.50425s^2 + 3.382317s - 10.2823689}$	$\frac{166s^2 - 0.06141s - 0.0001875}{s^2 + 0.0005765s + 7.203e - 008}$
G23	$\frac{-150.691621s^2 - 1.446301e + 12s + 1.682835}{-1.40965e + 08s^2 + 4.809870e + 06s + 0.000157}$	$\frac{8739s^2 + 7.32s + 0.0007719}{s^2 + 0.0005765s + 7.203e - 008}$
G24	$\frac{0.016439s^2 + 1.190322e + 08s + 1.31212}{2.249456e + 07s^2 + 4.0072857e + 06s + 0.044163}$	$\frac{8.171s^2 + 0.02981s + 2.14e - 006}{s^2 + 0.0005765s + 7.203e - 008}$
G31	$\frac{470334.30765s + 3.270734}{45.137649s^2 - 1.79416s - 8.741614e - 06}$	$\frac{2.322e005s^2 - 243.4s + 0.001665}{s^2 + 0.0005765s + 7.203e - 008}$
G32	$\frac{3.716764e - 07s^2 - 160.82437e - 07s + 2.548966}{3.470826s^2 + 3.7707s + 3.13062e - 04}$	$\frac{-375.9s^2 + 0.4627s + 0.0005865}{s^2 + 0.0005765s + 7.203e - 008}$
G33	$\frac{-3.1067766e + 05s + 2.683919}{-189.668841s^2 - 551.803398s + 0.005885}$	$\frac{2926s^2 - 3.597s + 3.286e - 005}{s^2 + 0.0005765s + 7.203e - 008}$
G34	$\frac{59112.633568s + 1.437609}{-2.6507907e + 05s^2 - 3511.145849s - 0.053218}$	$\frac{-0.4876s^2 - 0.0597s + 3.286e - 005}{s^2 + 0.0005765s + 7.203e - 008}$
G41	$\frac{-2.720125e + 07s + 2.501114}{2381.895719s^2 + 740.758982s - 2.580143e - 05}$	$\frac{3.296e005s^2 + 24.4s - 0.006982}{s^2 + 0.0005765s + 7.203e - 008}$
G42	$\frac{2.01188e - 03s^2 - 0.087054s + 3.7086}{1304.883s^2 + 1.097948e + 05s - 3.3664e - 04}$	$\frac{72.72s^2 - 0.4004s - 7.935e - 005}{s^2 + 0.0005765s + 7.203e - 008}$
G43	$\frac{5.456232e + 04s + 3.632051}{20.535311s^2 + 20.261304s + 0.000559}$	$\frac{1.258e004s^2 + 4.893s + 0.0004882}{s^2 + 0.0005765s + 7.203e - 008}$
G44	$\frac{1.2505305e + 04s + 1.333895}{-6380.258041s^2 - 613.370677s - 0.020417}$	$\frac{1.715s^2 - 0.02503s - 4.705e - 006}{s^2 + 0.0005765s + 7.203e - 008}$
Gd1	$\frac{1.427343e + 06s - 1.570372e + 06}{-8.105303e + 14s^2 - 2.361753e + 11s - 1.1699040e + 07}$	$\frac{0.1224s^2 - 2.013e - 005s + 9.669e - 009}{s^2 + 0.0005765s + 7.203e - 008}$
Gd2	$\frac{-2.7855e - 06s^2 + 0.122333e - 03s + 1.29029}{41901.74673s^2 + 41909.74673s + 0.3277079e - 03}$	$\frac{9.213e - 005s^2 + 3.538e - 007s + 2.198e - 010}{s^2 + 0.0005765s + 7.203e - 008}$
Gd3	$\frac{1.319354e + 11s + 2.217903}{2.175717e + 11s^2 - 3.741932e + 11s + 2.32512}$	$\frac{0.9534s^2 + 0.0005484s + 6.87e - 008}{s^2 + 0.0005765s + 7.203e - 008}$
Gd4	$\frac{2.3313028e + 04s + 1.298735}{2.220288e + 08s^2 - 1.65059e + 09s - 4.327659e + 04}$	$\frac{-3.39e - 005s^2 - 3.341e - 008s - 2.163e - 012}{s^2 + 0.0005765s + 7.203e - 008}$

Appendix C: Transfer function obtained of 0% load

Transfer function blocks	algebraic method	reduced order approximation
G11	$\frac{6412495.306104s + 59.515387}{81552.782731s^2 + 14.789253s + 8.741323e - 05}$	$\frac{3.828e004\, s^2 + 561.7\, s + 0.006739}{s^2 + 0.0002741\, s + 9.897e - 009}$
G12	$\frac{5.156462e - 6s^2 + 2.9510432e - 4s + 3.342709638}{-3.34349077s^2 - 3.3438166s - 8.398176626e - 05}$	$\frac{79.85\, s^2 - 0.955\, s - 0.0003939}{s^2 + 0.0002741\, s + 9.897e - 009}$
G13	$\frac{203764.661731s^2 - 10907532.587163s + 56.60632}{-2370.734866s^2 - 5009.316672s + 0.038167}$	$\frac{178.3\, s^2 + 1.338\, s + 1.467e - 005}{s^2 + 0.0002741\, s + 9.897e - 009}$
G14	$\frac{8.872773e + 40s^2 - 4.859238e + 42s + 1.656547}{-1.042016e + 42s^2 - 9.464071e + 40s + 0.032264}$	$\frac{3.845\, s^2 + 0.05121s + 5.082e - 007}{s^2 + 0.0002741\, s + 9.897e - 009}$
G21	$\frac{-2.966146e + 10s + 114.021444}{1762415.861223s^2 - 382735.79834s + 0.000915}$	$\frac{-4.377e005\, s^2 + 120.2\, s + 0.001232}{s^2 + 0.0002741\, s + 9.897e - 009}$
G22	$\frac{-2.21638386e - 03s^2 + 0.1268436s + 3.475304171}{-41.36659216s^2 + 9.7254335s + 3.685291183e - 04}$	$\frac{948.2\, s^2 + 0.117\, s - 9.333e - 005}{s^2 + 0.0002741\, s + 9.897e - 009}$
G23	$\frac{-1.783467s^2 - 4.004262e + 09s + 2.15279}{-392574.725689s^2 - 671071.290709s + 0.000361}$	$\frac{4701\, s^2 + 1.852\, s + 5.905e - 005}{s^2 + 0.0002741\, s + 9.897e - 009}$
G24	$\frac{-0.000362s^2 - 573649.046541s + 1.698815}{-81530.566687s^2 - 7775425361s + 0.028148}$	$\frac{11.85\, s^2 + 0.02519\, s + 5.973e - 007}{s^2 + 0.0002741\, s + 9.897e - 009}$
G31	$\frac{470334.30765s + 3.270734}{45.137649s^2 - 1.79416s - 8.741614e - 06}$	$\frac{5.378e005\, s^2 - 408.2\, s - 0.003703}{s^2 + 0.0002741\, s + 9.897e - 009}$
G32	$\frac{2.136783444e - 07s^2 - 122.288e - 07s + 3.334867}{4.423179153s^2 + 4.802113698s + 3.116106e - 05}$	$\frac{-1552\, s^2 + 0.2945\, s + 0.0003176}{s^2 + 0.0002741\, s + 9.897e - 009}$
G33	$\frac{67905.569056s + 3.379449}{407.551672s^2 - 1891.8753s - 0.007252}$	$\frac{1502\, s^2 - 0.8885\, s - 4.612e - 006}{s^2 + 0.0002741\, s + 9.897e - 009}$
G34	$\frac{77345.248573s + 1.937428}{-472193.214731s^2 - 1741.323615s - 0.024306}$	$\frac{-5.796\, s^2 - 0.05223\, s - 7.889e - 007}{s^2 + 0.0002741\, s + 9.897e - 009}$
G41	$\frac{-1.664227e + 08s + 3.266391}{10606.927662s^2 + 1483.255839s - 9.560048e - 06}$	$\frac{9.186e005\, s^2 - 25.51\, s - 0.003382}{s^2 + 0.0002741\, s + 9.897e - 009}$
G42	$\frac{1.52637013e - 03s^2 - 0.08735416s + 4.8185565}{-2490.9933s^2 - 247081.2502s - 15.27197e - 04}$	$\frac{45.08\, s^2 - 0.4357\, s - 3.123e - 005}{s^2 + 0.0002741\, s + 9.897e - 009}$
G43	$\frac{187977.323085s + 4.505946}{69.134727s^2 + 119.112126s + 0.001149}$	$\frac{6900\, s^2 + 1.238\, s + 3.885e - 005}{s^2 + 0.0002741\, s + 9.897e - 009}$
G44	$\frac{58954.826124s + 1.729767}{-26761.155753s^2 - 1437.956917s - 0.01438}$	$\frac{-4.063\, s^2 - 0.02662\, s - 1.191e - 006}{s^2 + 0.0002741\, s + 9.897e - 009}$
Gd1	$\frac{-1.344731e + 23s - 1475658.562121}{2.994947e + 31s^2 + 3.271975e + 26s + 2.050176e + 07}$	$\frac{-0.1992\, s^2 - 7.057e - 005\, s - 7.124e - 010}{s^2 + 0.0002741\, s + 9.897e - 009}$
Gd2	$\frac{-5.6523279e - 06s^2 + 0.323485e - 03s + 3.3032}{96936.4763s^2 + 179827.0165s + 754.357}$	$\frac{-0.0003971\, s^2 + 6.566e - 008\, s + 4.334e - 011}{s^2 + 0.0002741\, s + 9.897e - 009}$
Gd3	$\frac{-7.490644e + 11s + 2.88116}{-6.847024e + 11s^2 - 2.019168e + 12s + 2.921795}$	$\frac{0.9858\, s^2 + 0.0002702\, s + 9.76e - 009}{s^2 + 0.0002741\, s + 9.897e - 009}$
Gd4	$\frac{-40689.070969s + 1.677564}{-1.634099e + 08s^2 - 9828353.140255s - 264.279412}$	$\frac{-1.935e - 005\, s^2 - 6.714e - 009\, s - 6.282e - 014}{s^2 + 0.0002741\, s + 9.897e - 009}$

Appendix D: Lower order Transfer function reduction

Consider an nth higher order system represented by its transfer function

$$G(s) = \frac{N(s)}{D(s)} = \frac{\sum_{i=0}^{n-1} A_i s^i}{\sum_{i=0}^{n} a_i s^i}$$

$$= \frac{A_{n-1} s^{n-1} + A_{n-2} s^{n-2} + \dots + A_2 s^2 + A_1 s + A_0}{a_n s^n + a_{n-1} s^{n-1} + \dots + a_2 s^2 + a_1 s + a_0}$$

$$\text{First Order} = \frac{A_0}{a_1 s + a_0} \quad (5)$$

$$\text{Second order} = \frac{A_1 + A_0}{a_2 s^2 + a_1 s + a_0} \dots \quad (6)$$

$$\text{n-1 order} = \frac{A_{n-2} s^{n-2} + A_{n-3} s^{n-3} + \dots + A_2 s^2 + A_1 s + A_0}{a_{n-1} s^{n-1} + a_{n-2} s^{n-2} + \dots + a_2 s^2 + a_1 s + a_0} \quad (7)$$

Equations (5) through (7) gives the lower order model for higher order system G(s). For n higher order system, (n-1) lower order models can be formulated.

10. References

Asmar, B.N. WE Jones and Ja Wilson, A process engineering approach to the alsotm gasifier problem proc, Inst.Mech Eng.I, J.System control Eng., 2000,214 pp.441-452.

Dixon R Alstom Benchmark challenge II: control of Nonlinear Gasifier model, 2002. http://www.iee.org/omcomms/PN/controlauto/Specification_v2.pdf

Dixon R, Becnhmark challenge at control, 2004, Comput. Control Eng IEE vol 10 No3 pp 21-23 2005.

Dixon. R Advanced gasifier control, computing and control engineering journal IEE vol 10 N0 3 pp 93-96, 1999

Griffin, I.A., P. Schroder, AJ Chipperfield and PJ Fleming multiobjective optimization approach proc, Inst.Mech Eng.I, J.System control Eng., 2000, 214 pp.453-468.

Haryanto, A., P.siregar, D.Kurniadi and Keum-shik Hong, Development of Integrated Alstom gasification Simulator for implementation using DCS CS 3000 proceedings of the 17th world congress The international conference federation of automatic control, seoul, korea 2009

Liu, G.P., RDixon S Daley, multiobjective optimal tuning proportional integral controller design for the Alstom Gasifier problem proc, Inst.Mech Eng.I, J.System control Eng., 2000, 214 pp.395-404.

Mitchell, M., An introduction to Genetic Algorithm, Prentice-Hall of India, New Delhi, Edition: 2004

Munro, N., JM Edmunds, E. Kontogianees and St Impram A sequential loop closing approach to the Alstom gasifier problem proc, Inst.Mech Eng.I, J.System control Eng., 2000,214 pp.427-439.

Phillips, J. 2006, Different types of gasifiers and their integration with gas turbines. In: The Gas Turbine handbook.

http://www.netl.doe.gov/technologies/coalpower/turbines/refshelf/handbook/1.2.1.pdf

Pike A.W., Donne M.S and Dixon. R, Dynamic modeling and simulation of the air blown gasification cycle prototype integrated plant in proceedings of the international conference on simulation, IEE publication 457, York university 1998 pp 354-361.

Poongodi, P., S. Victor Genetic algorithm based PID controller design forLTI system via reduced order model, International Conference on Instrumentation, Control & Automation ICA2009 October 20-22, 2009, Bandung, Indonesia.

Prempain, E., I.Postlethwaite and XD sun Robust control of the gasifier using a mixed H_∞ approach proc, Inst.Mech Eng.I, J.System control Eng., 2000,214 pp.415-426.

Ramezan and Stiegel, 2006. Integrated coal gasification combined cycle. In: The Gas Turbine handbook.
http://www.netl.doe.gov/technologies/coalpower/turbines/refshelf/handbook/1.2.pdf

Rice M, Rosster. J and schurmans J An advanced predictive control approach to the Alstom gasifier prolem, proc, Inst.Mech Eng.I,J.System control Eng., 2000,214 pp.405-413

Seyab, R.K., Y. Cao and S.H Yang, The second alstom benchmark challenge on gasifier control predictive control for the ALSTOM gasifier problem IEE proceedings on control theory, vol153, N03 May 2006.

Sivakumar, L., Anithamary.X A low order transfer function model for MIMO ALSTOM gasifier, International conference on process Automation, Control and computing, IEEE 2011.

Sivanandam, S.N. S.N. Deepa, A Comparative Study Using Genetic Algorithm and Particle Swarm Optimization for Lower Order System Modelling International Journal of the Computer, the Internet and Management Vol. 17. No.3 pp 1 -10, (September - December, 2009)

Taylor, C.J., AP Mccabe, PC young and A chotai Proportional integral plus control of AlstomGasifier problem proc, Inst.Mech Eng.I, J.System control Eng., 2000,214 pp.469-480.

Wang, X., Ke Wu, Jianhong Lu, Wenguo Xiang, Non linear identification of Alstom gasifier based on wiener model International conference on sustainable power generation and supply pg 1-7 april 2009.

Sivakumar. L, "Performance analysis, diagnosis and and optimisation (PADO) for power plants", Seminar on Power Plant Automation Concepts & Applications –by The Instrumentation Systems and automation Society- Bangalore, April 22, 2006.

Sivakumar. L and Ganpathiraman. G "Performance analysis diagnostics and optimisation in generation", Conference on: IT Power- Improving Performance And Productivity,New Delhi ; Sep 2006.

Sivakumar, L, Reddy K.L and Sundararajan. N, "Detailed circulation analysis to determine the DNB margin in natural circulation boilers" Proceedings of International conference on Heat and Mass Transfer. Hyderabad), Feb 13- 16 1980, p 1-8.

Ponnusamy.P,Sivakumar, L. and Sankaran, S. V."Low-order dynamic model of a complete thermal power plant loop", Proceedings of the Power Plant Dynamics,Control and Testing Symposium, Vol. 1, 1983, p 10. 01-10

Sivakumar. L.andBhattacharya. R. K, "Dynamic analysis of a power boiler using a non-linear mathematical model", Proceedings of second symposium on Power Plant Dynamics and Controls, Hyderabad (Record of Proceedings), Feb 14-16, 1979, p 21-29.